MUSIC, SOUND, AND THE LABORATORY FROM 1750-1980

EDITED BY
Alexandra Hui, Julia Kursell, and Myles W. Jackson

O S I R I S | 28

A Research Journal Devoted to the
History of Science and Its Cultural Influences

Osiris

Series editor, 2013–2018

ANDREA RUSNOCK, *University of Rhode Island*

Volumes 28 to 32 in this series are designed to connect the history of science to broader cultural developments, and to place scientific ideas, institutions, practices, and practitioners within international and global contexts. Some volumes address new themes in the history of science and explore new categories of analysis, while others assess the "state of the field" in various established and emerging areas of the history of science.

28   ALEXANDRA HUI, JULIA KURSELL, & MYLES W. JACKSON, EDS., *Music, Sound, and the Laboratory from 1750 to 1980*

Series editor, 2002–2012

KATHRYN OLESKO, *Georgetown University*

17   LYNN K. NYHART & THOMAS H. BROMAN, EDS., *Science and Civil Society*
18   SVEN DIERIG, JENS LACHMUND, & J. ANDREW MENDELSOHN, EDS., *Science and the City*
19   GREGG MITMAN, MICHELLE MURPHY, & CHRISTOPHER SELLERS, EDS., *Landscapes of Exposure: Knowledge and Illness in Modern Environments*
20   CAROLA SACHSE & MARK WALKER, EDS., *Politics and Science in Wartime: Comparative International Perspectives on the Kaiser Wilhelm Institute*
21   JOHN KRIGE & KAI-HENRIK BARTH, EDS., *Global Power Knowledge: Science and Technology in International Affairs*
22   GREG EGHIGIAN, ANDREAS KILLEN, & CHRISTINE LEUENBERGER, EDS., *The Self as Project: Politics and the Human Sciences*
23   MICHAEL D. GORDIN, KARL HALL, & ALEXEI KOJEVNIKOV, EDS., *Intelligentsia Science: The Russian Century, 1860–1960*
24   CAROL E. HARRISON & ANN JOHNSON, EDS., *National Identity: The Role of Science and Technology*
25   ERIC H. ASH, ED., *Expertise and the Early Modern State*
26   JAMES RODGER FLEMING & VLADIMIR JANKOVIC, EDS., *Klima*
27   ROBERT E. KOHLER & KATHRYN M. OLESKO, EDS., *Clio Meets Science: The Challenges of History*

Cover Illustration:

Central office of the Paris théâtrophone, 1892, from Georges Mareschal, "Le théâtrophone," *La Nature* 20 (1892): 58. This service allowed listeners to enjoy musical concerts and theatrical performances in their own home over the telephone.

Acknowledgments

We three editors would like to thank the following for providing funds and support for two conferences, where preliminary drafts of the essays in this collection were discussed: Mitchell Memorial Library, Mississippi State University; College of Arts and Sciences, Mississippi State University; Department of History, Mississippi State University; Dibner Family Professorship in the History and Philosophy of Science and Technology of the Polytechnic Institute of New York University; the Max Planck Institute for the History of Science in Berlin (particularly Regina Held, Lorraine Daston, and Hans-Jörg Rheinberger). We would also like to thank Lyndsey Cockwell and Tessa Dunkel for their assistance in editing the various essays, and we are particularly grateful to copy editor Rachel Kamins for her editing and to Andrea Rusnock for her assistance in shepherding us through the process of putting this volume together. All of them have made our lives infinitely easier, and we thank them wholeheartedly. Finally, we dedicate this work to the memory of our dear friend and colleague, Elfrieda Franz Hiebert, who sadly passed away as we were working on the volume.

Music, Sound, and the Laboratory
from 1750 to 1980

by Alexandra Hui,* Julia Kursell,† and Myles W. Jackson‡

Music and science have a long common history dating back to the ancient Greeks. This shared history became uncoupled some time in the late nineteenth century, as music ceased to be the most important source of sound that could be subjected to investigation. Knowledge of sound always relies on some form of medium—that is, a way to store, transmit, transform, manipulate, and encode or decode sound. Until the second half of the nineteenth century, the techniques and practices of producing, describing, and reproducing sounds had been the most developed in music, and scientists, engineers, and musicians were engaged in similar, reciprocal projects. Music remained privileged as an object of acoustic investigation, driven in part by the bourgeois culture of music that flourished during the nineteenth century, especially in central Europe, and created a public of listeners and producers of musical sounds. Yet, at the same time, scientists and engineers broadened the scope of investigation beyond music, to sound, with new mathematical tools, new fields of experimental inquiry, and eventually new media technologies, most often located in laboratories. Music continued to frame the interaction among scientists, engineers, and musicians, and, though perhaps less immediately apparent, this reciprocal engagement has continued, over time becoming more intricate, intriguing, and historically informative.

Our volume illuminates the ongoing reciprocal relationship between scientific investigation, on one hand, and music and sound, on the other. We offer a series of historically and locally specified case studies, mostly between 1750 and 1980, devoted to the entanglement of music, sound, and the laboratory. The authors approach the topic from a variety of perspectives: history of science and technology, music history, science studies, and media studies. Starting in the 1980s, Anglo-American histories of science drew upon new trends being generated in the social sciences with a view to avoiding earlier distinctions between the context and content of the scientific enterprise. Any discipline or field that wants to tackle the role of sound must reflect on the media through which sound propagates. Historians, in particular, have explored in recent years the multiple ways in which we can glean information about sensory experiences in the past.[1] Although the project to find out how people lived

* Department of History, Mississippi State University, Box H, Mississippi State, MS 39762; ahui@history.msstate.edu.
† Capaciteitsgroep muziekwetenschap, Universiteit van Amsterdam, Nieuwe Doelenstraat 16, 1012 CP Amsterdam, Netherlands; j.j.e.kursell@uva.nl.
‡ NYU-Gallatin, 1 Washington Place, Room 405, New York, NY 10003; myles.jackson@nyu.edu.
¹ See, e.g., Mark M. Smith, *Sensing the Past: Seeing, Hearing, Smelling, Tasting, and Touching in History* (Berkeley, Calif., 2008).

© 2013 by The History of Science Society. All rights reserved. 0369-7827/11/2013-0001$10.00

OSIRIS 2013, 28 : 1–11   1

in different historical contexts is as exciting as it is methodologically difficult, the history of sound is not alone in its quest for uncovering sounds of earlier periods. Another discipline engaged in this quest, which until fairly recently was conspicuous in its absence from the history of science and science studies, is media studies.[2] As Bernhard Siegert's contribution to this volume suggests, media studies scholars are interested in studying the question of how scientific instruments can turn into media. A history of science that seeks insights into the concepts of sound that prevailed historically must include media history. We hope this volume provides an introduction to media scholars' use of concepts and methods relevant to the history of science, which can enhance our understanding of the scientific enterprise and its growth.

In the study of vision the interplay between questions of mediation and scientific inquiry has proven less problematic than in the case of sound. In the process of developing this volume, we have taken some cues from the work of science-studies scholars on the role of vision and visualization in the practice of science. Diagrams, sketches, equations, formulas, tables, graphs, photographs, and related vision-based analytical practices have been the focus of the work by science and technology studies (STS) scholars and historians of science and technology for the past quarter century. Many informative studies have shown how scientific instruments, in the form of inscription devices, render the invisible visible.[3] Michael Lynch has spoken of the age of the "externalized retina" to denote the importance of the visual culture to the study of science.[4] While our libraries are filled with the important studies relevant to the visual elucidation of science, the importance of listening and hearing to the scientific enterprise has, until recently, been ignored.[5]

This volume attempts to redress the imbalance. We do not wish to ignore the visual, as sound is often defined in comparison to it. Rather, we claim that studies focusing on sound and listening stress other critical aspects of experimental science. Music provided knowledge, techniques, instruments, and data necessary for research in physiology, thermodynamics, psychology, phonetics, communication technology, and many other disciplines that seem remote from the art of sound. At the same time laboratory noises and sounds yielded important information. To give one example: pitches generated by labial and reed pipes in nineteenth-century laboratories con-

---

[2] A notable exception was the Writing Science series edited by Timothy Lenoir and Hans Ulrich Gumbrecht at Stanford University Press. The journal *Grey Room* published a valuable introduction into "new German media studies" (vol. 29 [2007], ed. Eva Horn), with contributions by scholars who, after Friedrich Kittler, have given a new foundation to the field of media studies in Germany (see his *Discourse Networks, 1800/1900*, with a foreword by David E. Wellbery [Stanford, Calif., 1990] and *Gramophone, Film, Typewriter* [Stanford, Calif., 1999]) and have also taken into account the history of science. The journal recently took up this thread, also giving room to historians of science, in the volume *Audio/Visual* (43 [2011], ed. Mara Mills and John Tresch).

[3] The classic works are Bruno Latour and Steve Woolgar, *Laboratory Life* (London, 1979), and Latour, "Visualization and Cognition: Thinking with Our Eyes and Hands," *Knowledge in Society* 6 (1986): 1–40.

[4] Lynch, "Externalized Retina: Selection and Mathematization in the Visual Documentation of Objects in the Life Sciences," *Human Studies* 11 (1988): 201–34.

[5] This point has been made by a variety of scholars. See, e.g., Trevor Pinch and Karin Bijsterveld, "Should One Applaud? Breaches and Boundaries in the Reception of New Technology in Music," *Technology and Culture* 22 (2003): 536–59; Pinch and Frank Trocco, *Analog Days: The Invention and Impact of the Moog Synthesizer* (Cambridge, Mass., 2002); Emily Thompson, *The Soundscape of Modernity: Architectural Acoustics and the Culture of Listening in America, 1900–1933* (Cambridge, Mass., 2002); Jonathan Sterne, *The Audible Past: Cultural Origins of Sound Production* (Raleigh, N.C., 2003).

veyed the necessary data for the calculation of the speed of sound through air and other gases.[6] Though originating in organ building, pipes thus took part in the experimental confirmation of Laplace's constant for the ratio of a sound wave's density to its pressure and the study of adiabatic phenomena.[7] Another example: in the biological laboratories of the 1980s, listening skills were used to track radioactive probes.[8] As these examples show, scientific instruments not only generated transcriptions, they created aural information as well. Both the skill of hearing and the knowledge necessary for sound production were involved. Our volume tackles this reciprocity.

We believe, perhaps unsurprisingly, that this reciprocity is historically bound. The relation between music and science changed profoundly during the two centuries this volume seeks to investigate. We argue that this change cannot be ascribed to the history of music or the history of science alone, but points to the entwined histories of Western science and music that musicologist Emily Dolan and historian of science John Tresch envisage in their shared contribution to this volume. The interplay of research, technology, and music created new ways of listening. These new ways of listening in turn conferred status upon the sense of hearing in science. At the same time as audible data began to provide evidence, music led to detecting and listening to sounds created in the laboratory. This stimulated a circulation of material objects between science and music that forms an important theme of our volume. A brief gloss on this narrative follows.

By the beginning of the nineteenth century, both music and science had undergone radical transformations, which in turn colored their interactions. Music was no longer the main source of objects for acoustical experimentation. Natural scientists had turned to techniques of visualization and recording; the beautiful vibrating plates of Ernst Florens Friedrich Chladni are perhaps the best-known example. Sensory physiology, as well as the psychology of music and listening, emerged as empirical sciences that firmly integrated experimentation into their methodology. Seeking controlled phenomena for their investigations, these fields relied more and more on electronics for both the production and analysis of the investigated phenomena. A boom of acoustic technologies for recording, transmitting, and transforming sounds accompanied this development.[9] The upheavals in Western musical culture at the threshold of the twentieth century were equally profound. Music historians of the period noted the collapse of the tonal system, challenged both by compositions generated within the Western musical culture and exposure to the new sounds and harmonies of remote places.[10] Those in the emerging field of ethnomusicology asserted that the Western tonal system was culturally defined rather than given by

[6] E. F. F. Chladni, "Bemerkungen über die Töne einer Pfeife in verschiedener Gasarten," *Magazin für den neuesten Zustand der Naturkunde mit Rücksicht auf die dazu gehörigen Hülfswissenschaften* 1 (1798): 65–79.

[7] Myles W. Jackson, *Harmonious Triads: Physicists, Musicians, and Instrument Makers in Nineteenth-Century Germany* (Cambridge, Mass., 2006).

[8] This was a daily ritual for one of us (Jackson) as a graduate student in molecular biology laboring away in DNA-sequencing laboratories. On this, see Cyrus C. M. Mody, "The Sounds of Science: Listening to Laboratory Practice," *Science, Technology, and Human Values* 30, no. 2 (2005): 175–98.

[9] See, e.g., Thom Holmes, *Electronic and Experimental Music: Technology, Music, and Culture*, 3rd ed. (New York, 2008); Mark Katz, *Capturing Sounds: How Technology Has Changed Music* (Berkeley, Calif., 2004); Pinch and Trocco, *Analog Days* (cit. n. 5).

[10] See, e.g., George Perle, *Serial Composition and Atonality: An Introduction to the Music of Schoenberg, Berg, and Webern*, 6th ed. (Berkeley, Calif., 1991); Jim Samson, *Music in Transition: A Study in Atonal Expansion and Atonality, 1900–1920* (Oxford, 1995); Charles Rosen, *Arnold Schoenberg*

nature, as had been claimed up until the eighteenth century.[11] But what changed musical culture most profoundly was electricity, culminating with the current aesthetics of editing sounds.

The role of media in this development is critical. The demand for tools to address sounds was just as strong in science as in musical culture. While scientists began to create their own instruments to generate controlled phenomena, the musical culture of the nineteenth century saw an unparalleled role for the piano not only as a concert instrument for virtuosi, but also as the preferred tool to teach music and to reproduce music at home; it became the workbench of composers and the drawing board of theoreticians. As a consequence, the importance of the piano experienced an immense upswing. Although piano builders maintained the mystery of the good sound, they turned its production into a full-fledged industry. Individual craftsmen gave way to factories, which strove to keep up with ever-increasing demand through mass production of these instruments.

The piano contributed to a redistribution of knowledge that affected its users, the many bourgeois households among them most of all. This process, however, was not merely a social phenomenon, as it was deeply linked to the very production of sound itself and the way in which listening was affected. The extreme popularity of the piano was tied to the fact that this instrument's tone production requires no control by the ear. Learning music soon meant learning to press down a piano's keys and listen to the resulting sounds. Once a key was struck, the sound could no longer be changed. Players became listeners, who could embody musical notation in sound. The piano in turn shaped the players' listening abilities. Players trained their ears to get used to the piano's specific equal-tempered tuning system. Reconfiguring sound production and listening habits in these ways, the piano turned into a medium. By 1900, in addition to being a site not only of artistic but of industrial, social, and scientific activity, the piano was firmly established as the medium through which to convey knowledge of music.

Music did not remain removed from the technological developments, passively waiting to be changed by them. This is best illustrated in the French-born composer Edgard Varèse's appeal to his fellow musicians to integrate electrically produced sounds into music. Beginning in the 1920s, Varèse was particularly fascinated with the science of sound and timbre and relentlessly sought out new musical instruments.[12] He turned to physicists and engineers, among them such prominent researchers as Harvey Fletcher, the acoustical research director of Bell Telephone Laboratories, to produce new sounds, and in 1922 he explained that "the composer and the electrician will have to labor together to get" what Varèse would later call "a new musical expression."[13] In his compositions, we clearly hear sounds that were unthinkable in the centuries before. The siren in his "Amériques," the ensemble consisting of percussion instruments developed exclusively for his piece "Ionisation," and, eventu-

---

(1975; repr. with a new preface, Chicago, 1996); Oliver Neighbour, *The New Grove Second Viennese School: Schoenberg, Webern, Berg,* New Grove Series (New York, 1983).

[11] See, e.g., Bruno Nettl and Philip V. Bohlman, eds., *Comparative Musicology and Anthropology of Music* (Chicago, 1991); Helen Myers, ed., *Ethnomusicology: Historical and Regional Studies,* Norton/Grove Handbooks in Music (New York, 1993).

[12] Fernand Ouellette, *Edgard Varèse,* trans. Derek Coltman (New York, 1968), 103.

[13] Quoted in Chou Wen-Chung, "Varèse: A Sketch of the Man and His Music," *Musical Quarterly* 52 (1966): 151–70, on 165. Originally published in the *Christian Science Monitor* (1922).

ally, the tape-recorded electronic sounds of "Déserts" had entered the realm of music. In music, noise thus took the place of tone—that is, of periodic sound waves.

Correspondingly, in technology, music was employed in the analysis of sound. For example, the harmonic series, which through the eighteenth and nineteenth centuries formed the basis of the Western tonal system, contributed to the analysis, compression, and synthesis of speech sounds. Engineers' incorporation of the harmonic series as a model made their references to music explicit. Such was the case, for instance, with the harmonic telegraph for telephone engineers in the early twentieth century. The harmonic telegraph broke the voice into pieces in order to analyze, compress, and synthesize speech.

In examining the reciprocity between musicians and scientists, this volume aims to demonstrate that a history of music and science can introduce new perspectives on the evolution of scientific methodology during the nineteenth and twentieth centuries. It will substantiate this claim with regard to three interdependent subdivisions of the contributed articles. The first section of our volume will investigate techniques of evidence. Of course, when it comes to providing an object of reference for science, hearing and listening never had the same status as seeing. Sounds fade, and research in sound and hearing needed some form of mediation to demonstrate their erstwhile existence. Yet in many historical constellations, sounds helped to demonstrate the existence of even more remote or fugitive objects. Indeed, the temporal nature of sound could be an asset or an object of study as well. Scientists employed—and still employ—all of their senses to comprehend nature.

The second section reconstructs what we call a phenomenotechnique of sound.[14] While music provided objects of investigation for experimental natural philosophers up to the eighteenth century, acoustics in the nineteenth century created its own sounds, which then conformed to physico-mathematical definitions and the conditions of psychophysiological experimentation rather than following the rules of music. The sounds of the siren, the tuning fork, and the resonator, and especially the electronically generated sinusoidal sound, revolutionized the investigation of hearing. In the twentieth century these new sounds entered music and changed the very definitions of music and sound.[15]

The third section is devoted to the circulation of sound objects in and beyond the scientific community. During the nineteenth century sounding objects were simultaneously stabilized and mobilized. Standardization enabled the flow of sound to permeate the laboratory as well as to enter broad arenas of society. Circulation reshaped sounds and their function, as the circulating objects were exposed to different regimes of use. Thus, both standardization and the circulation it facilitated did not necessarily progress in an uncontested way. Though the worlds of music and science could share an interest in standardizing objects, bodily movements, and even sounds, when the circulating objects passed through the hands and minds of musicians and scientists, artisans and technicians, they could also motivate the participants of this exchange to distinguish themselves from the alleged other side. For example, the implementation of equal temperament, while embraced by many members of the musical and

[14] On this notion, coined by Gaston Bachelard, see Hans-Jörg Rheinberger, *An Epistemology of the Concrete: Twentieth-Century Histories of Life* (Durham, N.C., 2010).
[15] Olivia Mattis, "Varèse's Multimedia Conception of *Déserts*," *Musical Quarterly* 76 (1992): 557–83; Felix Meyer and Heidy Zimmermann, eds., *Edgard Varèse: Komponist, Klangforscher, Visionär* (Mainz, 2006).

scientific communities, had detractors among music critics and scientists alike. As a result, piano builders, for instance, had to negotiate the competing goals of standardization and the flexibility necessary to accommodate musicians' desires to create their own identity on a single instrument.

This three-part division of our volume has the advantage of both highlighting themes that have been previously examined by historians of science and introducing new ones. For example, the history of standardization has been of interest to historians of science and STS scholars for the past twenty years.[16] The standardization of music and sound, although sharing the inextricable links among science, technology, economics, and politics highlighted in these earlier accounts, was rather unique. Aesthetic issues were at stake.

This volume also sheds light on the increasing specialization and separation of scientific disciplines in the nineteenth and twentieth centuries. Indeed, the isolation and alienation felt by the early nineteenth-century British scholar Thomas Young, as detailed by Peter Pesic's contribution, and by the twentieth-century American physicist Sheridan Speeth, as described by Axel Volmar, hint at the frustration generated by the lack of appreciation for interdisciplinary approaches. Interestingly, such nineteenth- and twentieth-century German-speaking physicists as Hermann von Helmholtz, Max Planck, and Ernst Mach rarely experienced or articulated this frustration as they moved easily between the realms of music and science.

This ease of movement begins to explain the plethora of articles on nineteenth-century German science and music. Many of these scientists possessed an impressive musical training as well as extensive contact with individuals populating the music world—musicians, composers, music critics, musical-instrument makers, and musicologists. That was part and parcel of nineteenth-century *deutsche Kultur*. In addition, the German-speaking countries were, of course, home to some of the world's leading composers and musicians. Given the importance of music to the persona of the *Bildungsbürgertum*, perhaps it is not all that surprising that German and Austrian savants could pass in and out of that sphere rather seamlessly. In short, much of the elaboration of the scientific properties of music and musical-instrument manufacturing occurred during the second half of the nineteenth century, precisely the time when the German-speaking lands were enjoying a privileged status in the physical sciences, similar to what France had enjoyed earlier in the century. This is not, of course, to say that this mixture of science and music is merely a German-speaking (and, later, an American) phenomenon. We hope that the preponderance of work focused on the German-speaking lands in this volume will inspire other scholars specializing in different geographical regions to apply questions similar to those ad-

[16] See, e.g., Tore Frängsmyr, John L. Heilbron, and Robin E. Rider, eds., *The Quantifying Spirit in the Eighteenth Century* (Berkeley, Calif., 1990); Simon Schaffer, "Late Victorian Metrology and Its Instrumentation: A Manufactory of Ohms," in *Invisible Connections: Instruments, Institutions, and Science*, ed. Robert Bud and Susan E. Cozzens (Bellingham, Wash., 1992), 35–56; Ken Alder, "A Revolution in Measure: The Political Economy of the Metric System in France," 39–71, Schaffer, "Accurate Measurement Is an English Science," 135–72, and Kathryn M. Olesko, "The Meaning of Precision: The Exact Sensibility in Early Nineteenth-Century Germany," 103–34, in *The Values of Precision*, ed. M. Norton Wise (Princeton, N.J., 1995); Olesko, "Precision, Tolerance, and Consensus: Local Cultures in German and British Resistance Standards," *Archimedes: New Studies in the History and Philosophy of Science and Technology* 1 (1996): 117–56; Graeme J. M. Gooday, *The Morals of Measurement: Accuracy, Irony, and Trust in Late Victorian Electrical Practice* (Cambridge, 2004); Peter Galison, *Einstein's Clocks and Poincaré's Maps* (London, 2003), 84–155.

dressed here to their regions. More scholarship can only further illuminate the extent and nuance of the reciprocal relationship between music and science.

The volume also opens up new social spaces for the investigation of sound: the cafés of nineteenth-century Vienna; nineteenth-century artisanal workshops; twentieth-century acoustical laboratories investigating the interfaces of computer science and music, as discussed in Cyrus Mody and Andrew Nelson's piece, and even of computer science and earthquakes, as detailed in Volmar's contribution; and of course concert halls. In a sense, this volume historicizes the work recently tackled in social studies of science, and we would like to think of this collection of articles as a complementary work.[17] Much is to be gained through the symmetry of historically informed social studies of science and a history of science that is devoted to understanding the culture of science in the very broadest sense possible. We hope this volume motivates further efforts of such mutual illumination.

Our volume discusses sound in general, although there is a bias toward music. Epistemologically, it is extremely difficult to differentiate between sound and music. Indeed, their definitions are historically contingent. Perhaps Pierre Schaeffer, one of the originators of musique concrète in the 1950s, summed it up best: "For all that, traditional music is not denied any more than the theater is supplanted by the cinema. Something new has been added, a new art of sound. Am I wrong in still calling it music?"[18] Rather than attempt to address the differences between sound and music ourselves, we have allowed our authors to draw upon their actors' categories in order to differentiate between the two. In so doing, it becomes clear that experimental agendas in these categories opened up new ways of understanding sound, music, and nature itself.

The first group of articles describes the techniques of evidence of engineers' and scientists' work on sound. Peter Pesic's article, "Thomas Young's Musical Optics: Translating Sound into Light," details the acoustical work of the British physician and savant Thomas Young, who is best known for his optical research, and establishes the foundation for this section. Pesic shows that Young's commitment to the unity of nature led him to argue for a wave theory of light propagation based on his acoustical research. Sound traveled in waves; so too must light. Such unity of natural phenomena in Young's eyes was not of a reductionist, top-down sort. Rather, it was a lateral kind of unity, much in line with romantic depictions of nature.

Roland Wittje's contribution, "The Electrical Imagination: Sound Analogies, Equivalent Circuits, and the Rise of Electroacoustics, 1863–1939," traces the transformation of acoustics into electroacoustics in the early twentieth century. The introduction of electroacoustic technologies into the laboratory fueled the interest of electrical engineers in the field of acoustics, which had hitherto been understood as concerning fundamentally mechanical phenomena. The resulting electrification of the acoustics laboratories brought about new representations of acoustic phenomena, which, in turn, also made it easy to design real circuits for the electrical generation and manipulation of sound. Wittje demonstrates how electroacoustics became much more than a research technology and evolved from a laboratory practice into a new way of thinking about acoustics.

[17] Trevor Pinch and Karin Bijsterveld, eds., *The Oxford Handbook of Sound Studies* (Oxford, 2012).
[18] Quoted in Herbert Russcol, *The Liberation of Sound: An Introduction to Electronic Music* (Englewood Cliffs, N.J., 1972), 83, and Holmes, *Electronic and Experimental Music* (cit. n. 9), 58.

We can understand the major shift underlying this change as rooted in hearing. Sound could be conceived in terms of energy, as Helmholtz first demonstrated—a conception in which sound quality was detached from a material sound source. The listener could then determine the form and features of the sound object. This path to sonification, where auditory data were considered a legitimate form of evidence, required a series of steps to acquire some reliability in the ears of the beholder. While "evidence" literally points to vision as its major source, listeners first had to trust their ears and, in order to do so, train their hearing (a process that suggests right and wrong or scientific and unscientific ways of hearing). Daniel Gethmann's article, "The Aesthetics of the Signal: Noise Research in Long-Wave Radio Communications," explores such conditions of evidence in his examination of two experiments on radio reception where the source of the received data was unknown. The first experiment laid the foundation for radio astronomy, with a view to assisting the improvement of radio communication. The second experiment can be seen as leading to a new source of media, as it went beyond the initial euphoria of detecting cosmic signals and generated a new understanding of the relation between noise and communication. Karl Guthe Jansky's categorization of "star noise" in his search for the causes of strong and regular interferences in long-distance radio telephony introduced noise as a definitive factor in communication, one that was a constant feature rather than a temporary distortion. Gethmann's historical point: once it was standardized and the human psychological component was removed, noise itself became a form of communication.

Axel Volmar's article, "Listening to the Cold War: The Nuclear Test Ban Negotiations, Seismology, and Psychoacoustics, 1958–1963," similarly examines the sonification of evidence resulting from collaborations among seismologists, radio engineers, and musicians. Experimenters transposed seismograms into the audible range for interpretation by cellists, who were chosen for their training and trustworthiness in judging auditory data. Volmar highlights this rendering of evidence—that was never meant to be heard at all—into audible rather than visible form as part of a more general examination of the use of sound and listening as tools of scientific study.

The second section, organized around the introduction of new phenomena of sound and techniques of sound generation and measurement, presents a series of episodes in which physicists and physiologists reexamined musical topics, such as tonality, tonal systems, consonance and dissonance, and time. While these topics had been of central interest to acousticians up to the eighteenth century, nineteenth-century scientists realized that they were only using the concepts for ease of description. To those writing about music, it became clear that any sound that was newly created in the laboratory eluded straightforward description. New sounds required new characterizations. Acousticians developed a new interest in music as the field gradually moved toward an interest in human hearing. Music regained its role of providing objects to which physiologists and psychologists could easily refer. Their new approaches to the study or use of sound in the laboratory informed new understandings of hearing.

The first article in this section proffers approaches from the domain of media theory. Bernhard Siegert's contribution deals with long-term processes in which the creation of the investigated phenomena is constantly adjusted to the state of the art in science, and in this way contributes to both cultural history and the history of science. The example of the bell in Siegert's article, "Mineral Sound or Missing Fundamental: Cultural History as Signal Analysis," links the history of science and sound with semiotics. The sounding of bells remained a mystery to acousticians until the 1980s, especially the so-called strike note of a bell, which defied measurement. It was only during the second half of the twentieth century that acousticians began to realize that human hearing itself creates the strike note from the bell's partials; musical object and listener are inextricably intertwined. Jan F. Schouten eventually offered a new terminology, which described the strike note as the sound's "residuum." What the listener perceived was from that moment on understood as "virtual pitch," undetectable by measuring devices. Indeed, it was only in connection with such devices that this conclusion could be reached. Juxtaposing the history of sound analysis with the cultural semiotics of the bell, the article is also intended to demonstrate one way in which media studies resonates (pun intended) with historians of science interested in acoustics and sound. We hope that it inspires further communication between fields.

Alexandra Hui's article, "Changeable Ears: Ernst Mach's and Max Planck's Studies of Accommodation in Hearing," traces the growing sensibility of the physicist toward the psychophysiological implications of such a reevaluation of tonal systems. Both Mach and Planck, leading turn-of-the-century physicists, devoted significant effort to understanding the phenomenon of accommodation in hearing, in which an individual could alter his or her experience of sound by changing the focus of his or her attention. Both believed that accommodation in hearing was the arbiter of meaning in music and was bound to the origins and development of musical systems. Accommodation in hearing was also beyond direct physical measurement, located at a psychophysiological junction seemingly only accessible through the subjective experience of generating or listening to music. Yet, as Hui argues, this phenomenon directly informed both physicists' ideas about the historicity of sound sensation, the universality and supremacy of the nineteenth-century Western musical aesthetic, and, potentially, the nature of knowledge itself.

In "The Audiovisual Field in Bruce Nauman's Videos," Armin Schäfer explores the limits of possibility of the subjective experience of generating music. In the 1960s, the artist Bruce Nauman explored the unconsciously acquired gestures we adopt in our daily lives. In his studio, he exposed himself to various common situations, which he then defamiliarized in numerous ways. A number of these explorations dealt with the playing of musical instruments. The practices of instrumental playing involved scientific knowledge about both the body and the economy of work, as investigated by fin-de-siècle physiologists, psychiatrists, and applied psychologists. This research created a notion of "fatigue" that had already long been understood by pedagogues of instrumental playing. Students, they warned, must repose between phases of exercise in order to avoid exhaustion. Nauman, however, explored the limits of fatigue in the video *Violin Tuned D.E.A.D.* by continuously playing a violin for an hour, intentionally pushing himself to exhaustion. In this aesthetic experience, music took part as a bodily technique that had been informed by pedagogy, the life sciences, and medicine. In drawing this connection between bodily discipline and scientific practice and highlighting the critical role of music, Schäfer extends a lasting interest among historians of science into the realms of music and media theory.

Henning Schmidgen's article, "*Camera Silenta*: Time Experiments, Media Networks, and the Experience of Organlessness," offers this section a paradigmatic case: in the nineteenth century, silence was thought to be the condition of sound. As detailed in Schmidgen's study of the material culture of the *camera silenta* and similar structures, the initial motive behind the construction and scientific use of soundproof

rooms was to avoid the disturbance of test subjects involved in reaction-time experiments and the resulting measurement errors. The physiologists and psychologists
eventually realized that the greatest acoustic distraction for the experimental subjects
came from the subjects themselves. John Cage would later turn this into an insight
about the conditions of music making: there is no such thing as silence. Silence is but
a condition of change. In dealing with the material and acoustical boundaries of the
laboratory, Schmidgen's article serves as a segue to the final collection of contributions, which investigate the transgression of these boundaries.

The third section of the volume focuses on the circulation of sounding objects both
within and beyond the boundaries of the scientific community and the consequences
of these peregrinations for shared material practices and attitudes as well as a collective aesthetic forged by natural philosophers, musicians, composers, and instrument makers. Julia Kursell's contribution, "Experiments on Tone Color in Music and
Acoustics: Helmholtz, Schoenberg, and *Klangfarbenmelodie*," traces the relationship between Arnold Schoenberg's music and Helmholtz's physiology. In his seminal
work of 1863, *On the Sensation of Tone as a Physiological Basis for the Theory of
Music*, Helmholtz proffered a new definition for the term *Klangfarbe* that was based
on his experiments on the physiology of hearing. According to previous definitions,
Klangfarbe only referred to those aspects of a tone that allowed of formal description. In this article, Kursell argues that Helmholtz's definition of *Klangfarbe* as the
difference in the calculable properties of sound is unique and critical to twentieth-
century music, encouraging both artists and scientists to come to terms with those
qualities of sound that had formerly escaped both musical theorizing and practice.
In addition, the experiment in sound synthesis he made to demonstrate his new definition became a model for composing with sounds, in two respects. On the one hand,
it embodied theory in sound. On the other hand, it created a new sound, the audible
sinusoidal wave, which embodied sound's smallest element. Kursell traces this development of sound synthesis to Schoenberg's 1909 composition *Farben*, discussing
it as an example of how orchestral composition transformed experimental synthesis
back into an orchestral synthesis of sound. Both synthetic sound and orchestral synthesis became standard tools in the composition of orchestral music in the second
half of the twentieth century.

Two contributions concerning the paradigmatic instrument of the nineteenth century are next in this section. Elfrieda Hiebert and Sonja Petersen investigate the piano
from related but different angles. Here, plain scientific knowledge is the object of circulation; in both articles, education and pedagogy is the main field where this object
is circulated. The two authors, however, bring in quite different modes of circulation,
both very important for this volume. In "Craftsmen-Turned-Scientists? The Circulation of Explicit and Working Knowledge in Musical-Instrument Making, 1880–
1960," Petersen demonstrates how the tools and sensory-aesthetic capabilities of
piano makers, which were also important for the production of high-quality musical
instruments, circulated through their writings and drawings. She explores the tension
between practical, hands-on knowledge and theoretical knowledge relevant for piano
manufacture. This differentiation tackles the question of how aesthetic criteria pro-
nounced by instrument makers can mediate between these two types of knowledge.

Hiebert's contribution, "Listening to the Piano Pedal: Acoustics and Pedagogy in
Late Nineteenth-Century Contexts," brings the issue of pedagogy back to the initial
question of the volume; namely, How does hearing shape the relation between music

and science anew? She focuses on instrument makers and analyzes the vital role the pedal has played in changing the sound of the evolving piano, shaping a new "acoustical" model of piano sound. To attain this sound three things were necessary: a new technique of pedaling, insight into contemporary acoustics, but also and most significantly, a deliberate and more conscious effort by listeners to control the ear than had ever been demanded in previous piano pedagogy.

In their article, "'A Towering Virtue of Necessity': Interdisciplinarity and the Rise of Computer Music at Vietnam-Era Stanford," Cyrus Mody and Andrew Nelson discuss a different power of music. Scholars at Stanford's Center for Computer Research in Music and Acoustics responded to the call for academic scientists and engineers to reconvert from military and space applications to research on "human problems" through efforts to develop the intersections between music and engineering, despite ongoing struggles of legitimacy. Though every bit of electroacoustic engineering had once been entangled with warfare, its reconversion into a peaceful application can claim to be a success. While in the other articles of this section, heard evidence is at stake, this article deals with iconicity, turning it into an auditory feature.

John Tresch and Emily Dolan's article, "Toward a New Organology: Instruments of Music and Science," concludes the volume by offering a comparative overview of the discourses and practices of musical and scientific instrumentality. Examining closely related examples taken from the science and music of the Renaissance through the twentieth century, they focus on the relations perceived among instruments, the senses, aesthetics, and nature. These case studies reveal the entwined histories of Western science and music: their shared material practices, aesthetics, and attitudes toward technology, as well as their parallel and complementary views of natural order and human agency.

* * *

To sum up, this volume covers questions on the relations among music, technology, and science and the permeable boundaries of the laboratory, sense perception, its investigation, and the reintegration of knowledge of sensory capacities into laboratory practice. Music is unique in combining aspects of science and learning, handicraft and technology, theory and calculus, material and immaterial objects, communication, art, and ambience. We hope that this volume will provide a collection of exemplary studies to open the field to more general history of science questions for which the study of music and science provides a fertile ground.

TECHNIQUES OF EVIDENCE

Thomas Young's Musical Optics:
Translating Sound into Light

by Peter Pesic*

ABSTRACT

Thomas Young's interest in music affected his scientific work throughout his career. His 1800–1803 papers interrelate musical, acoustic, and optical topics to translate the wave theory from sound to light, as does his synoptic *Lectures on Natural Philosophy* (1807) in justifying his discoveries to a larger audience. Returning to optics in 1817, in the aftermath of the work of Fresnel, Young grounded in musico-acoustical studies his suggestion that light may be a transverse (rather than longitudinal) wave, along with its paradoxical implications for the ether, which he discussed in 1823. Young's decipherment of Egyptian hieroglyphs also rested on phonology.

Thomas Young (1773–1826) made crucial interventions in the development and application of wave theory to light. Throughout his career, he used studies of music and sound to advance the theory of wave motion, especially the concept of interference, which he worked out in sound and then applied to light. Sir John Herschel singled out Young's insight into sound interference as "the key to all the more abstruse and puzzling properties of light, which would alone have sufficed to place its author in the highest ranks of scientific immortality, even were his other almost innumerable claims to such a distinction disregarded."[1]

Young's accomplishment should be placed in the context of Newton's skepticism about the wave theory of light, which shadowed the century after his *Opticks* (1704).[2] In general, Continental writers adopted wave theories, following Descartes's picture of an all-pervasive fluid continuum, whose vortices moved the planets and whose vibrations were visible as light. In contrast, British scholars preferred Newton's particle theory. They followed him in considering Francesco Grimaldi's experiments as not fully sufficient to prove the wave theory, hence leaving the particle theory as the

* St. John's College, 1160 Camino de la Cruz Blanca, Santa Fe, NM 87505; ppesic@sjcsf.edu.

I thank the John Simon Guggenheim Memorial Foundation for its support. I am grateful to Alexandra Hui, Myles Jackson, and Julia Kursell for inviting me to contribute to this issue and thank my fellow contributors for stimulating discussions, comments, and advice. I also thank Jed Buchwald for his very helpful comments. I would like to dedicate this essay to Gerald Holton, mentor and friend, who first directed my attention to the signal importance of Young's investigations of sound as well as light.

[1] Quoted in George Peacock, *Life of Thomas Young, M.D., F.R.S.* (London, 1855), 128.
[2] Henry Steffens, *The Development of Newtonian Optics in England* (New York, 1977); Geoffrey N. Cantor, *Optics after Newton: Theories of Light in Britain and Ireland, 1704–1840* (Manchester, 1983); Jed Z. Buchwald, *The Rise of the Wave Theory of Light: Optical Theory and Experiment in the Early Nineteenth Century* (Chicago, 1989).

© 2013 by The History of Science Society. All rights reserved. 0369-7827/11/2013-0002$10.00

"simpler," more obviously "mechanical" explanation for light, absent more powerful evidence to the contrary.

The crucial arguments did not emerge until around 1800 in the work of an amazingly multitalented individual who, though unique in his constellation of abilities, manifests the fruitful breadth of scope so important in the advances made by other contemporary natural philosophers. Among the rich diversity of his serious interests, music occupied a special place for Young, helping him formulate and advance his advocacy of the wave theory as he translated it into the realm of light. I will show how musical concerns emerged in each phase of his work, first in the overt context of vibrating bodies and pipes, then applied by analogy to light. To illuminate how his thinking carried forward these acoustical archetypes into progressively more elaborate optical formulations, I shall follow the intertwined development of these musical and optical themes chronologically through Young's writings, with close attention to their timing and arrangement. The specific order and placement he gave these issues reflects their interplay and influence as he took these analogies ever deeper, for music functioned both as an "external" and an "internal" force in the detailed development of Young's work on acoustics and optics. I will argue that, point by point, his optical innovations were prepared by his prior studies in sound and music, which began as external artistic studies that soon became embedded in the internal development of Young's natural philosophy.

Accordingly, my treatment will be guided by the detailed sequence of his arguments. After discussing Young's musical background, I will give a close reading of his papers during 1800–1803, which illuminate the role of musical concerns in his unfolding optical discoveries. Then I will discuss Young's retrospective account of these same developments as he justified them to a broader public in his lectures at the Royal Institution. A decade later, Young's sonic analogy affected his final thoughts on light as a transverse wave. Throughout, his interest in translation and linguistics bore on his rendition of music and sound into optics and light. Conversely, his concern for sound also affected his pioneering work in deciphering Egyptian hieroglyphics.

YOUNG'S MUSICAL BACKGROUND

Young's intellectual development should be considered against his background as the eldest of ten children in a pious Society of Friends (Quaker) family. His uncle was an eminent physician and member of the Royal Society. Young showed a prodigious early talent for languages. Though basically self-taught (his family could not afford the elite public schools and their specialized tutoring), by age nineteen he was fluent in Latin and Greek, had a good command of the principal European living languages, could read biblical Hebrew, and had even studied Chaldean, Syriac, and Arabic.[3] For instance, his youthful rendition of a speech from Shakespeare into

---

[3] See Peacock, *Life of Thomas Young* (cit. n. 1, on 12), which remains a valuable source from a near-contemporary (though Peacock did not know Young personally), along with Hudson Gurney, *Memoir of the Life of Thomas Young* (London, 1831). Both are ultimately reliant on Young's own autobiographical notes; see Victor L. Hilts, "Thomas Young's 'Autobiographical Sketch,'" *Proceedings of the American Philosophical Society* 122 (1978): 248–60. The most complete modern biography is Alexander Wood and Frank Oldham, *Thomas Young, Natural Philosopher, 1773–1829* (Cambridge, 1954), which notes that Young's father and grandmother "were not merely nominal Quakers, but active members of the Society" and adduces "a certain affinity between the Quaker pursuit of truth, with its emphasis on verification in personal experience, and the scientific method" (3). More re-

classical Greek (fig. 1) exemplifies his lifelong interest in the problem of translation, a recurrent theme in his later work, along with general issues of phonology and comparative linguistics.

During those years, Young also taught himself mathematics and developed an interest in science. He read Newton's *Principia* by himself, a notable feat, for this formidably difficult book (the bible of contemporary science) was usually treated at the university level as a work of extraordinary difficulty, only accessible through specialized commentary and tutoring. Nor were his studies purely theoretical; he ground colors, studied drawing, and constructed scientific instruments. After leaving one of the local schools, he devoted himself "almost entirely to the study of Hebrew and to the practice of turning and telescope-making."[4] Yet despite his amazing breadth and depth of learning, Young was quite unaware of popular literature; his Quaker upbringing removed him from the ordinary activities of his contemporaries.

Whatever may have been his personal preferences, his family's finances dictated that he take up a career in medicine, following his uncle's lead. This he did without complaint, seemingly considering it a continuation of his interests in physics and mathematics, now extended to a physiological sphere. Following the practices of the time, Young first served an apprenticeship in London as a pupil in St. Bartholomew's Hospital and showed extraordinary abilities in anatomy. In 1793, at age twenty, he made a major discovery about the function of the lens in accommodation, the process through which the eye adjusts its focus from near to distant objects.[5] In studying the eye of an ox, Young thought he found evidence of fibers inside the lens that could plausibly act as focusing muscles, which earlier anatomists had conjectured but not seen definitively. Through the good offices of his uncle, Young read a paper on his discovery to the Royal Society, which led to his being elected a fellow at age twenty-one, though this accolade was overshadowed by controversy. Young's discovery was claimed by an eminent anatomist, John Hunter, as his own, while another anatomist asserted that he could find no such muscular structures in the lens. At that point, Young withdrew his discovery, in deference to this authority, though he later reasserted it in light of further research.

Young's medical apprenticeship led him next to Edinburgh, where many Quakers chose to study, excluded from Oxford and Cambridge on account of their faith.[6] Still, at Edinburgh Young began to play the flute and to take dancing lessons, which disobeyed Quaker precepts, as did his incipient experiments in theatergoing.[7] Not

---

cently, see Andrew Robinson, *The Last Man Who Knew Everything: Thomas Young, the Anonymous Polymath Who Proved Newton Wrong, Explained How We See, Cured the Sick, and Deciphered the Rosetta Stone, among Other Feats of Genius* (New York, 2006). Regarding the Quaker background, see Elizabeth Allo Isichei, *Victorian Quakers* (London, 1970); Geoffrey Cantor, "Real Disabilities? Quaker Schools as 'Nurseries' of Science," in *Science and Dissent in England, 1688–1945*, ed. Paul Wood (Aldershot, 2004), 147–66; Cantor, *Quakers, Jews, and Science: Religious Responses to Modernity and the Sciences in Britain, 1650–1900* (Oxford, 2005), 64, 82–3, 111; Genevieve Mathieson, "Thomas Young, Quaker Scientist," (MA thesis, Case Western Reserve Univ., 2007), available at http://etd.ohiolink.edu/view.cgi?acc_num=case1196288181 (accessed 7 Nov. 2012).

[4] Peacock, *Life of Thomas Young* (cit. n. 1), 7.

[5] Ibid., 35–41; Robinson, *Last Man Who Knew Everything* (cit. n. 3), 36–40.

[6] Cantor, "Real Disabilities?" (cit. n. 3), 147–9.

[7] Regarding a Quaker doctor of the generation before Young, it was noted that "music, dancing, the theatre, the opera, wine, women and song, gambling, attendance at cock-fights, bull-baitings, race meetings, all the rough hearty joys of the Englishman of the time were incompatible with the Quaker costume he wore." Wood and Oldham, *Thomas Young* (cit. n. 3), 35.

ΟΥΛΣΙΟΥ ΜΟΝΟΛΟΓΙΑ.

χαίροις ἂν ἤδη μακρὰ πᾶσ᾽ ἐνδοξία·
χαίροιτε δυνάμεις, αἳ πισωρεύεσθέ μοι.
οὕτως ἔχει δὲ τἀνθρώπεια· σήμερον
ἀνὴρ τὰ χλωρὰ φύλλα τἀλπίδος φύει·
αὔριον ἀκμάζει, πορφυρέοις τ᾽ ἐπ᾽ ἄνθεσι
τιμῶν ὅσων περ εὗχε, πολλ᾽ ἀβρύνεται·
τριταῖον αὖτε ῥῖζος ἐμπίπλει βαρύ,
κἀπεὶ πεποιθὼς κάρτα δ᾽ ἐλπίζει τάλας
καρπὸν μεγίστων ἐκπεπαίνεσθαι καλῶν,
ῥίζῃ πρὸς αὐτῇ δύσμορος ξηραίνεται,
κἄπειτα πίπτει δειλός, ὡς ἐγὼ τὰ νῦν.
ὁποῖα παῖδες νήπιοι παράφρονες,
ἐπὶ κύςεσιν νεῖν ἐν θέρει πειρώμενοι,
ὅπως τὰ πολλὰ δ᾽ εἰ κεκινδύνευκ᾽ ἐγὼ
δόξης θαλάσσῃ, πρὸς βάθος μηδὲν σκοπῶν·
κόμπος δ᾽ ἀραιὸς ὃν ἐπεφυσήκειν ἄδαν
ἐσχισμένος λέλοιπέ μ᾽ ἐν κλυδωνίῳ,
γέροντα, μόχθῳ καὶ χρόνῳ κεκμηκότα,
κἀνταῦθα λάβροις κύμασι βυθισθήσομαι.
ὦ λαμπρότητος καὶ τρυφῆς κενὴ σκιά!
ἀπεχθὲς ὄνομα! νῦν δὲ καρδίαν ἐμὴν
αὐταρχίας τυχοῦσαν εὖ δ᾽ ἐπίσταμαι.
φεῦ δυστάλαιναν τοῦ τρισαθλίου τύχην
χάριτος τυράννων ὅστις ἐκκρεμάννυται!

Figure 1. Young's translation into classical Greek verse of a speech given by Cardinal Woolsey in Shakespeare's Henry VIII, *in Young's handwriting; reprinted from Peacock,* Life of Thomas Young *(cit. n. 1), facing 23.*

surprisingly, the experience of new places and people helped Young break away from the doctrinal limitations of his upbringing. His succeeding stay in Göttingen further broadened his horizons. His doctoral dissertation, "De Corporis Humani Viribus Conservatricibus" (1796), concerned the physiology of the human voice and included an alphabet of forty-seven letters intended to convey every sound of which the voice is capable.[8] In this work, his interests in sound directly address his ongoing linguistic and phonological concerns.

Young's disorientation in adjusting to foreign customs paradoxically intensified his pursuit of the social and artistic activities excluded from his Quaker upbringing. He began to take dancing lessons five or six times every week, as he wrote an English friend, nor was he "very punctual in some of the medical courses." George Peacock, Young's early biographer, noted that he had been precluded from the pursuit of the "personal accomplishments" that he now followed so avidly. "It was in vain that his fellow-students, whether in banter or in earnest, told him that his musical ear was not good, and that he would fail to acquire ease and grace as a dancer. A difficulty thus presented to him as insuperable was a sufficient motive to attempt to conquer it; and though different opinions have been expressed with respect to the entire success of the experiment, there is no doubt that the mastery of those arts, which he really attained, was another triumph of his unconquerable perseverance."[9]

Precisely because it was a relatively late interest that emerged in his formative years and spoke to a side of his nature that had been underdeveloped, the musical side of Young deserves special attention as his bridge to a common social life with others, whose previous absence he may have felt acutely. By the time he left Germany, he could dance the complexities of a cotillion and ride a horse with ease, passions he cultivated the rest of his life.[10] He noted in Germany "the love of new inventions singularly combined with a pedantic habit of systematizing the old," so that Young felt that "the general spirit of the country rather tended to confirm than to correct the habits of his earlier education."[11] Young remained critical of German thought; though Schiller and Goethe were "rare luminaries among an infinite number of twinkling stars and obscure nebulae," he thought Goethe's *Wilhelm Meister* "vanishes in comparison with some of our English novels."[12] In his later work, Young never used the phrase "unity of nature," so dear to German *Naturphilosophie*, and he wrote a strong critique of Goethe's color theory as "a striking example of the perversion of the human faculties."[13] Young's own efforts to bring the wave theory of sound to bear on light may be compared with his interest in translating between different languages, rather than with a prior commitment to the unity of nature.

Young's final stage of medical apprenticeship led him to matriculate at Cambridge in 1797 because the Royal College of Physicians would only admit as fellows those who had attended Oxford or Cambridge. Thus, though already a fellow of the Royal

---

[8] Ibid., 49–50.

[9] Ibid., 49.

[10] Peacock, *Life of Thomas Young* (cit. n. 1), 114. For comparison of Young's experience in Germany with that of other contemporaries, see Linde Katritzky, "Coleridge's Links with Leading Men of Science," *Notes and Records of the Royal Society* 49 (1995): 261–76.

[11] Hilts, "Thomas Young's 'Autobiographical Sketch'" (cit. n. 3), 252.

[12] Peacock, *Life of Thomas Young* (cit. n. 1), 109.

[13] Young, "Zur Farbenlehre: On the Doctrine of Colours; By Goethe," *Quarterly Review* 10 (1814): 427–8. See also Frederick Burwick, *The Damnation of Newton: Goethe's Color Theory and Romantic Perception* (Berlin, 1986), 30–3.

Society, Young had still years to wait before he completed the statutory requirements
for a full medical degree. His Cambridge matriculation required him to profess An-
glican orthodoxy, abjuring Quaker nonconformism. Though the Westminster Quaker
meeting formally disowned him in 1798, Young still struck his Cambridge classmates
as having "something of the stiffness of the Quakers"; he did not much associate with
the other young men, who called him "Phænomenon Young," indicating both their
respect and their disdain.[14] One of them recalled that "he read little, and though he
had access to the college and university libraries, he was seldom seen in them. There
were no books piled on his floor, no papers scattered on his table, and his room had all
the appearance of belonging to an idle man. I once found him blowing smoke through
long tubes [though Young never smoked tobacco], and I afterwards saw a representation of the effect in the *Transactions of the Royal Society* to illustrate one of his
papers upon sound; but he was not in the habit of making experiments."[15] We will
shortly return to this scene. Young himself noted, shortly after arriving in Cambridge,
that, starting with his Göttingen thesis on "the various sounds of all the languages
that I can gain knowledge of," he had "of late been diverging a little into the physical
and mathematical theory of sound in general. I fancy I have made some singular observations on vibrating strings, and I mean to pursue my experiments."[16]

YOUNG'S PIPES AND ORGANS

In 1797, Young's uncle died, leaving generous bequests to his friends (and patients)
Edmund Burke and Samuel Johnson, as well as to Young himself, who now was free
to follow his own interests without financial concerns.[17] The following year, after an
accident and broken bone that kept him from his usual exercise, Young devoted himself to what he called "observations of harmonics," by which he meant experimental
studies of wave motion in sound.[18] During his recovery, he also read contemporary
French and German mathematics and noted that "Britain is very much behind its
neighbours in many branches of the mathematics; were I to apply deeply to them I
would become a disciple of the French and German school; but the field is too wide
and too barren for me."[19] His choice not to engage further with Continental mathematics had lasting consequences, as we shall see.

 As he thought through the problem of sound during his work on harmonics, Young
thereby prepared himself to apply the very same physical models and mathematical
description to light. As shall emerge, his famous two-slit experiment for light was a
translation of parallel experiments for sound, rather than a direct transcription, reflecting the differences between the "languages" of light and sound. Yet Young's awareness of sonic and musical phenomena prepared the ground for his work on light, down
to the precise details of the experiment that would finally satisfy Newton's demand

---

[14] Peacock, *Life of Thomas Young* (cit. n. 1), 115–20, on 118, 120. For Young's expulsion, see Westminster Monthly Meeting Minutes, 15 Feb. 1798, Library of the Society of Friends, London (reference: 11 b 7), and Mathieson, "Thomas Young, Quaker Scientist" (cit. n. 3), 15–6.
 [15] Peacock, *Life of Thomas Young* (cit. n. 1), 121.
 [16] Wood and Oldham, *Thomas Young* (cit. n. 3), 50.
 [17] Young himself attributed "the ultimate extent of his uncle's protection" to Burke's "friendly and
indulgent" interest and his "good offices"; Hilts, "Thomas Young's 'Autobiographical Sketch'" (cit.
n. 3), 251.
 [18] Peacock, *Life of Thomas Young* (cit. n. 1), 129.
 [19] Wood and Oldham, *Thomas Young* (cit. n. 3), 65.

that light be shown positively bending around obstacles. As Olivier Darrigol has emphasized, "Young realized that the fruitful development of the analogy between sound and light required a prior improvement of acoustic knowledge."[20]

The course of Young's work in the years after his early paper on the accommodation of the eye clearly shows the interweaving of music, sound, and light. Three essays he published in the year 1800 show the remarkable overlay and simultaneity of his thinking in these domains. In January 1800, while still at Emmanuel College, Cambridge, he read to the Royal Society his "Outlines of Experiments and Inquiries Respecting Sound and Light," which in essence lays out the fundamental premise of his ensuing research and whose title emphasizes the yoking of these two fields.[21] Young begins with nine topics in acoustics, presenting a series of experiments that measured the quantity of air discharged through an aperture, the direction and velocity of the airstream, the velocity of sound, its degree of spatial divergence, and the harmonic sounds of pipes and the decay of their sounds, ending with a general discussion of the vibration of various elastic fluids. He often connects his work with those who preceded him, especially Leonhard Euler, whose arguments about the wave theory in sound and light he had studied closely.[22] Here, and throughout the later works we will discuss, Young often interweaves musical references very naturally, as if he clearly expected his audience to find them familiar and congenial. Apparently, such connections between music and more general scientific topics were not wholly idiosyncratic but were well known to learned writers and their educated readers. Thus, our account goes beyond Young alone to describe this larger current of thought as it emerged in his work.

Young first studied how pipes make harmonic sounds. His mystified classmate had observed him measuring the flow of air through a pipe, recording the varying pressures required to sound various overtones by "overblowing" it, exciting higher overtones with greater air pressure, a familiar technique to him as a flute player (see fig. 2).[23] With these acoustic investigations in mind, Young's tenth section addresses "the analogy between light and sound," first listing the evidence that light is a wave, including "Newton's rings," the pattern of concentric rings between two glass surfaces compressed against each other (one curved, the other flat) that very nearly moved Newton himself to accept a wave theory.[24] Young notes the difficulty and complexity of Newton's putative "fits of transmission and reflection" and adds that the recurrence of the same color in Newton's rings is "very nearly similar to the production

[20] Darrigol, "The Analogy between Light and Sound in the History of Optics from Malebranche to Thomas Young [Part 2]," *Physis* 46 (2009): 111–217, on 114. The present article aims to extend Darrigol's outstanding work on this analogy by deepening the musical background and indicating some facets of Young's work beyond Darrigol's treatment.

[21] Young, "Outlines of Experiments and Inquiries Respecting Sound and Light," in *Thomas Young's Lectures on Natural Philosophy and the Mechanical Arts*, 4 vols. (Bristol, 2002), 4:531–54.

[22] See Euler, *Letters on Different Subjects in Natural Philosophy: Addressed to a German Princess*, 2 vols. (New York, 1837), 1:34–56, 83–7, which first appeared in English in 1795, and Cantor, *Optics after Newton* (cit. n. 2), 117–23. On Euler, see Darrigol, "Analogy between Light and Sound [Part 2]" (cit. n. 20), 169–80; Peter Pesic, "Euler's Musical Mathematics," *Mathematical Intelligencer* 35, no. 2 (2013): 35–43.

[23] For the sound of an overblown c″ on an alto recorder, performed by the author in Santa Fe, N.M., 2013, hear audio 1 (123 KB; MP3) in the electronic version of this article.

[24] For the preceding history of this analogy, see Olivier Darrigol, "The Analogy between Light and Sound in the History of Optics from the Ancient Greeks to Isaac Newton: Part 1," *Centaurus* 52 (2010): 117–55; see also Darrigol, "Analogy between Light and Sound [Part 2]" (cit. n. 20), which treats Young on 185–217.

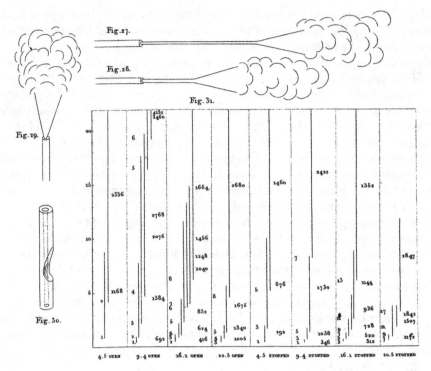

Figure 2. *Plate 3 from Young's "Outlines of Experiments and Inquiries" (cit. n. 21), which he captioned: "Fig. 27. The appearance of a stream of smoke forced very gently from a fine tube. Fig. 28 and 29, the same appearance when the pressure is gradually increased. Fig. 30. A mouth piece for a sonorous cavity. Fig. 31. The perpendicular lines over each division of the horizontal line show, by their length and distance from that line, the extent of pressure capable of producing, from the respective pipes, the harmonic notes indicated by the figures placed opposite the beginning of each, according to the scale of 22 inches parallel to them. The larger numbers, opposite the middle of each of these lines, show the number of vibrations of the corresponding sound in a second."*

of the same sound, by means of a uniform blast, from organ pipes which are different multiples of the same length," showing his knowledge of pipe organs as well as of the recurrent overtones in woodwind overblowing.[25] Young considers this "very nearly similar" to the fact that "the same colour recurs, whenever the thickness [of the pressed glass plate or lens used to show Newton's rings] answers to the terms of an arithmetical proportion," such as 1, 2, 3, . . . He also notes "the analogy between the colours of a thin plate and the sounds of a series of organ pipes, which, indeed, Euler adduces as an argument in favour of [the wave theory of light], although he states the phenomena very inaccurately."[26] Young leaves his exact analogy unclear; he does not seem bothered that Newton's rings follow an arithmetic proportion (1, 3, 5, . . .), whereas the organ pipes are governed by a geometric proportion (2, 4, 8, . . .). Even

[25] Young, "Outlines of Experiments and Inquiries" (cit. n. 21), 543. For instance, pipes in eight-foot and four-foot organ stops can each sound middle C. See the section titled "Of the Harmonic Sounds of Pipes," ibid., on 539–40, esp. table 11. Cf. Darrigol, "Analogy between Light and Sound [Part 2]" (cit. n. 20), 188–90, which does not discuss overblowing.
[26] Young, "Outlines of Experiments and Inquiries" (cit. n. 21), 543.

so, he considers the recurrence of colors in Newton's rings to be precisely comparable to the "recurrence" of pitches produced by organ pipes, which is only comprehensible on the grounds of a wave theory.[27]

In his later writing about various musical instruments, Young notes that the "various compoundings of the stops" give the organ its particular "quality of sound, sometimes called its tone, register, colour, or *timbre*." This "fourth component part of music" Young esteems highly, for "much of the pleasure derived from music depends on it; but as it is capable of little diversity on the same instrument, it is seldom considered in treating of the theory of music." This sound-color "depends on the law by which the sounding body, and the particles of the air, are governed with respect to the velocity of their progress and regress in each vibration, or in different successive vibrations." Young considers the true appreciation of timbre to be a question for natural philosophy, though "all this relates to the quality of sound, and whoever adequately relishes the works of the great modern masters, will be fully competent to judge of its practical importance."[28]

Thus, when Young compares Newton's rings to an organ, we realize the full appropriateness of his application of timbre or sound-color to visual color, for Newton himself had noted the importance of the recurrent pattern of the coloration in his rings. Young "hears" Newton's rings as resembling an organ's cyclical structure of pitches, overtones, and stops, in which the pressure of the airstream can excite recurrent harmonic pitches as the pressure exerted on the glass can evoke the recurrent colors of the rings. Note also that, implicitly, Young here translates a temporal phenomenon (the frequencies of the organ pipes) to a spatial one (the varying lens thicknesses producing Newton's rings).[29]

ADVANCING THE MUSICO-OPTICAL ANALOGY

Having established this fundamental analogy between music and light, Young then turns to a musical phenomenon that will provide a crucial insight into light. His point of departure is a troubling assertion by Robert Smith, the eminent Cambridge astronomer, in his *Harmonics, or, The Philosophy of Musical Sounds* (1749) that "the vibrations constituting different sounds should be able to cross each other in all directions, without affecting the same individual particles of air by their joint forces." On the contrary, Young notes, "undoubtedly they [the vibrations] cross, without disturbing each other's progress; but this can be no otherwise effected than by each particle's partaking of both motions." As proof, he instances "the phenomena of beats" as observed by the violinist Giuseppe Tartini and discussed by Smith himself.[30] To illustrate them, Young devises a kind of thought experiment, supposing "what

[27] Newton's rings appear even with incoherent light, thus allowing Young's analogy with coherent musical tones to go forward, whereas other optical setups would depend on the issue of coherence (the correlation between different waves in space or time). I thank Jed Buchwald for drawing my attention to this point.

[28] All quotations are from Young, "An Essay on Music," in *Young's Lectures* (cit. n. 21), 4:563–72, on 565; the history of the organ is in *A Course of Lectures on Natural Philosophy*, vols. 1–3 in *Young's Lectures*, on 1:404.

[29] For the relation of spatial and temporal interference, see also Darrigol, "Analogy between Light and Sound [Part 2]" (cit. n. 20), 197.

[30] For an example of the beats between two tones that are close in frequency (262 and 272 Hz), generated electronically by the author using the music-notation software Finale, hear audio 2 (273 KB; MP3) in the electronic version of this article.

24   PETER PESIC

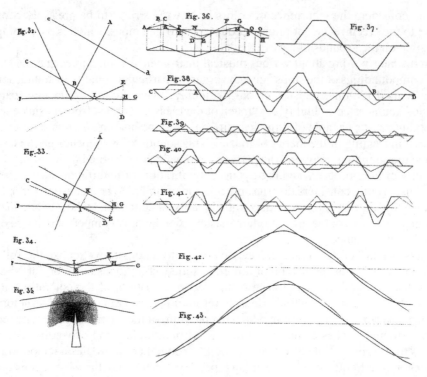

Figure 3. *Plate 4 from Young's "Outlines of Experiments and Inquiries" (cit. n. 21). Figs.
32–5 show "affections of light": its reflection (32), refraction (33), total reflection (34), and
light passing "near an inflecting body" (35); figs. 36–43 show the waveforms of various
combinations of two sounds: an octave (37), a major third (38), a major tone (39), a minor
sixth (40), a fourth "tempered about two commas" (41), a fourth further tempered by "sub-
ordinate vibrations of the same kind in the ratios of 3, 5, and 7" (42), and a vibration "corresponding with the motion of a cycloidal pendulum" (43).*

probably never precisely happens, that the particles of air, in transmitting the pulses
[of sound], proceed and return with uniform motions," and in a series of figures he
draws their motion along the horizontal axis, their displacement along the vertical
(see fig. 3).[31] Young includes a number of different cases in which, "by supposing
any two or more vibrations in the same direction to be combined, the joint motion
will be represented by the sum or difference of the ordinates." Thus, two sounds
of nearly the same strength and pitch will produce a "joint sound" called a "beat"
that reaches its maximum (the sum of the maximum of each component) on a slow
rhythm determined by the difference between their respective frequencies or pitches.
Young's sequence of cases shows the graphic difference between the joint sounds
produced by different components, and he notes that "the greater the difference in
the pitch of two sounds, the more rapid the beats, till at last, like the distinct puffs
of air in the experiments already related, they communicate the idea of a continued

---

[31] Young, "Outlines of Experiments and Inquiries" (cit. n. 21), 544. For discussion of the controversy with Smith, see also Darrigol, "Analogy between Light and Sound [Part 2]" (cit. n. 20), 188–93.

sound; and this is the fundamental harmonic described by Tartini."[32] His diagrams show "snapshots" of the vibrating string, translating its temporal motion into instantaneous spatial waveforms.

At this point, Young's description breaks free from the presumption that sound is a vibrating *body* by noting that sufficiently frequent puffs of air by themselves "communicate the idea of a continued sound."[33] Thus, the locus of the investigation of sound has been shifted to the vibrating air, away from the body no longer needed to produce it. We now realize that, in his student rooms, Young had been producing not just puffs of smoke but a sound of very low frequency, as if he had slowed the phenomenon of musical sounds emitted by a pipe down to an immensely slower time scale on which it could be carefully observed and thoroughly compared with the flowing air that caused it.

Young immediately draws a musical corollary from his description of beats. Returning to the addition of two almost equal sounds, "the momentum of the joint sound is double that of the simple sound only at the middle of the beat, but not throughout its duration." Therefore, "the strength of sound in a concert will not be in exact proportion to the number of instruments composing it." Young has reached this counterintuitive result from his thought experiment, rather than any actual observation, but he now realizes its possible significance as evidence of the wave theory. "Could any method be devised for ascertaining this by experiment, it would assist in the comparison of sound with light" by demonstrating the palpable reality of beats in waves, whether of sound or light.[34] Young will seek evidence of the "beating" of light waves that will be as clear as that of the beating of sound; to do so, he will arrange a "concert" of light.

Indeed, his whole plate of diagrams (in fig. 3) richly illustrates the way he juxtaposes light and sound. Where the diagrams on the right illustrate various possible sound forms, those on the left show "the affections of light," its behavior in reflection, refraction, and passing "near an inflecting body," perhaps a string or knife's edge. The very layout of the plate invites us to contemplate sound and light together. To that end, he returns to the problem of determining the frequency of vibrations, shape, and state of motion of a "chord," a stretched string. Here the visual appearance of a sounding body illuminates its vibrations.

In fact, Young may have been among the first to use the piano, a rather recent arrival among musical instruments, as a *scientific* instrument. He used "one of the lowest [wire-wrapped] strings of a square piano forte" to make an optical experiment:

> Contract the light of a window, so that, when the eye is placed in a proper position, the image of the light may appear small, bright, and well defined, on each of the convolutions of the wire [due to its wrapping]. Let the chord be now made to vibrate, and the luminous point will delineate its path, like a burning coal whirled round, and will present to the eye a line of light, which, by the assistance of a microscope, may be very accurately observed. According to the different ways by which the wire is put in motion, the form of this path is no less diversified and amusing, than the multifarious forms of the quiescent lines of vibrating plates, discovered by Professor Chladni.[35]

[32] Young, "Outlines of Experiments and Inquiries" (cit. n. 21), 544.
[33] Ibid.
[34] Ibid.
[35] Ibid., 546–7.

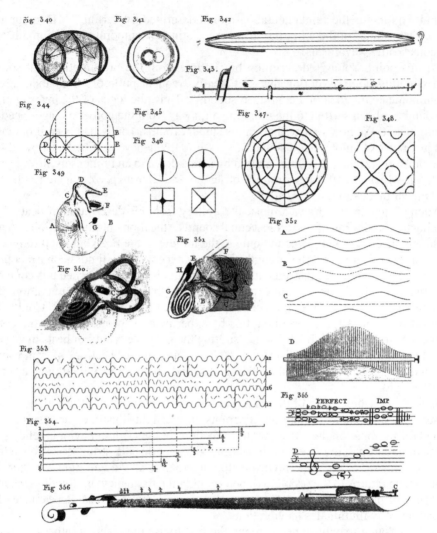

Figure 4. *Plate 25 from Young's* Course of Lectures *(cit. n. 28), showing three-dimensional contours of sound waves (figs. 340, 341), speaking and hearing trumpets (342), a bow exciting a violin string (343), Chladni patterns of sand on vibrating plates (346–8), the mechanism of the human ear (350, 351), Young's own temperament (355), and the "trumpet of Marigni" or tromba marina (356), a kind of bowed monochord, here illustrating overtones.*

Young positions himself in relation to Chladni, whose striking demonstration of standing waves made visible the sonorous modes of vibrating plates (see Young's figs. 346–8 in my fig. 4), a translational artifice Young considers the direct antecedent of his own work. Though his primary object was to gauge the shape of the vibrating string, the details of Young's experimental arrangement are, in fact, very close to what will turn out to be his crucial demonstration of light interference: a thin string illuminated by a small, well-defined light source. Young's own illustration of light passing "near an inflecting body" (in fig. 3) gives evidence that he was aware of this parallelism, even though in this paper he does not take the next step, to allow the vibrating string to come to rest and then to see the vibrations of light surrounding it, as if that were silence made visible.

Young connects his studies of pipes with the problem of the human voice, "the object originally proposed to be illustrated by these researches." This recalls the physiological and medical aspects of his Göttingen dissertation, though here Young seems more interested in purely musical aspects of timbre and resonance. He connects the voice with his smoke pipes by noticing that, analogous to his rhythmic pipe puffs, the human glottis can produce a slow vibration "making a distinct clicking sound" that can be made more continuous "but of an extremely grave pitch: it may, by a good ear, be distinguished two octaves below the lowest A of a common bass voice, consisting in that case of about 26 vibrations in a second." Young connects this glottal clicking with the methods used by ventriloquists to "throw" their voices and also (at still higher pitches) with falsetto singing. Though he refers to anatomy and physiology, he more often relies on "a good ear"; he tells us he can hear four harmonics above the fundamental sung by "a loud bass voice."[36]

The finale of this remarkable paper returns to one of the oldest musical conundrums. Young, like so many before him, became fascinated with the question of temperament and here offers his own solution to its age-old problems in his astutely practical variant of well temperament, which has been revived in recent performances of late eighteenth-century music that emphasize authenticity.[37] Young illustrates his own temperament in a diagram comparing various systems of tuning (see fig. 5), using spatial visualization to illustrate sonic issues. His wide-ranging comparative musical investigations closely resemble, in scope and structure, his concurrent comparative work on languages, as if they were various possible "temperaments" of living speech.

Only four months later (April 1800), Young published "An Essay on Music," giving important evidence of his ongoing interest in music during the height of his optical researches. He begins this essay by acknowledging "the agreeable effect of melodious sounds, not only on the human ear, but on the feelings and on the passions," yet he considers music far more than "delicate titillation" or even than "giving expression to poetical and impassioned diction," which Coleridge and other romantic thinkers emphasized. Contra Kant, Young argues that the study of music is not "amusement only" but reveals a science "scarcely less intricate or more easily acquired than the most profound of the more regular occupations of the schools." Those who show "superior brilliancy" in music "seem almost to require the faculties of a superior order of beings." Young's essay shows considerable familiarity with the history and theory of music, as well as the importance he ascribed to it. He emphasizes the role of harmonics or overtones for the common triads and scales of contemporary musical practice. Finally, he discusses the terminology of musical tempo and gives a detailed table of

[36] Ibid., 549–50. For Young's example of the 26 Hz low A as generated electronically in Finale by the author, hear audio 3 (97 KB; MP3) in the electronic version of this article.

[37] Myles W. Jackson, *Harmonious Triads: Physicists, Musicians, and Instrument Makers in Nineteenth-Century Germany* (Cambridge, Mass., 2006), 172–6. For performances in Young's temperament, hear Enid Katahn (piano), *Beethoven in the Temperaments*, recorded Peterborough, N.H., 1997, Gasparo GSCD-332, and *Six Degrees of Tonality*, recorded Peterborough, N.H., 2000, Gasparo GSCD-344, compact discs. Compare Katahn's performance in Young's temperament of Beethoven's Sonata in C Major, op. 53, introduction to the second movement (reproduced courtesy of Gasparo), with the same passage played in equal temperament, recorded live by the author, Santa Fe, N.M., 2003, in audio 4 (5 MB; MP3) and audio 5 (4.1 MB; MP3), respectively, in the electronic version of this article.

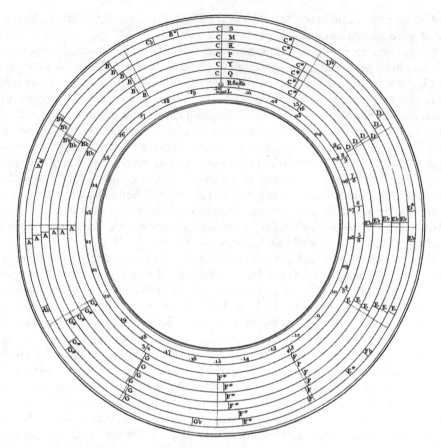

Figure 5. *Young's comparison of different schemes of musical temperament (plate 6 from "Outlines of Experiments and Inquiries" [cit. n. 21]), including his own temperament (the ring labeled Y); the entire circle spans an octave around C, shown at the top.*

the number of measures per minute used in various tempi and meters by composers such as Handel, Haydn, and Mozart.[38]

Seven months later, in November 1800, Young presented his paper "On the Mechanism of the Eye" to the Royal Society.[39] Revisiting his maiden discovery about the accommodation of the eye, Young argues that he had been fundamentally right that changes in the shape of the lens were responsible for accommodation, not the cornea nor the length of the eyeball, as had been suggested by others. To measure these changes, Young pressed instruments against his sclera, the white of his own eye, as Newton had inserted a bodkin behind his own eyeball.[40] Though these excruciating measurements and Young's ensuing physiological deductions make up the bulk of his paper, he first lays his groundwork on another extended comparison of sound and

[38] See Young, "Essay on Music" (cit. n. 28), here quoted at 562, 565–7; Peter Pesic, "Thomas Young and Eighteenth Century Tempi" (unpublished manuscript, St. John's College).

[39] Young, "On the Mechanism of the Eye," in *Young's Lectures* (cit. n. 21), 4:573–606.

[40] Peter Pesic, *Sky in a Bottle* (Cambridge, Mass., 2005), 167–9.

sight. He judges that the ear is "the only organ that can be strictly compared" with the eye, for the other senses operate through more immediate contact of their objects with the nerves.[41]

For Young, contrast with the ear illuminates the eye's functioning. He calculates the quantitative difference between the ear's ability to discriminate the angular direction from which sounds are coming (only within about 5°) and the eye's far sharper directional abilities (90,000 times finer). On the other hand, the eye's "field of perfect vision, for each position of the eye, is not very great," whereas "the sense of hearing is equally perfect in almost every direction." Using these comparisons between eye and ear as an initial point of reference, Young then goes on to devise what he calls a new optometer that will allow precise measurement of the eye's focal distances, as well as the other parameters needed to make his argument about accommodation fully detailed and complete.[42] Thus, all three papers of Young's *annus mirabilis* of 1800 invoke sound, hearing, and music in fundamental ways that inform and shape his arguments about seeing and light.

HEARING COLORS

In August 1801, Young published a letter reaffirming his account of sound and his new musical temperament against the criticisms of a Professor Robinson in Edinburgh. In November, his paper "On the Theory of Light and Colours" juxtaposed excerpts from Newton's writings with Young's own series of new propositions, presented in Euclidean-style hypotheses and demonstrations.[43] Young's rhetoric enlists Newton on the side of the wave theory of light, defusing Newton's objections to it by juxtaposing them with the many passages in which he recognized its merits.

As the essential background for his argument in favor of an ether carrying the vibrations of light, Young assumes the prior case of air as the medium for sound vibrations. "Every experiment, relative to sound, coincides with the observation already quoted from Newton, that all undulations are propagated through the air with equal velocity," which Young thought a capital point in favor of the wave theory of light that Euler himself did not seem to understand when he maintained incorrectly that waves of higher frequency travel faster. Here and throughout, Young uses the wave theory of sound to establish the essential results he will apply to light; returning to his earlier arguments against Smith, he notes that "it is obvious, from the phenomena of elastic bodies and sound, that the undulations may cross each other without interruption" by "uniting their motions," though different frequencies of wave will not intermix. Likewise, he relies on the example of sound to establish that waves expand spherically through a homogeneous medium.[44]

Though Young claims not to "propose any opinions which are absolutely new," he offers an important suggestion that color vision relies on only "three principal colours, red, yellow, and blue," which he chooses because their "undulations are related in magnitude nearly as the numbers 8, 7, and 6," whose integral ratios recall those of

---

[41] Young, "On the Mechanism of the Eye" (cit. n. 39), 574.

[42] Ibid., 574–5; for his optometer, see 575–7.

[43] For his "Letter to Mr. Nicholson . . . Respecting Sound and Light," see *Young's Lectures* (cit. n. 21), 4:607–12; "On the Theory of Light and Colours," ibid., 4:613–31.

[44] Young, "On the Theory of Light and Colours" (cit. n. 43), 618–20; Geoffrey Cantor, "The Changing Role of Young's Ether," *British Journal for the History of Science* 5 (1970): 44–62.

music theory.[45] Thus, green light, whose frequencies are about 6.5 in terms of these ratios, "will affect equally the particles in unison with yellow and blue, and produce the same effect as a light composed of those two species: and each sensitive filament of the nerve may consist of three portions, one for each principal colour."[46]

Young continues to follow what Newton called "the analogy of nature" closely, noting that, on the basis of his own argument, "any attempt, to produce a musical effect from colours, must be unsuccessful, or at least . . . nothing more than a very simple melody could be imitated by them" because the ratios of the primary colors limit the range of any such "color melody" to less than an octave, for anything wider would go "wholly without [outside] the limits of sympathy of the retina, and would lose its effect; in the same manner as the harmony of a third or a fourth is destroyed, by depressing it to the lowest notes of the scale." That is, musical melodies would not translate directly to colors because musical intervals become indistinguishable when transposed to the extreme limits of audible frequencies. The analogy between the ear and the eye guides Young's hypothesizing even when he becomes aware of their important differences, which are no less significant to him than their similarities. "In hearing, there seems to be no permanent vibration of any part of the organ," implying its greater simplicity and unity, compared to the eye as a two-dimensional field of sensors that, at every point, cannot possibly have the range of vibrations available to the ear in its single canal. His three-color hypothesis emerges under the direct pressure of the pitch-distinguishing capabilities of the ear.[47]

Young goes on to offer additional evidence in favor of the wave theory of light, drawing especially on the arguments about the superposition of waves he had earlier made against Smith, and culminating in his proposition VIII: "*When two Undulations, from different Origins, coincide either perfectly or very nearly in Direction, their joint effect is a Combination of the Motions belonging to each.*" Young notes that he had earlier "insisted at large on the application of this principle to harmonics; and it will appear to be of still more extensive utility in explaining the phenomena of colours." He applies it now to "Mr. Coventry's exquisite micrometers; such of them as consist of parallel lines drawn on glass, at the distance of one five hundredth of an inch," what we now call diffraction gratings.[48]

From proposition VIII, Young derives a simple mathematical criterion for the light waves of a given monochromatic wavelength (coming from a point source of red light, say) to combine constructively and yield a bright red spot whenever the sine of the angle of that spot is an integral multiple of the ratio of the spacing between lines on the grating and the wavelength of light. Because the incident red light can reflect constructively off the grating at a whole series of angles, we will see not one but a series of red spots, each corresponding to a different integer in Young's formula. He notes that the particle theory of light would not produce any such periodic and recurrent spots, so that "it is impossible to deduce any explanation of it from any hypothe-

[45] Young, "On the Theory of Light and Colours" (cit. n. 43), 617; in his next paper, "An Account of Some Cases of the Production of Colours," Young will change these three primaries to red, green, and violet, whose ratios are as 7, 6, and 5, to meet William Wollaston's corrections of the spectral ratios.
[46] Ibid.
[47] The Newton quote about "the analogy of nature" is cited ibid.; the following quotes come from 618 (emphasis in the original).
[48] Ibid., 624–6. For the development of the technology of these gratings, see Myles W. Jackson, *Spectrum of Belief: Joseph Von Fraunhofer and the Craft of Precision Optics* (Cambridge, Mass., 2000).

sis hitherto advanced; and I believe it would be difficult to invent any other that would account for it. There is a striking analogy between this separation of colours, and the production of a musical note by successive echoes from equidistant iron palisades; which I have found to correspond pretty accurately with the known velocity of sound, and the distances of the surfaces." Once again, music gives the point of departure for his optical analogy. As he contemplates the lines of the grating, he analogizes them as "echoing" the light, as if audition and vision had merged.[49] Here again, a sonic, temporal phenomenon translates into a spatial, optical one.

Young's account of his sound experiment also suggests that he could have used it to connect the speed of sound with its wavelength and the spacing between the iron palisades. Though Young was quite aware of the significance of determining the wavelength of light experimentally, he does not do it here, reserving it for his reconsideration of Newton's rings, which (as noted above) Young had earlier instanced as the linchpin of his analogy with the recurrent frequencies of organ pipes. In "On the Theory of Light and Colours," Young obviously attaches special significance to determining the wavelength of light from Newton's own data, as if seeking Newton's support even in the process of overthrowing his conclusions.

Newton had framed his spectral colors by assuming that they formed an octave; he did not seem to recognize that his own ring data contradicted such a 2:1 ratio.[50] But now Young corrects Newton's musical mistake: "The whole visible spectrum appears to be comprised within the ratio of three to five, which is that of a major sixth in music; and the undulations of red, yellow, and blue, to be related in magnitude as the numbers 8, 7, and 6; so that the interval from red to blue is a fourth."[51] Thus, Young specifically returns to the same musical analogy that Newton had used, though Newton had mistakenly substituted the octave for the major sixth. By getting right what Newton had mistaken, Young is able to retrieve the accurate wavelengths of the optical spectrum, which he goes on to state in musical terminology:

> The absolute frequency [of light] expressed in numbers is too great to be distinctly conceived, but it may be better imagined by a comparison with sound. If a chord [vibrating string] sounding the tenor c, could be continually bisected 40 times, and should then vibrate, it would afford a yellow green light: this being denoted by c^{41}, the extreme red would be a^{40}, and the blue d^{41}.[52]

Even the identity of these colors is "better imagined" by giving their musical-note names, as if Young preferred to "hear" than to see them, though the "pitches" involved are enormously higher than any audible sound. The resultant synesthesia goes far beyond our normal senses: Young concludes that C is "yellow-green" and D is "blue," as if we were able to hear forty octaves above middle C. He also provides a table stating the "absolute length and frequency of each vibration" of different colors of light, thereby reminding us of their sheer physical reality in space and time.

[49] Young, "On the Theory of Light and Colours" (cit. n. 43), 626.
[50] Peter Pesic, "Isaac Newton and the Mystery of the Major Sixth: A Transcription of His Manuscript 'Of Musick' with Commentary," *Interdisciplinary Science Reviews* 31 (2006): 291–306. See also Alan E. Shapiro, "The Evolving Structure of Newton's Theory of White Light and Color," *Isis* 71 (1980): 211–35.
[51] Young, "On the Theory of Light and Colours" (cit. n. 43), 627.
[52] Ibid.

Against the background of these 1801 musico-optical results, the following July Young distilled his proposition VIII into "a simple and general law":

> Wherever two portions of the same light arrive at the eye by different routes, either exactly or very nearly in the same direction, the light becomes most intense when the difference of the routes is any multiple of a certain length, and least intense in the intermediate state of the interfering portions; and this length is different for light of different colours.[53]

Using this law, Young returns to simple experiments mentioned by Newton and Francesco Maria Grimaldi, from which he now can deduce the exact wavelengths they themselves did not calculate. Recounting an experiment in which he observed the "fine parallel lines of light which are seen upon the margin of an object held near the eye," Young notes "that they were sometimes accompanied by coloured fringes, much broader and more distinct." To make these fringes more distinct still, he observed a horse hair, then a wool fiber, then a single strand of silk, which gave the clearest, broadest pattern. Young made a rectangular hole in a card and bent the card's edges to support a hair parallel to the sides of the hole, a stabilizing mounting that allowed him to measure the deviations of the various colored fringes, which coincided with those he had measured in Newton's rings.[54]

In November 1803 Young took these experiments a step further in his final paper before the Royal Society, which begins by noting "that fringes of colour are produced by the interference of two portions of light," proving "*the general Law of the Interference of Light*" and hence the wave theory in a "decisive" way.[55] His new experiment was even simpler: making a small hole in a window shade, on which a mirror directed the sun's light, he used his artificial sunbeam to illuminate "a slip of card, about one thirtieth of an inch in breadth, and observed its shadow, either on the wall, or on other cards held at different distances." Young now proves that the fringes were the joint effects of light passing on both sides of the card, not just one. He used "a little screen" to block the light coming on one side of the card and notes that "all the fringes which had before been observed in the shadow on the wall immediately disappeared, although the light inflected on the other side was allowed to retain its course." Therefore the fringes could only be produced by the joint action of light "passing on each side of the slip of card, and inflected, or rather diffracted, into the shadow."[56] He goes on to show that his results are quantitatively consistent with his "general law" and that the distances between the dark lines in his fringed shadows agree accurately with analogous distances that he calculates from Newton's own observations of the shadow of a knife's edge and of a hair.[57]

---

[53] Young, "An Account of Some Cases of the Production of Colours Not Hitherto Described," in *Young's Lectures* (cit. n. 21), 4:633–38, on 633.

[54] Ibid. He also adduces "coloured atmospherical halos" and supernumerary rainbows as meteorological examples of his colored fringes, writ large in the heavens; ibid., 634–5, 643–5.

[55] Young, "Experiments and Calculations Relative to Physical Optics," in *Young's Lectures* (cit. n. 21), 4:639–48, on 639; emphasis in the original. See also J. D. Mollon, "The Origins of the Concept of Interference," *Philosophical Transactions of the Royal Society of London* 360 (2002): 807–19, and especially Naum S. Kipnis, *History of the Principle of Interference of Light* (Basel, 1991).

[56] Young, "Experiments and Calculations Relative to Physical Optics" (cit. n. 55), 639–40.

[57] Oddly, Young does not calculate the value of the incident wavelength of light for any of these cases, as he had done in his 1801 paper for Newton's rings and for the diffraction grating. Though some have therefore questioned whether he really performed the measurements, the table shown seems perfectly

Young concludes that light "is possessed of opposite qualities, capable of neutralising or destroying each other, and of extinguishing the light, where they happen to be united," so that light plus light may yield darkness. As he emphasizes, this paradoxical-seeming conclusion is the essence of the wave theory, which gives it the power to explain the recurrences, fringes, and inner rainbows he identified. The concert of light is now complete; Young's conclusion takes him full circle, back to the musical hypotheses with which he began:

> But, since we know that sound diverges in concentric superficies [surfaces], and that musical sounds consist of opposite qualities, capable of neutralising each other, and succeeding at certain equal intervals, which are different according to the difference of the note, we are fully authorized to conclude, that there must be some strong resemblance between the nature of sound and that of light.[58]

YOUNG'S SYMPHONIC LECTURES

In 1801, in the midst of this series of papers, Young became professor of natural philosophy at the Royal Institution, where he delivered the talks that were later published in his *Course of Lectures on Natural Philosophy and the Mechanical Arts* (1807), one of the first attempts at general synthesis in the aftermath of Newton.[59] Addressing a broad audience, including women and others excluded from the universities, Young presented a general picture, emphasizing the leading concepts and omitting mathematical details. His 1800–1803 papers showed the importance of music and sound as he discovered his new insights; his Royal Institution lectures show how he continued to rely on sound and music in the context of their public justification and popularization.[60]

The fifteen hundred quarto pages that gather Young's lectures integrate natural philosophy with practical arts such as machinery, carpentry, and shipbuilding, as well as drawing, engraving, printing, and even "the art of writing" (here including his linguistic concerns).[61] Music occupies a special place in his encyclopedic edifice as the core of his central series of lectures on hydrodynamics, even though acoustics "has usually been considered as exceedingly abstruse and intricate."[62] This is an understatement; by the end of the eighteenth century, acoustics had become a quiet backwater of natural philosophy, not a center of controversy (in contrast to optics). As part of his symphonic synthesis, Young revived the study of sound by connecting it to the larger issues of wave motion.

---

definite, unless one doubts that the numbers listed there really were observed by Young (rather than cooked up after the fact). See John Worrall, "Thomas Young and the 'Refutation' of Newtonian Optics: A Case-Study in the Interaction of Philosophy of Science and History of Science," in *Method and Appraisal in the Physical Sciences*, ed. Colin Howson (Cambridge, 1976), 107–80; cf. Kipnis, *History of the Principle* (cit. n. 55), 118–24. Young may have thought it sufficient to show the consistency of his new experiment with those of Newton, relying on his 1801 determination of wavelength from Newton's rings and diffraction gratings to establish that number's value.

[58] Young, "Experiments and Calculations Relative to Physical Optics" (cit. n. 55), 645.

[59] Bence Jones, *The Royal Institution, Its Founder and Its First Professors* (New York, 1975).

[60] Regarding Young's work at the Royal Institution, see Peacock, *Life of Thomas Young* (cit. n. 1), 134–7, G. N. Cantor, "Thomas Young's Lectures at the Royal Institution," *Notes Rec. Roy. Soc. Lond.* 25 (1970): 87–112, and Robinson, *Last Man Who Knew Everything* (cit. n. 3), 85–94.

[61] Young, *Course of Lectures* (cit. n. 28); Robinson, *Last Man Who Knew Everything* (cit. n. 3), 120–1.

[62] Young, *Course of Lectures* (cit. n. 28), 1:367.

Young interweaves his own successive discoveries with his account of sound waves. His presentation of the overtones characterizing various orchestral instruments leads him to speculate that the human ear is a musical instrument composed of fibers ready to respond sympathetically to external sounds (see Young's figs. 350 and 351 in my fig. 4). A close link between physics and physiology also characterizes his work on the eye. At the same time, Young proceeds without complete knowledge of the central mechanism by which the ear (or the eye) functions. Even so, his use of musical instruments allowed him access to other organs, made of pipes whose structure was fully open to inspection, and thus helped bridge over the central lacunae in his analogies, the unknown mechanisms of hearing and vision themselves.

Young treats harmony in considerable detail, adducing the phenomenon of beats as an example of rhythmic recurrence:

> The most barbarous nations have a pleasure in dancing; and in this case, a great part of the amusement, as far as sentiment and grace are not concerned, is derived from the recurrence of sensations and actions at regular periods of time. Hence not only the elementary parts of music, or the single notes, are more pleasing than any irregular noise, but the whole of a composition is governed by a rhythm, or a recurrence of periods of greater or less extent.[63]

He surveys the sound quality of every common instrument, including the human voice, and the shapes of organ pipes that might sound the vowels, such as the *vox humana* stop he describes as part of the modern organ (see fig. 6).[64] Here his discourse circles back to language, as if the study of music could somehow generate speech itself.

His next chapter turns to optics, showing how the study of sound leads naturally to the study of light. He systematically takes the scientific insights he grounded in musical experience and applies them to solve the enigmas of light. Young's treatment of the nature of light and color forms the climax of the second part of his *Course of Lectures*, from which he then builds the case for the wave theory; as in his 1800–1803 papers, his lecture figures also rhetorically juxtapose sound with light (fig. 6).[65] He uses arguments about the constancy of the speed of sound to justify the constancy of the speed of light.[66] Likewise, Young compares phosphorescent substances, which reradiate light earlier shone on them, to the sympathetic vibration of strings "which are agitated by other sounds conveyed to them through the air."[67]

In his climactic lecture, Young expresses the general principle of interference as emerging from "the case of the waves of water, and the pulses of sound," in which "the beating of two sounds has been explained from a similar interference." Young seals his case by presenting the "beating" of two light sources, exactly as he had shown the beating of two sounds, including his precise determination of the wavelengths of red and violet light.[68] He learned from Robert Hooke that "red and blue

[63] Ibid., 392.
[64] For the sound of this stop in the Möller organ at the Culver Academies Chapel, Culver, Ind., played in 2009 by John Gouwens (and reproduced with his permission), hear audio 6 (203 KB; MP3) in the electronic version of this article.
[65] Ibid., 457.
[66] Ibid., 459–60.
[67] Ibid., 462; Young calls this phenomenon "solar phosphori."
[68] Ibid., 464–5.

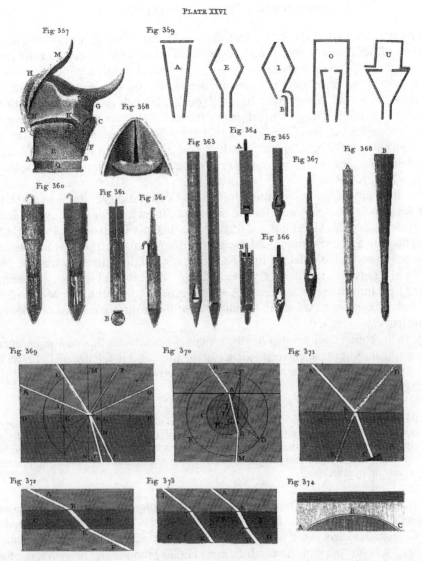

Figure 6. Plate 26 from Young's Course of Lectures *(cit. n. 28); the top registers compare the human glottis (figs. 357, 358) with organ pipes of various kinds, including the* vox humana *(360); the bottom registers illustrate various optical phenomena, such as reflection and refraction.*

differ from each other in the same manner as the sound of a violin and of a flute," a hypothesis then refined by Newton to mean "that the difference of colours, like that of tones in music, depends on the different frequency of the vibrations constituting light."[69] For Young, the full realization of the wave theory of light rested on the power of the musical analogy he had grasped better than Newton.

[69] Ibid., 475 (Hooke), 479 (Newton).

LONGITUDINAL VERSUS TRANSVERSE WAVES

After 1803, Young left the Royal Institution and active research in optics, discouraged by vitriolic attacks by Lord Brougham, an immovable adherent of the particle theory of light. Young subsequently wrote on medical subjects and increasingly worked on the decipherment of hieroglyphics. Later, he was greatly encouraged by the recognition and praise given him by younger French researchers in optics, especially Dominique Arago and Augustin Fresnel. The "Young-Fresnel theory," as it came to be called, prevailed by the 1820s, having converted all except for a few stubborn partisans of Newtonian orthodoxy (such as Brougham). In 1817, Young surveyed these confirmations in a magisterial review article for the *Encyclopaedia Britannica* on chromatics.[70]

The discovery of the polarization of light by Étienne-Louise Malus in 1807, however, seemed to raise problems for the undulatory theory. Gazing through an Iceland spar (calcite) crystal, Malus noticed that the two images of the reflected sunlight from a neighboring glass window would alternately disappear and appear as he rotated the crystal. Somehow, the reflected light had some kind of directionality that the crystal could only transmit when correctly oriented. The crystal would split the incoming reflected light into two separate beams, each "polarized" differently, as Malus phrased it. If indeed light was a wave, how could it exist in the different states of orientation Malus had discovered?[71]

By 1815, Young doubted that his theory could account for this new phenomenon, as he wrote in his private correspondence at the time. But in a letter of 1817, he himself proposed a solution that both used and reversed the analogy with sound. Writing to Arago, he noted that

> it is a principle in this [wave] theory, that all undulations are simply propagated through homogenous mediums in concentric spherical surfaces like the undulations of sound, consisting simply in the direct and retrograde motions of the particles in the direction of the radius [i.e., the direction of propagation of the wave], with the concomitant condensation and rarefactions.[72]

That is, sound is a *longitudinal* wave, causing fluctuations of density of the air along the direction of propagation. In his 1807 *Course of Lectures*, Young had noted that "Dr. Chladni has discovered that solids, of all kinds, are capable of longitudinal vibrations," though "the vibrations which most bodies produce are, however, not longitudinal but lateral."[73] Thus, Chladni's vibrating plates showed Young visible evidence of both longitudinal and lateral (transverse) motion. In 1817, though Young clearly

[70] Young's penchant for encyclopedism led him to contribute articles not just on optics but also on Egypt (a seminal work in the beginnings of Egyptology), bridges, and tides, among many others; see Robinson, *Last Man Who Knew Everything* (cit. n. 3), 179–88, which discusses the reception of Young by the French school on 165–78. See also F. Arago, *Éloge historique du Docteur Young* (Paris, 1832); Eugene Frankel, "Corpuscular Optics and the Wave Theory of Light: The Science and Politics of a Revolution in Physics," *Social Studies of Science* 6 (1976): 141–84; Frank A. J. L. James, "The Physical Interpretation of the Wave Theory of Light," *Brit. J. Hist. Sci.* 17 (1984): 47–60.

[71] For Malus and polarization, see Pesic, *Sky in a Bottle* (cit. n. 40), 84–9. See also David Park, *The Fire within the Eye: A Historical Essay on the Nature and Meaning of Light* (Princeton, N.J., 1997), 252–3, 273–4.

[72] Young, *Miscellaneous Works of the Late Thomas Young*, 3 vols. (London, 1855), 1:383.

[73] Young, *Course of Lectures* (cit. n. 28), 1:380.

understood the force of the example of sound, he now realized that light waves might operate in an importantly different manner: "And yet it is possible to explain in this theory a transverse vibration, propagated also in the direction of the radius, and with equal velocity, the motions of the particles being in a certain constant direction with respect to that radius: and this is a *polarization*."[74]

If the vibrations of the light wave are transverse (perpendicular) to their direction of propagation, they can then be polarized in the plane transverse to that direction. The two split beams transmitted by Iceland spar turned out to exemplify the two orthogonal directions in that plane: Malus's images appeared and disappeared as the crystal was rotated, first transmitting the polarized light, then not.[75] Thus, Young suggested, as did André-Marie Ampère, Arago, and Fresnel independently, light could be a transverse wave, compared to sound waves as longitudinal.[76] Though several of Young's biographers assert at this point that he and Arago had been "blinded" by the analogy with sound, Young's letter suggests the opposite, for he says that he was led to his new suggestion precisely by sound itself.[77] Note that he speaks, in both the case of transverse and of longitudinal waves, of "this theory" in the singular, indicating that the general characteristics of undulatory theory are shared by both, including the concepts of wavelength, frequency, velocity, and direction of propagation.

Returning to this issue in 1823, Young again represents himself as "strongly impressed with the analogy of the properties of sound," but now notices that the possibility of transverse light waves leads to a "perfectly *appalling*" consequence: because they had always been formulated in terms of the vibrations of a solid, "it might be inferred that the lumeniferous ether, pervading all space, and almost all substances, is not only elastic, but absolutely solid!!!"[78] Though Young's biographers take this as even stronger evidence of his blinding by the analogy to sound, his objection indicates the very difficulties with the ether that loomed so large by the end of the nineteenth century. In 1878, James Clerk Maxwell noted the "difficulties we may have in forming a consistent idea of the constitution of the aether" as both dilute and rigid, "certainly the largest, and probably the most uniform body of which we have any knowledge."[79] He and Young both accepted that mysterious body, but Young had realized its profoundly paradoxical character long before. As with the earlier issue of transversality, Young credited this final contribution to optics to his consideration of the "undulations of sound."

This concluding example confronts us with the full richness of Young's translation of sound vibrations into light waves. His youthful rendition of Shakespeare into classical Greek surely involved his awareness of both the possibilities and the perils

---

[74] Young, *Miscellaneous Works* (cit. n. 72), 1:383.

[75] For detailed discussion, including the work of Fresnel and Arago, see Buchwald, *Rise of the Wave Theory of Light* (cit. n. 2), 205–32.

[76] Ibid., 203–14.

[77] See Wood and Oldham, *Thomas Young*, 186, quoted and echoed by Robinson, *Last Man Who Knew Everything*, 173 (Both cit. n. 3). Darrigol, "Analogy between Light and Sound [Part 2]" (cit. n. 20, on 114 n. 2), also notes that "it could even be argued that the analogy blocked the understanding of polarization."

[78] A supplement to the *Encyclopaedia Britannica* entitled "Theoretical Investigations Intended to Illustrate the Phenomenon of Polarisation," reprinted in Young, *Miscellaneous Works* (cit. n. 72), 1:412–7, on 414, 415.

[79] Maxwell, *The Scientific Papers of James Clerk Maxwell*, 2 vols. (Cambridge, 1890), 2:763–75, on 775.

Figure 7. *Young's illustration of the translation between Egyptian hieroglyphics and classical Greek, from his article on Egypt for the* Encyclopaedia Britannica *(1819), reprinted as plate 5 in Young,* Miscellaneous Works *(cit. n. 72), 3: facing 197.*

of such translation.[80] In the present case, his translation yielded both the possibility of transverse light waves but also the attending paradox of the ether. Young was content to follow this translation from sound to light far enough to contemplate these new, "appalling" implications; characteristically, he left to Fresnel and Arago the detailed mathematical exploration of the new terrain.[81] Similarly, in his subsequent work on Egyptian hieroglyphics, Young discovered that the language was phonetic and cor-

[80] For Young's attitude toward this translation, see Peacock, *Life of Thomas Young* (cit. n. 1), 20–3.
[81] For Fresnel's final understanding of transversality, see Buchwald, *Rise of the Wave Theory of Light* (cit. n. 2), 228–31.

rectly identified many characters on the Rosetta stone, such as the cartouche of the pharaoh Ptolemy, leaving to Champollion the full decipherment of the text and the attendant *réclame* (fig. 7).[82] As with his work on light, Young's great linguistic discovery essentially involved sound.

Ironically, French acclaim for Young's light theories was accompanied by British neglect; conversely, the British magnified and the French minimized his achievements in hieroglyphics, compared to those of Champollion. In the tumult of the Napoleonic era, Young experienced the frustrations of a cosmopolitan polymath traversing the British-French divide. The crucial moment of breakthrough in translation may have been more satisfying for Young than the subsequent labor to fill in the gaps and continue the work to the bitter end. Ultimately, he may have been most hampered by his aversion to the "too wide and too barren" mathematical language Fresnel used so powerfully. Though admired for knowing so many tongues, Young may have known one too few, insofar as he eschewed the Continental mathematical language. Perhaps his disinclination may reflect his education, steeped in Newton's intentionally archaizing, anti-Cartesian geometrical language, rather than the algebraic symbology associated with Gottfried Wilhelm Leibniz. This may have been not mere imitation of Newton, though, but rather a reflection of Young's (and Newton's) deep respect for antiquity, their shared curiosity about *prisca sapientia*, strongly manifest in Young's work on hieroglyphics, Newton's on ancient chronology.

However one reads his own wide-ranging quest, Young himself thought that "it is probably best for mankind that the researches of some investigators should be conceived within a narrow compass, while others pass more rapidly through a more extensive sphere of research."[83] Though this elegant statement does not make explicit the difficulties and frustrations involved, Young was the exemplar of this second path, poised between languages in ways that parallel his fundamental role in translating the wave theory between sound and light. As he pursued these multiple projects, his experience with music at many points affected not only his approach to acoustics but the way he then deployed its analogy with light. His sensitivity to sound clearly affected his approach to the problem of translating Egyptian hieroglyphs. No less richly did the successive stages of his acoustical and optical work show a keen interplay between the force of his musical experience and the ensuing dialectic of translation that characterizes the emergent innovations he brought to the theory of interference and its application from sound to light.

[82] Cyrus Herzl Gordon, *Forgotten Scripts: Their Ongoing Discovery and Decipherment* (New York, 1982), 27–30: "Young established the principle of homophony" (28). See also Jed Z. Buchwald and Diane Greco Josefowicz, *The Zodiac of Paris: How an Improbable Controversy over an Ancient Egyptian Artifact Provoked a Modern Debate between Religion and Science* (Princeton, N.J., 2010), 316–27.

[83] Hilts, "Thomas Young's 'Autobiographical Sketch'" (cit. n. 3), 254.

The Electrical Imagination:

Sound Analogies, Equivalent Circuits, and the Rise of Electroacoustics, 1863–1939

*by Roland Wittje**

ABSTRACT

The transformation of acoustics into electroacoustics in the early twentieth century was brought about by at least two significant changes in the mechanical world of acoustics. Electrical technologies entered the acoustics laboratory and profoundly changed the research practices therein. At the same time, electrodynamic theory and electric circuit design advanced rapidly to replace mechanical conceptions as the explanatory basis for the physical sciences. Equivalent-circuit diagrams facilitated a reductionist representation as well as the design of real circuits for electric generation and manipulation of sound by translating acoustic problems into electric systems. Consequently, electroacoustics became more than a research technology and evolved from a laboratory practice into a new way of thinking and talking about sound.

INTRODUCTION

The transformation of acoustics into electroacoustics is central to the reshaping of the science of sound in the first half of the twentieth century. In addition to science, electric recording, transmission, manipulation, and amplification of sound have defined a large part of our common aural experience ever since. The history of electroacoustics has thus far been understood predominantly as a history of technologies such as the telephone, microphones, loudspeakers, and electric amplification. These were coproduced with new technologies of mass media, particularly radio broadcasting and sound motion pictures, and then entered the research laboratory.[1] I argue that this transformation of acoustics into electroacoustics reached far beyond electrical technology and led to a conceptual redefining of our understanding of sound. The new electrical understanding of sound was represented by means of equivalent-

* History of Science Unit, University of Regensburg, D-93040 Regensburg, Germany; roland.wittje @psk.uni-regensburg.de.

I would like to thank the editors, participants in the Berlin workshop in Aug. 2011, Christine Nawa, and two anonymous reviewers for valuable comments.

[1] See Frederick Vinton Hunt, *Electroacoustics: The Analysis of Transduction, and Its Historical Background* (Cambridge, Mass., 1954); Robert T. Beyer, *Sounds of Our Times: Two Hundred Years of Acoustics* (New York, 1999), 177–86; Emily Thompson, *The Soundscape of Modernity: Architectural Acoustics and the Culture of Listening in America, 1900–1933* (Cambridge, Mass., 2002), 229–93; and Thompson, "Dead Rooms and Live Wires: Harvard, Hollywood, and the Deconstruction of Architectural Acoustics, 1900–1930," *Isis* 88 (1997): 596–626.

© 2013 by The History of Science Society. All rights reserved. 0369-7827/11/2013-0003$10.00

circuit diagrams and other types of electrical analogies that became firmly established in the 1930s.

Scientific knowledge and technology are developed locally but within a transnational network of exchange of ideas and practices. That local and national development can only be understood in international context is particularly true with regard to the rapid expansion of acoustic knowledge and technologies of mass media. Acoustic knowledge and technologies became highly relevant for warfare during the Great War, as well as for the national and international markets of the electrical and media industries of the interwar period. Being internationally well informed and able to adapt became a matter of competitiveness, if not survival. Thus, the history of electroacoustics must be understood as a transnational history that had its local and national implementations. My story draws upon examples and developments during the interwar period mainly from Germany and Great Britain, but also from Norway and the United States, where Bell Telephone Laboratories in particular took an international lead in electroacoustics research and development.

Scientists in the nineteenth as well as in the twentieth century have emphasized the specificities of acoustics within the physical sciences. According to Hermann von Helmholtz, "*physical acoustics* is essentially nothing but a section of the theory of the motions of elastic bodies."[2] For Helmholtz and other scientists, it was precisely the human sensation of hearing that made acoustics an interesting chapter in the branch of mechanics. John William Strutt, third Lord Rayleigh, specified that "we shall confine ourselves to those classes of vibrations for which our ears afford a ready made and wonderfully sensitive instrument of investigation. Without ears we should hardly care much more about vibrations than without eyes we should care about light."[3] Comparing the sensation of hearing to the sensation of vision, Helmholtz emphasized that "music stands in a much closer connection with pure sensation than any other [including visual] arts," thereby arguing for a distinctive relationship between the human perception of musical tones and aesthetics.[4]

In the 1920s, the industrial physicist Ferdinand Trendelenburg argued in similar ways; it was the close relationship of acoustics with other branches of knowledge and its importance for general cultural issues that justified a special treatment of acoustics next to the problems of mechanics. Just like Rayleigh, Trendelenburg confined his treatment of acoustics to the physical processes that act upon the human sense of hearing.[5] But acoustics had been transformed in the interwar period. Next to the importance of acoustics for "general cultural issues," Trendelenburg highlighted "recent technical issues" that had significantly reshaped the scientific treatment of acoustics.[6] In the nineteenth century, acoustics was primarily perceived and presented as occupying a domain in between the physical world of mechanics and the cultural world of the music of the educated upper middle class.[7] From this position between mechanics and music as high culture, acoustics was gradually relocated into electrodynamics,

---

[2] Helmholtz, *On the Sensations of Tone as a Physiological Basis for the Theory of Music* (London, 1875), 4.

[3] Rayleigh, *The Theory of Sound*, 2 vols. (London, 1877–8), 1:vi.

[4] Helmholtz, *On the Sensations of Tone* (cit. n. 2), 3.

[5] Trendelenburg, ed., *Akustik*, vol. 8 of *Handbuch der Physik* (Berlin, 1927), 1.

[6] Ibid. See also Erich Waetzmann, ed., *Technische Akustik*, vol. 1 of *Handbuch der Experimentalphysik* (Leipzig, 1934), v.

[7] See Myles W. Jackson, *Harmonious Triads: Physicists, Musicians, and Instrument Makers in Nineteenth-Century Germany* (Cambridge, Mass., 2006).

electrical technology, and mass media. With the rise of electroacoustics, the way music was produced, consumed, and understood was transformed as well.[8] In order to understand these developments in electroacoustics, we must consider both the development of electrical technologies as well as the changing understanding of electrodynamics, especially electric oscillations and electric circuit design.

The electroanalog field became a new language, a new way of thinking and talking about sound in the twentieth century. Its practices were shaped by the circuit diagram, rather than linguistic or mathematical expressions. Circuit diagrams belonged to an array of abstract visual concepts, such as flowcharts and thermodynamic cycles, developed by scientists and engineers. Within the framework of these visual concepts, technical problems could be formulated, analyzed, and solved.[9] Circuit diagrams, just like Swedish scientist and engineer Christopher Polhem's mechanical alphabet and German mechanical engineer Franz Reuleaux's machine grammar, conveyed their own sign language and grammar. These consisted of standardized symbols for standardized electric components and a network of idealized electric connections.[10] Even though engineers and scientists could easily envision a real electric circuit from the diagram and produce it in the workshop, the relationship between the drawing and the material circuit was not straightforward. Circuit diagrams were not about the materiality of the circuit but about its operation. In contrast to mechanical drawings, circuit diagrams were meant to show not spatial arrangements but functional relations.[11]

This abstract and conceptual nature of circuit diagrams was especially prominent in electroacoustics. Circuit drawings could represent actual electric circuits such as amplifier circuits, microphone and loudspeaker wiring, or electric filters. They could also represent acoustic systems that were partly or entirely nonelectric. This is evinced by an example from research in sound insulation in buildings at Norges Tekniske Høgskole (the Norwegian Institute of Technology) in the 1930s. In figure 1 an equivalent-circuit diagram illustrates the transmission of sound through a double wall. While the upper drawing shows the wall construction with the incoming and

[8] See Joachim Stange, *Die Bedeutung der elektroakustischen Medien für die Musik im 20. Jahrhundert* (Osnabrück, 1988); Peter Donhauser, *Elektrische Klangmaschinen—die Pionierzeit in Deutschland und Österreich* (Vienna, 2007); and Thompson, *Soundscape of Modernity* (cit. n. 1).

[9] Precious little has been written on the history of circuit design. See Edward Jones-Imhotep, "Icons and Electronics," *Historical Studies in the Natural Sciences* 38, no. 3 (2008): 405–50, for debates in the postwar period, and Eugene S. Ferguson, *Engineering and the Mind's Eye* (Cambridge, Mass., 1992), 11, for abstract visual concepts in engineering. Ferguson argues for these concepts being specific to engineers while scientists tend to use mathematical concepts. In the history of electroacoustics, however, it becomes impossible to always separate scientists from engineers, especially in the case of technical and industrial physicists, many of whom crossed the line and felt at home in both communities—science and electrical engineering.

[10] The mechanical alphabet consisted of a collection of seventy-nine wooden mechanical models, used as a pedagogical instrument by Polhem (1661–1751) in the Laboratorium Mechanicum, a school of mechanics that he founded in 1697. Reuleaux (1829–1905) formulated his machine grammar in his 1875 monograph *Theoretische Kinematik: Grundzüge einer Theorie des Maschinenwesens* (published in English as *The Kinematics of Machinery* [1876]). This system uses a kinematic notation to unite pairs of elements into kinematic chains. For more on Polhem and Reuleaux, see Ferguson, *Engineering and the Mind's Eye* (cit. n. 9), 137–47.

[11] Jones-Imhotep, "Icons and Electronics" (cit. n. 9), 416. Engineers, scientists, and technicians could build material circuits from a circuit diagram, thanks in large part to standardized electric components. Radio amateurs in particular built their own equipment using circuit diagrams published in journals such as *Wireless World* and components from radio-supply shops.

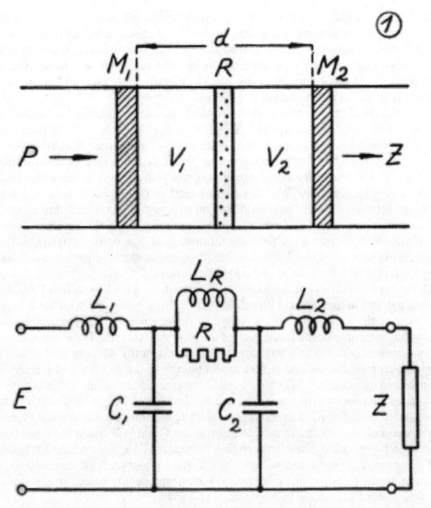

Figure 1. *The electric equivalent-circuit diagram of a double wall as an oscillating system with damping; Berg and Holtsmark, "Schallisolation von Doppelwänden" (cit. n. 12), 75.*

transmitted sound, the lower equivalent-circuit diagram represents the wall as an oscillating system with damping.[12]

What was truly remarkable about the representation of the double wall by an equivalent-circuit diagram was that there was nothing electrical about the wall or the sound traveling through it. By the 1930s electric circuit diagrams, even in the case of representations of nonelectric systems, had become a lingua franca of the new acoustics. Moving from the acoustic to the electric system was an act of transla-

[12] Reno Berg and Johan P. Holtsmark, "Die Schallisolation von Doppelwänden I: Holzwände," *Det Kongelige Norske Videnskabers Selskabs Forhandlinger* 8, no. 23 (1935): 75–8. Holtsmark was professor of physics at the Norwegian Institute of Technology in Trondheim, where he established an acoustic laboratory in 1929. His assistant Berg was an electrical engineer. See Roland Wittje, *Acoustics, Atom Smashing and Amateur Radio: Physics and Instrumentation at the Norwegian Institute of Technology in the Interwar Period* (Trondheim, 2003), 69–142.

tion. Electric oscillations with the same waveforms as acoustic vibrations could be described by the same mathematical equations by substituting the equivalent electric variables for the acoustic variables. Acoustic variables such as force, speed, displacement, mass, and elasticity were replaced by electric variables such as tension, current, charge, self-induction, and capacity. The electric systems then became mathematically identical representations of the acoustic systems.[13]

Acousticians of the interwar period frequently pointed out that acoustics was perceived as a dormant and antiquated discipline from the 1890s until the outbreak of the Great War.[14] On the industrialized battlefields of the Great War, acoustics became an important body of knowledge, and music ceased to be its main frame of reference. But instead of emphasizing discontinuity, the history of acoustics before, during, and after the Great War can also be read as one of continuity. The use of analogies between different physical phenomena was rather common throughout the nineteenth century. Analogies served didactic purposes and linked apparently separate phenomena in order to achieve the ultimate goal of a unified physical science. Analogies between electrical and acoustic phenomena were frequently used, but so were analogies with light, heat, and other phenomena.

The use of the term *electroacoustics* itself can be traced back to around 1900, when it appeared first in the German literature.[15] The timing coincided with the rise of an electromagnetic worldview, which sought to replace mechanical conceptions with electromagnetic ones as the foundation of the physical sciences. The electromagnetic regime eventually failed as a worldview, but it brought the development of electromagnetic representations and electrical technology to the forefront of the physical sciences and engineering. Scientists studied the electric arc as an oscillating system for wireless telephony. They also invented the amplifier tube for long-distance telephony. Both were regarded as research in electricity rather than acoustics. Precisely this work in electricity in the decade prior to the Great War made the rapid development and application of artillery ranging, submarine detection, wireless telephony, and other electroacoustic technologies for warfare possible. At the same time, the tradition of nineteenth-century acoustics, which was deeply rooted in the study of classical music and musical instruments, seems to have stagnated.

As the first industrial war, the Great War was an important turning point for acoustics research through the introduction of new themes and technologies to sound studies, an experience that scientists on both sides of the Atlantic shared. In the immediate interwar period, actors in the electrical and media industry became the main driving forces in the production and consumption of the new electroacoustic knowledge.[16]

[13] See, e.g., Bjørn Trumpy, *Akustikk—utvalgte forelesninger for elektroavdelingens, 4. årskurs (linje for svakstrøm)* [Acoustics—selected lectures for the Department of Electrical Engineering, fourth year (Program of Low-Current Engineering)] (Trondheim, 1930), pt. 1, 13–4, 52–3.
[14] Waetzmann, *Technische Akustik* (cit. n. 6), v; Erwin Meyer and Waetzmann, "Die Bedeutung der Akustik im Rahmen der gesamten Physik und Technik," *Zeitschrift für technische Physik* 17, no. 12 (1936): 508–12, on 508; Bruce Lindsay, preface, in John William Strutt, Baron Rayleigh, *The Theory of Sound*, vol. 1, 2nd ed. (1894; repr., New York, 1945), xxviii; Beyer, *Sounds of Our Times*, 177; Thompson, *Soundscape of Modernity*, 59 (Both cit. n. 1).
[15] Robert Hartmann-Kempf, *Über den Einfluß der Amplitude auf Tonhöhe und Decrement von Stimmgabeln und zungenförmigen Stahlfedern: Elektroakustische Untersuchungen* (Frankfurt am Main, 1903).
[16] William H. Eccles, "The New Acoustics," *Proceedings of the Physical Society* 41 (1929): 231–9.

Frederick Vinton Hunt began his extensive historical introduction to electroacoustics with the proclamation that the field was as old as thunder and lightning, which was, for the scientist, a natural electroacoustic phenomenon.[17] For our aims, however, it will be sufficient to trace the history of the entanglement of acoustics with electrical technology and electrodynamic theory back to the middle of the nineteenth century and its two most influential works, those by Helmholtz and Rayleigh.

THE ACOUSTICS OF HELMHOLTZ AND RAYLEIGH: THE *SENSATIONS OF TONE* AND THE *THEORY OF SOUND*

> The first and principal difference between various sounds experienced by our ear, is that between *noises* and *musical tones*. . . . We can easily compound noises out of musical tones. . . . This shews us that musical tones are the simpler and more regular elements of the sensations of hearing, and that we have consequently first to study the laws and peculiarities of this class of sensations.[18]

> Before proceeding further we must consider a distinction, which is of great importance, though not free from difficulty. Sounds may be classified as musical and unmusical; the former for convenience may be called *notes* and the latter *noises*. . . . We are thus led to give our attention, in the first instance, mainly to musical sounds.[19]

The two most influential works in acoustics in the nineteenth century were Helmholtz's *Lehre von den Tonempfindungen als physiologische Grundlage für die Theorie der Musik* (On the sensations of tone as a physiological basis for the theory of music) of 1863 and Rayleigh's *Theory of Sound*, which was in two volumes, published in 1877 and 1878, then revised and enlarged in the 1894 and 1896 second editions. Helmholtz's *Sensations of Tone* and Rayleigh's *Theory of Sound* were both very similar as well as remarkably different in their scientific aims, their representations of sound phenomena, and their targeted readership.

Helmholtz's involvement with acoustics, like most of his work during the period when he was professor of physiology at the University of Heidelberg (1858–71), was informed by his approach to physiology: to reduce all phenomena to the simple laws of mechanics. Helmholtz's acoustics must also be viewed in the context of his investigations in physiological optics, carried out during the same period. While both treated issues of sensation and perception, Helmholtz's acoustics went further than his optics in its aim to place the aesthetic perception of music on scientific grounds. In addition to physicists and physiologists, Helmholtz sought out musicologists and aestheticians as the main readership of his *Sensations of Tone*.[20] Rayleigh targeted a very different readership of mathematically trained scientists. He aimed at laying "before the reader a connected exposition of the theory of sound, which should include the more important of the advances made in modern times by Mathematicians and Physicists."[21] Helmholtz limited the use of higher mathematics, especially differential calculus, in order to make his volume accessible to readers from musicol-

[17] Hunt, *Electroacoustics* (cit. n. 1), 1.

[18] Helmholtz, *On the Sensations of Tone* (cit. n. 2), 11–2, "Noise and Musical Tone."

[19] Rayleigh, *Theory of Sound* (cit. n. 3), 1:4.

[20] Hermann von Helmholtz, *Die Lehre von den Tonempfindungen als physiologische Grundlage für die Theorie der Musik* (Brunswick, 1863), 1.

[21] Rayleigh, *Theory of Sound* (cit. n. 3), 1:v.

ogy or physiology, who did not necessarily undergo special training in mathematics or physics. Rayleigh's book, in contrast, was filled with differential calculus; it was intended as a handbook to provide the trained reader with comprehensive tools for mathematical treatment of all kinds of acoustic phenomena. While Helmholtz used many of the visual representations of sound, including inscriptions from tuning forks, Chladni figures, and Lissajous figures, such representations are remarkably absent in Rayleigh's books.[22]

We should not, however, lose sight of the similarities between Helmholtz's and Rayleigh's approaches to acoustics. Both treated sound fundamentally as a mechanical phenomenon, which was explained by oscillations in solids, liquids, and gases. Both used musical sounds as a main reference for their investigations. The very fact that acoustic oscillations were understood as audible frequencies was the main argument, for Helmholtz as well as Rayleigh, for keeping acoustics as a field of inquiry separate from mechanical vibrations. The ear remained the most sensitive sound detector for both, despite visual representations and Rayleigh's measurements of sound intensities by a disc suspended on a torsion thread (later known as the Rayleigh disc). Rayleigh's *Theory of Sound*, though conceptualized as a mathematical and physical treatise, contained a section on audition and discussed Helmholtz's theory of combination tones in the second edition of its volume 2 in 1896. By discussing anatomical and physiological issues that were outside the scope of his original work, Rayleigh paid tribute to Helmholtz and the subsequent debate over his theory of combination tones.[23]

THE ELECTRIFICATION OF SOUND: ELECTRODYNAMIC THEORY, CIRCUIT DESIGN, AND ELECTRICAL INSTRUMENTATION

Both Helmholtz and Rayleigh drew upon emerging electrical technologies and electrical analogies in their work on acoustics. Helmholtz used telegraph technology in his acoustic investigations, employing electromagnets and interrupters to drive the tuning forks for his tuning-fork resonator and his vibration microscope. Helmholtz also employed telegraph wires as an analogy, explaining that the nerves conveyed sensation as wires conveyed telegraph pulses. Helmholtz's representations of sound were, however, visual, not electrical or electroanalog. He did not draw upon an analogy between electric oscillations and acoustic vibrations; the analogy between perceptions of sound and perceptions of light was nevertheless important. This mirrors the understanding of electric oscillation as well as communication technology at the time. Telegraph signals were understood as pulses, or electric shocks, which could drive Helmholtz's tuning forks, and not as electric waves, which could be translated into equivalent acoustic waves by means of transducers.

Physicists' understanding of electric oscillations rapidly changed with the advance

---

[22] Among the few exceptions are the Lissajous figures in vol. 1; ibid., 29.

[23] This debate was a major dispute arising from Helmholtz's *Sensations of Tone*. See David Pantalony, "Rudolph Koenig's Workshop of Sound: Instruments, Theories, and the Debate over Combination Tones," *Annals of Science* 62, no. 1 (2005): 57–82, on 80–1, and Julia Kursell, "Wohlklang im Körper: Kombinationstöne in der experimentellen Hörphysiologie von Hermann v. Helmholtz," in *Resonanz: Potentiale einer akustischen Figur*, ed. Karsten Lichau, Viktoria Tkaczyk, and Rebecca Wolf (Paderborn, 2009), 55–74, on 73–4. See also Erich Waetzmann, *Die Resonanztheorie des Hörens* (Brunswick, 1912).

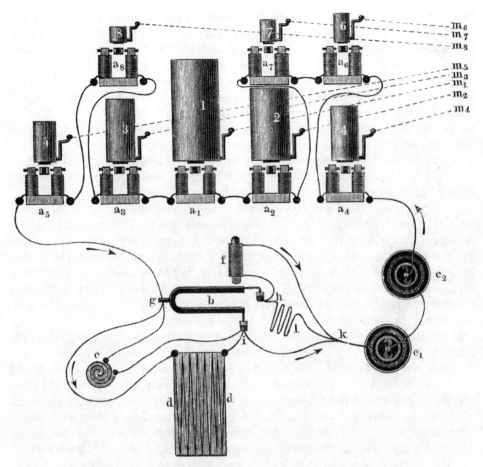

Figure 2. Helmholtz's wiring diagram for his electromagnetically driven tuning-fork synthesizer; Helmholtz, Lehre von den Tonempfindungen *(cit. n. 20), 584. The diagram follows very different conventions than twentieth-century circuit diagrams. Components* a_1 *to* a_5 *represent the electromagnets driving the tuning fork;* b *is the interrupting fork driven by the coil* f; e_1 *and* e_2 *represent two Grove cells;* c *is a condenser; and* d *is a "very great" resistance.*

of electrodynamic theory, the design of electric circuits, and the development of the telephone as the first reversible electric sound transducer. These were all developments in which Helmholtz himself played an important role. Like his work in acoustics, Helmholtz's interest in electricity was initially informed by his work in physiology. In the 1850s he conducted a number of experiments on nervous excitation that led him to the first quantitative treatment of self-induction and the extension of Ohm's law to variable currents.[24] He was an early proponent of what is today known as the voltage-source equivalent, a foundation for the equivalent-circuit concept, which he established while working on the measurement of voltages and currents in

[24] Olivier Darrigol, *Electrodynamics from Ampère to Einstein* (Oxford, 2000), 220.

muscle tissue.[25] Helmholtz was mainly interested in open circuits and oscillatory discharge. In the 1870s he published a series of papers on the theory of electrodynamics, in which he compared the assumptions and theories of Wilhelm Weber and Franz Neumann with those of Michael Faraday and James Clerk Maxwell.[26] The industrialist Werner Siemens presented his work on the electromagnetic telegraph to the Berlin Physical Society around 1850, the same time period in which Helmholtz presented his work on nerve transmission. Helmholtz's adaptation and use of telegraph technology and analogy in his physiological investigations comes as no big surprise given that he and Siemens were close acquaintances.[27]

It was, however, not Prussia but Great Britain that was the undisputed leader in establishing a worldwide network of cable telegraphy, which became the "nervous system" of the British Empire, in the second half of the nineteenth century.[28] Enlarging and improving the telegraph system created a large demand for electrical knowledge. William Thomson, professor of natural philosophy at the University of Glasgow, became chief adviser to the first transatlantic telegraph cable project in 1857 and developed electrical measurement instruments for the telegraph industry. The testing laboratories of the cable industry, rather than university laboratories, were the most sophisticated and best-equipped electrical laboratories of the time. The requirements of cable telegraphy led to the development of standard electric units, precision measurement instruments, electric circuit design, and electrodynamic theory.[29]

In his *Treatise on Electricity and Magnetism* (1873), Maxwell acknowledged the importance of telegraphy to the understanding of electrodynamics.[30] In his electromagnetic theory, Maxwell applied analogies not between electric oscillations and acoustics, but between the propagation of electric and magnetic forces in the ether and the propagation of light. This was to support his theory that the electromagnetic ether and the light ether were the same medium, and that light was an electromagnetic disturbance in this medium. The theory of hydromechanics nevertheless established a link between the imponderable ether and the ponderable media where sound propagated. Maxwell made use of mechanical models as vivid representations of electrical phenomena, declaring that illustrative mechanical models provided the best way to translate complex mechanical relationships into a concrete and readily grasped form without a loss of rigor.[31] Among the Maxwellians, a group of physicists who interpreted and developed Maxwell's theory after his death in 1879, George Francis FitzGerald and Oliver Lodge were the ones who adopted the use of mechanical models to explain the mode of operation of electromagnetic phenomena within the ether. According to FitzGerald these models embodied only analogies, or likenesses, and not a true representation of how the ether was supposed to be imagined.[32]

[25] Hermann Helmholtz, "Ueber einige Gesetze der Vertheilung elektrischer Ströme in körperlichen Leitern mit Anwendung auf die thierisch-elektrischen Versuche," *Annalen der Physik und Chemie* 89, no. 6 (1853): 211–33; Don H. Johnson, "Origins of the Equivalent Circuit Concept: The Voltage-Source Equivalent," *Proceedings of the IEEE* 91 (2003): 636–40.
[26] Darrigol, *Electrodynamics* (cit. n. 24), 223–33.
[27] Timothy Lenoir, "Helmholtz and the Materialities of Communication," *Osiris*, 2nd ser., 9 (1994): 184–207, on 187–8. See also Christoph Hoffmann, "Helmholtz' Apparatuses: Telegraphy as Working Model of Nerve Physiology," *Philosophia Scientiae* 7, no. 1 (2003): 129–49.
[28] Bruce J. Hunt, *The Maxwellians* (Ithaca, N.Y., 1991), 54.
[29] Ibid., 55.
[30] Ibid., 55–6.
[31] Ibid., 75.
[32] Ibid., 83.

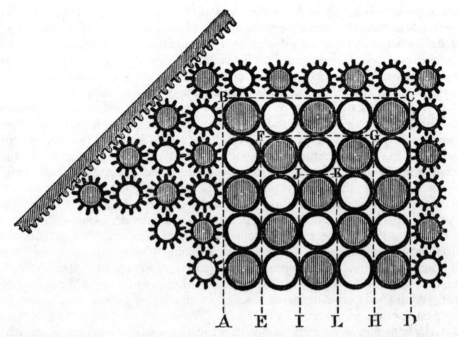

Figure 3. Model illustrating a current induced in metal by an increasing magnetic field; Oliver Lodge, Modern Views of Electricity *(London, 1889), 194. Some models of electricity and magnetism, such as Lodge's hydrostatic model of the Leyden jar, were actually built as lecture demonstration devices; see Max Kohl,* Physical Apparatus *(cit. n. 45), 836, fig. 60638.*

The analogy between sound waves and electromagnetic waves became important when the Maxwellians discussed the electromagnetic production of light. By 1880, FitzGerald had come to the conclusion that the electromagnetic production of light was impossible. But then, after carefully studying Rayleigh's *Theory of Sound*, he had to reconsider this viewpoint. Rayleigh's solution of the equation that FitzGerald dealt with, which took the same form for acoustics and electromagnetism, suggested that "a simply periodic current would originate wave disturbances such as light."[33] It was not the British Maxwellians but Helmholtz's former student Heinrich Hertz who was the first to produce electric waves that propagated in free space and to establish their affinity to light waves in 1887 and 1888. Hertz used the analogy between his electric oscillations and acoustics in order to describe the nature of the oscillations that he produced.[34] However, it was once again the analogy with light waves, not acoustic waves, that Hertz had to establish in order to identify his electric waves as light of very long wavelengths.

Hertz's experiments gave a boost to the understanding and further study of electrodynamic theory. Equally important for the development of the electroacoustic agenda was telephony. Oliver Heaviside, a telegraph operator who quit his job in 1874 to pursue private research, became a member of the inner circle of the Maxwellians and

[33] Ibid., 38–42, on 39.
[34] Hertz, *Electric Waves* (London, 1893), 49, 135–6.

one of the leaders in further developing Maxwell's theory. Between 1885 and 1887, Heaviside applied Maxwell's theory to an examination of how electric signals traveled along wires and how their distortion could be reduced or eliminated. Heaviside thereby conceived electric signals in a fundamentally new way, not as pulses in wires, but as trains of electromagnetic waves on wires. Heaviside's discovery of inductive loading in 1887 would be extremely valuable for long-distance telephony.[35]

Rayleigh had used analogies between acoustic phenomena and electric currents in his first edition of 1877–8 to introduce the concept of acoustic conductivity.[36] When the second edition of volume 1 of his *Theory of Sound* appeared in 1894, electrodynamic theory and telephone technology had advanced to such a degree that he found it necessary to add a chapter on electric vibrations (chap. XB). It was rather disconnected from the other chapters, and Rayleigh did not use the term *electro-acoustics*. He nevertheless clearly realized the growing importance of electrodynamic theory, instruments, and practice for the future development of acoustics.

Equally as important as electrodynamic theory was the rise and emancipation of electrical engineering for the advance of electroanalog thinking and the development of standardized circuit diagrams as a language to represent physical phenomena. In the 1880s electrical engineering was institutionalized as a new discipline at the German Technische Hochschulen (Institutes of Technology). Its main drivers were the rapid advance of electrification and the heavy-power electrical industry; telegraphy and telephony played virtually no role. Electrical engineering emerged as an amalgam of physics and mechanical engineering, oriented toward practice in the industry.[37] As important as Maxwell's electrodynamic theory was for the understanding of electric oscillations and their propagation both on wires and in free space, it was too abstract to be taught to engineers and impractical for designing heavy-current machinery. Electrical engineers needed more visual and less mathematical tools to design calculations of electric devices.[38]

One of the pioneers in developing these new tools for designing calculations of heavy-current machinery was Carl Proteus Steinmetz. He was one of the leading figures in the theory of heavy-current electrical engineering and a forceful promoter of alternating current.[39] Steinmetz introduced graphical methods and complex numbers into heavy-current engineering and made ample use of analogies between electricity and mechanics. In his textbook on alternating-current phenomena of 1897, Steinmetz applied equivalent-circuit diagrams in explaining alternating-current transformers.[40]

[35] Hunt, *Maxwellians* (cit. n. 28), 129–51. Heaviside's work led to the "British electrical debate," a conflict between the Maxwellians on one side, who proposed loading the telephone lines with additional self-induction, and William Henry Preece of the Post Office Telegraph Department and other practitioners on the other side, who opposed such loading.

[36] Rayleigh, *Theory of Sound* (cit. n. 3), 2:159.

[37] Wolfgang König, *Technikwissenschaften: Die Entstehung der Elektrotechnik aus Industrie und Wissenschaft zwischen 1880 und 1914* (Chur, 1995).

[38] See Ronald Kline, *Steinmetz: Engineer and Socialist* (Baltimore, 1992), 108–12.

[39] Ibid. Steinmetz had studied mathematics, physics, and electrical engineering in Breslau (now Wrocław) and Zurich before he emigrated to the United States in 1889 and started working in the expanding electrical industry.

[40] Steinmetz and Ernst J. Berg, *Theory and Calculation of Alternating Current Phenomena* (New York, 1897), 183–5. For a history of the origins of the equivalent-circuit concept, see also Johnson, "Voltage-Source Equivalent" (cit. n. 25), and Johnson, "Origins of the Equivalent Circuit Concept: The Current-Source Equivalent," *Proceedings of the IEEE* 91 (2003): 817–21.

These representations of alternating-current transformers in equivalent-circuit dia-grams later inspired equivalent-circuit representations in electroacoustics.

ELECTROACOUSTICS AS MODERN PHYSICS: THE ELECTRIC ARC AS AN EXPERIMENTAL SYSTEM FOR *SCHWINGUNGSFORSCHUNG*

With the rise of electromagnetic theory, electric machinery, and the identification of cathode rays as electrons around 1900, an electromagnetic worldview swiftly sup-planted mechanical conceptions, replacing the fundamental laws of mechanics to which all physical phenomena should be reduced with fundamental laws of electro-magnetism.[41] As Bruce Hunt has put it, "Physicists ceased to feel a need to look for a mechanism behind the electromagnetic laws or to believe that their understanding would be improved by one."[42] The understanding of oscillations as the basis of a prin-ciple that transcended and interlinked all subdisciplines of the physical sciences would become equally important in the formation of such new research fields as quantum theory.[43] With the rise of the electromagnetic worldview and electrical technology, electric oscillations increasingly replaced mechanical models for oscillation theory.

Around the same time, the notion of electroacoustics made its appearance. In 1903, Robert Hartmann-Kempf added the subtitle *Elektroakustische Untersuchun-gen* (Electroacoustic investigations) to his dissertation, which he had worked on under Wilhelm Wien at the University of Würzburg.[44] In his dissertation, Hartmann-Kempf investigated the effect of amplitude on the resonance frequency and damping of tuning forks and steel springs driven by electric oscillations. Despite the new term, Hartmann-Kempf's experimental regime remained related to the one laid out for acoustics by Helmholtz forty years earlier. Around 1900, the telephone was classified and understood as an electrical, not an acoustic, instrument. Electrically driven tun-ing forks were generally classified as acoustic instruments.[45] Investigations in acous-tics were still mainly related to music and musical instruments and dominated by the debate about Helmholtz's theory of combination tones. The collection on technical acoustics at the Deutsches Museum in Munich, which opened in 1906, comprised only musical instruments.[46]

But a "new acoustics" emerged in this electric age. Amplifier tubes were devel-oped to solve the problem of long-distance telephone transmission. For the scien-

[41] The electromagnetic worldview reached its zenith around 1905 and lost its appeal around 1914 due to the rise of quantum theory and relativity. See Helge Kragh, *Quantum Generations: A History of Physics in the Twentieth Century* (Princeton, N.J., 1999), 103–19.

[42] Hunt, *Maxwellians* (cit. n. 28), 104.

[43] Wave formalism allowed for the transfer of formalism from acoustic wave theory to quantum mechanics, as Johan Holtsmark and Hilding Faxén, for example, transferred Rayleigh wave scatter-ing to electron scattering in 1927; see Faxén and Holtsmark, "Beitrag zur Theorie des Durchgangs langsamer Elektronen durch Gase," *Zeitschrift für Physik* 45 (1927): 307–24, and Wittje, *Acoustics, Atom Smashing and Amateur Radio* (cit. n. 12), 145–6.

[44] See Hartmann-Kempf, *Über den Einfluß der Amplitude* (cit. n. 15). Hartmann-Kempf was the son of Wilhelm Eugen Hartmann, one of the founders of the electrical instrument company Hartmann & Braun. Hartmann-Kempf used Hartmann & Braun instruments in the investigations for his disserta-tion, and later joined the company.

[45] See Max Kohl, *Physical Apparatus, Price List No. 50*, vols. 2 and 3 (Chemnitz, n.d. [1911]). Elec-trically driven tuning forks also served other purposes than to produce sounds; see, e.g., the tuning-fork chronograph.

[46] Franz Fuchs, *Der Aufbau der technischen Akustik im Deutschen Museum* (Munich, 1963), 1.

tists and engineers working with electrical communication technologies, the acoustic problems of sound propagation were intrinsically tied to the electrical problems of signal propagation. A similar relation was established in the sound laboratory by the rapid dominance of electrical measurement instrumentation.[47] As a transducer, the telephone connected the otherwise still separate electrical and acoustic worlds. The telephone was employed as an instrument in investigating the origin and nature of combination tones.[48]

In contrast to the telephone, the speaking or singing electric arc, an apparatus whose components are nowadays known as the plasma loudspeaker and plasma microphone, has nearly fallen into oblivion as an electroacoustic communication technology. Around 1900, however, the electric arc was a widespread technology with a promising future for electric lighting, for binding nitrogen from the air for fertilizers and explosives, and for transmitting wireless telegraphy and telephony. As an oscillating system it could broadcast both electromagnetic waves and the human voice in free space without using a membrane like the telephone. The arc could also detect electromagnetic waves as well as sound. What could be a more ideal experimental system for acoustics in the electric age?

William du Bois Duddell in Britain and Hermann Theodor Simon (known as Theodor) in Germany both developed and investigated the singing or speaking arc.[49] Their device was a carbon arc lamp in which a variable resistor or a microphone was used to alternate the sound produced by the arc (figs. 4 and 5). The speaking arc became a popular demonstration device in physics classes that could be used with a microphone or as an "electric piano," playing a simple melody (see fig. 5). In the instrument catalogs as well as in textbooks and teaching collections of the pre–World War I period, the speaking arc was classified as an electrical instrument, in the section of electric oscillations and wireless telegraphy along with other electroacoustic instruments like the telephone and magnetic sound-recording devices.

From 1901 Simon was professor of applied electricity (*angewandte Elektrizität*) at the University of Göttingen. The creation of the professorship was part of mathematics professor Felix Klein's efforts to establish applied sciences in the university.[50] Simon and his students systematically analyzed the electrical and acoustic properties of the arc as an oscillating system in the years before the Great War. They conducted both theoretical as well as experimental investigations that covered all different aspects of the arc's discharge. Several of Simon's students would become

[47] Thompson, "Dead Rooms and Live Wires," 596–626; Thompson, *Soundscape of Modernity*, 229–93 (Both cit. n. 1).

[48] See Karl L. Schaefer, "Über die Erzeugung physikalischer Kombinationstöne mittels des Stentor-telephons," *Annalen der Physik* 322 (1905): 572–83, and Waetzmann, *Resonanztheorie des Hörens* (cit. n. 23), 120–3.

[49] Duddell, "On Rapid Variations in the Current through the Direct-Current Arc," *Journal of the Institution of Electrical Engineers* 30 (1900): 232–67; Simon, "Akustische Erscheinungen am electrischen Flammenbogen," *Ann. Physik* 300, no. 2 (1898): 233–9; Simon, "Zur Theorie des selbsttönenden Lichtbogens," *Physikalische Zeitschrift* 7, no. 13 (1906): 433–45; Simon, *Der elektrische Lichtbogen—Experimentalvortrag* (Leipzig, 1911).

[50] Hermann Th. Simon, "Das Institut für angewandte Elektrizität der Universität Göttingen," *Physikalische Zeitschrift* 7, no. 12 (1906): 401–12; Adelheid Wein, *Heinrich Barkhausen und die Anfänge der wissenschaftlichen Schwachstromtechnik* (master's thesis, Philosophy Faculty I, Univ. of Regensburg, 2011); about Simon, see Theodor des Coudres, "Hermann Th. Simon †," *Physikalische Zeitschrift* 14 (1919): 313–20.

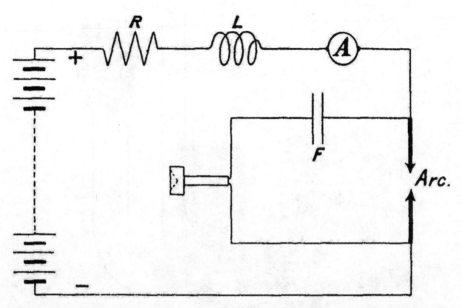

Figure 4. *Circuit diagram of the speaking arc as a telephone transmitter; Duddell, "On Rapid Variations" (cit. n. 49), 242.*

important figures in the formation of the electroacoustics research agenda in the interwar period. Karl Willy Wagner would be the founder and first director of the Heinrich Hertz Institut für Schwingungsforschung (Heinrich Hertz Institute of Oscillation Research). Hugo Lichte, who studied the sound produced by the electric arc, became head of the acoustics division of the Allgemeine Elektricitäts-Gesellschaft Forschungsinstitut (AEG Research Institute).[51] Heinrich Barkhausen became professor of *Schwachstromtechnik* (electrical communication engineering) at the Technische Hochschule Dresden. Barkhausen had compiled a comparative study on the generation of oscillations in electric and mechanical systems with a special emphasis on rapid electric oscillations generated by the arc.[52] His thesis, *Das Problem der Schwingungserzeugung*, was a point of origin of the program of *Schwingungsforschung* (oscillation and vibration research), which both Barkhausen in Dresden and Wagner at the Heinrich Hertz Institut followed in the interwar period.

In 1911 Simon summarized the different properties and potential capabilities of the arc as an electric system, including its heat- and light-producing powers, its capacity

[51] The AEG Forschungsinstitut was founded in 1928 with Carl Ramsauer as its first director. Lichte and the acoustics department worked mainly on a German system for sound motion pictures. For the AEG Forschungsinstitut, see Ernst Brüche, *Zehn Jahre Forschungsinstitut der AEG* (Berlin-Reinickendorf, 1938). For Lichte, see *Neue Deutsche Biographie*, vol. 14 (Berlin, 1985), s.v. "Lichte, Hugo," by Helmut Mielert; for his research on sound produced by the electric arc, see Lichte, "Über die Schallintensität des tönenden Lichtbogens," *Ann. Physik* 347, no. 14 (1913): 843–70. For Wagner, see Ernst Lübke, "Karl Willy Wagners Beiträge zur akustischen Forschung," *Akustische Zeitschrift* 8 (1943): 78–80, and Marianne Peilstöcker, "Professor Dr. Karl Willy Wagner [1883–1953]," *Jahrbuch Hochtaunuskreis 2003* (2002): 96–103.

[52] Barkhausen, *Das Problem der Schwingungserzeugung mit besonderer Berücksichtigung schneller elektrischer Schwingungen* (Leipzig, 1907).

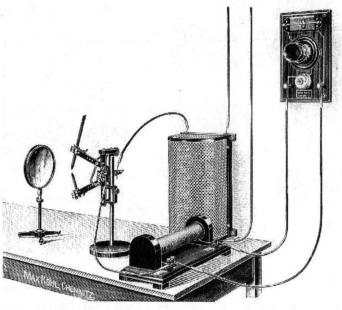

Max Kohl A. G. Chemnitz, Germany.

63 395, 50 892, 63 386, 63 388, 63 389. 1 : 8.

Complete Apparatus for Experiments with the Speaking Arc Lamp after Weinhold (W. D., £ s. d. Fig. 601), consisting of:

Arc Light Hand Regulator and Resistance: see Nos. 50,892, 63,386 and 63,387, p. 1057.

63,390. **Induction Coil,** F i g u r e 3. 4. 0

63,391. **Microphone with Switch and Regulating Resistance, and 1 Fuse with simple current
indicator for same** 3. 0. 0

Additional Apparatus for above so as to be able to demonstrate the automatically singing
(whistling) arc lamp also, consisting of:

63,392. **Small Induction Coil** (W. D., Fig. 604), with aluminium ring for Thomson's Experiment 0. 16. 0

61,122. **Switch** 0. 3. 6

63,393. **Paper Condenser** in simple wood box, approx. 8 mfd. 2. 0. 0

63,394. **Staged Paper Condenser,** with four steps, for use instead of No. 63,393; this permits
of obtaining a simple melody with the arc (electric piano) since the pitch varies with
the capacity cut in 6. 0. 0

*Figure 5. Apparatuses for singing- and speaking-arc experiments from the Max Kohl cata-
log, around 1911; Max Kohl, Physical Apparatus (cit. n. 45), 1058. In the same section of the
catalog, entitled "Telephony and Microphony," Max Kohl advertised sets for photophonic
apparatuses (used in wireless light telephony), employing acetylene light and selenium cells.*

to produce and consume sound, and finally its capacity to produce and detect rapid
electric oscillations in free space.[53] Its capacity to produce rapid electric oscillations
combined with its function as a reversible transducer made the carbon arc a perfect
candidate for wireless telephony based on electric waves, especially after the problem of producing a constant arc had been solved by Duddell. Simon did not stop with
sketching the fascinating technical prospects of the electric arc as a simple yet al-

[53] Simon, *Elektrische Lichtbogen* (cit. n. 49).

mighty wireless electroacoustic communication system. Emphasizing his theoretical and scientific interest, Simon placed the arc firmly in the electromagnetic worldview comprising electrons, atoms, and the all-penetrating world ether.[54] Through their research program on the electric arc as an oscillating system, Simon and his students placed acoustics, perhaps for the first time, in an electrical frame rather than a mechanical one.

ACOUSTICS IN THE CHEMISTS' WAR

Simon's investigations into the electric arc came to a halt with the outbreak of the Great War in 1914, when he and his students started to work on underwater acoustics related to submarine warfare for the German Imperial Navy in Kiel. The underwater world was an important battlefield but by no means the only one for acousticians. The Great War is sometimes called the War of the Chemists, emphasizing the importance of the chemical military-industrial-scientific complex in producing substitutes for scarce raw materials and, of course, in chemical warfare. But the notion belies the role of physicists and mathematicians in the mobilization and self-mobilization of scientists.[55] The Great War was the first large-scale industrial war where weapon systems like the submarine, the airplane, and the tank were used, and where large numbers of scientists were deployed in warfare research and development.[56] The field of acoustics became important for telephony, but even more so for artillery ranging, aircraft ranging, submarine detection, and the detection of tunnels under trenches.[57] In 1936 Erwin Meyer and Erich Waetzmann identified the Great War as the main driving force in the revival of acoustics research in the early twentieth century.[58] Meyer, born in 1899, was drafted as a soldier in 1917 before he started his studies with Waetzmann, professor of physics at Breslau. Waetzmann based his report on his own participation in scientific warfare. He had been a central actor in the German development of sound ranging of artillery and aircraft.[59]

Collaboration between scientists and the military in wireless telegraphy predated the Great War. Simon had initiated the Radioelektrische Versuchsanstalt für Marine und Heer (Radioelectric Testing Laboratory for the German Navy and Army) in Göttingen in 1908. His assistant Max Reich, who had been director of the Versuchsanstalt, became in 1915 head of a group at the Inspektion des Torpedowesens (German Command for Technical Inspection of the Torpedo Sector) in Kiel to develop

[54] Ibid., 32–5.

[55] Arne Schirrmacher, "Von der Geschossbahn zum Atomorbital? Möglichkeiten der Mobilisierung von Kriegs- und Grundlagenforschung füreinander," in *Mit Feder und Schwert: Militär und Wissenschaft—Wissenschaftler und Krieg*, ed. Matthias Berg, Jens Thiel, and Peter Th. Walther (Stuttgart, 2009), 155–75, on 153.

[56] Helmuth Trischler, "Die neue Räumlichkeit des Krieges: Wissenschaft und Technik im 1. Weltkrieg," *Berichte zur Wissenschaftsgeschichte* 19 (1996): 95–103; Guy Hartcup, *The War of Invention: Science in the Great War, 1914–18* (London, 1988).

[57] Hartcup, *War of Invention* (cit. n. 56), 68–80, 129–35, 164; Willem D. Hackmann, *Seek and Strike: Sonar, Anti-submarine Warfare, and the Royal Navy, 1914–54* (London, 1984); Schirrmacher, "Von der Geschossbahn zum Atomorbital?" (cit. n. 55).

[58] Meyer and Waetzmann, "Bedeutung der Akustik" (cit. n. 14), 508; see also Beyer, *Sounds of Our Times* (cit. n. 1), 197.

[59] Meyer, "Erich Waetzmann zum Gedächtnis," *Akustische Zeitschrift* 3 (1938): 241–4; Lothar Cremer, "Erwin Meyer, 21. Juli 1899 – 6. März 1972," *Jahrbuch der Akademie der Wissenschaften in Göttingen* 1972, 179–85.

and investigate a system for sound ranging of submarines, while Simon carried out complementary experiments at the University of Göttingen. Other students of Simon who were drafted for the Inspektion des Torpedowesens in 1915 to work on underwater acoustics were Barkhausen and Lichte. The group of physicists who worked on problems of underwater acoustics in Kiel during the Great War included Hans Riegger, who worked for Siemens & Halske, and Alard du Bois-Reymond (son of Emil du Bois-Reymond), Karl Heinrich Hecht, and Walter Hahnemann, who worked for the Signal Gesellschaft, a subsidiary of Hanseatische Apparatebaugesellschaft Neufeldt & Kuhnke.[60]

During the Great War it became clear to scientists and engineers that it was not the electric arc but the triode vacuum tube that held the potential to become the dominant, overarching component for generating, amplifying, modifying, and broadcasting electric oscillations. Electromechanically, electromagnetically, or electrostatically driven loudspeakers became the norm. The sound-producing and sound-consuming properties of the electric arc, as fascinating as they were, remained only a curiosity.

After the Great War, scientists who were involved in sound ranging and sound signaling published their war research, including Lichte, Barkhausen, and Du Bois-Reymond.[61] Hahnemann and Hecht of the Signal Gesellschaft in Kiel published a series of papers about theoretical properties of sound transmitters and sound receivers, which were structured as a generalized theory of electroacoustic transmitters.[62] In order to conceptualize the theory of the electromechanical transformer, Hahnemann and Hecht fell back on the well-understood theory of the alternating-current transformer, which Steinmetz had explained in 1897 using equivalent-circuit diagrams.[63] Hahnemann and Hecht's approach was to replace the mechanical movement of the coil driving the membrane of a sound transmitter with a secondary circuit that was equivalent to the mechanical system. Instead of driving a coil and membrane, the primary circuit would induce an equivalent oscillation in the secondary circuit (see fig. 6). Hahnemann and Hecht thereby replaced a coupled system, which contained both electric and mechanical components, with a purely electrical representation of the system. The electrical representation of coupled systems became the backbone of the theory of electroacoustic transducers and the use of equivalent-circuit diagrams,

[60] For Riegger, see Hans Gerdien, "Hans Riegger †," *Zeitschrift für technische Physik* 7 (1926): 321–4; for Hecht, see Hugo Lichte, "Dr. Heinrich Hecht zum 60. Geburtstag," *Zeitschrift für technische Physik* 21, no. 2 (1940): 25–7, and *Neue Deutsche Biographie*, vol. 8 (Berlin, 1969), s.v. "Hecht, Karl Heinrich," by Erhard Ahrens; for Hahnemann, see Jonathan Zenneck, "W. Hahnemann," *Hochfrequenztechnik und Elektroakustik* 53, no. 6 (1939): 214.

[61] Heinrich Barkhausen and Hugo Lichte, "Quantitative Unterwasserschallversuche," *Ann. Physik* 367 (1920): 485–516; Alard du Bois-Reymond, "Englische U-Boot-Abwehr," *Zeitschrift für technische Physik* 2, no. 9 (1921): 234–8.

[62] Walter Hahnemann and Heinrich Hecht, "Schallgeber und Schallempfänger I" and "Schallgeber und Schallempfänger II," *Physikalische Zeitschrift* 20 (1919): 104–14 and 245–51; Hahnemann and Hecht, "Der mechanisch-akustische Aufbau eines Telephons," *Ann. Physik* 365 (1919): 454–80. Hecht later published two monographs, on equivalent circuits and differential equations of mechanical and electrical oscillating systems in 1939, and on the theory of electroacoustic transducers in 1941: Hecht, *Schaltschemata und Differentialgleichungen elektrischer und mechanischer Schwingungsgebilde* (Leipzig, 1939); Hecht, *Die elektroakustischen Wandler* (Leipzig, 1941).

[63] Steinmetz and Berg, *Theory and Calculation of Alternating Current Phenomena* (cit. n. 40), 184–5; Hahnemann and Hecht, "Schallgeber und Schallempfänger I" (cit. n. 62). Hahnemann and Hecht did not make any reference to Steinmetz, but to Gisbert Kapp's 1907 textbook on transformers; see "Schallgeber und Schallempfänger I," 107, and Kapp, *Transformatoren für Wechselstrom und Drehstrom* (Berlin, 1907), 140, 227, 232. It is noteworthy that Kapp used a hydraulic analogy to explain magnetic scattering in his textbook (4).

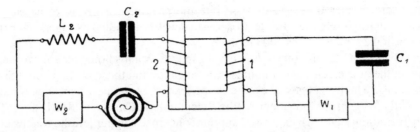

Figure 6. *Electric-circuit diagram of an electromagnetic transducer as a sound receiver. The circuit of the transducer was equivalent to that of an electromagnetic transformer. The circuit 2 represented the mechanical movement of the speaker, with* L_2 *and* C_2 *being equivalent to the elasticity of the membrane, the mass of the oscillating parts, and the mechanical and acoustic radiation resistance; Hahnemann and Hecht, "Schallgeber und Schallempfänger II" (cit. n. 62), 245.*

as presented in Frederick Vinton Hunt's monograph in 1954. But the equivalent-circuit regime could achieve even more. In the course of the interwar years, the equivalent-circuit diagram came to be mobilized to represent acoustic systems that did not have an electric component, such as walls in a house, automobile sound silencers, and even the human voice.

THE ADVANCE OF THE EQUIVALENT CIRCUIT:
TRANSLATING SOUND INTO ELECTRICAL NETWORKS

On 22 March 1929, the physicist William Henry Eccles, who had made important contributions to circuit design, gave a presidential address to the Physical Society of London, bringing to the attention of British physicists the novelty and importance of the "new acoustics" that had arisen in the previous decade.[64] In addition to the prominence that electroacoustic measurement technology had gained in acoustics research and the rise of electroacoustic media technology, Eccles paid special attention to the conceptual aspects that had transformed electroacoustics into a new language of sound:

> Besides these tangible adjuncts to the technique of experimental acoustics, electrical science has brought subtle assistance to the more theoretical aspects of the subject. This comes about because vibration phenomena of all kinds approximately satisfy the same linear differential equations. Inasmuch as the study of electrical vibrations in well-defined electrical circuits is easier and has been more cultivated (for practical purposes) than that of air vibrations, acoustic science profits from electrical by a free exchange of ideas about vibrations. Many acoustical problems can be translated into problems concerning electrical networks, and as there exists a great body of knowledge of such networks, the problem is often solved in the act of translation. Further, by adopting the phraseology of the electrician into acoustics, so that translation of the acoustic problem into the electrical problem becomes automatic, a language for thinking and talking becomes available and is found to clear the mind and assist reasoning.[65]

[64] Eccles, "New Acoustics" (cit. n. 16). For Eccles, see John Ashworth Ratcliffe, "William Henry Eccles, 1875–1966," *Biographical Memoirs of Fellows of the Royal Society* 17 (1971): 195–214. In 1918, Eccles and Frank Wilfred Jordan had patented a trigger circuit based on vacuum tubes, which later became known as the first flip-flop circuit.

[65] Eccles, "New Acoustics" (cit. n. 16), 233.

Furthermore, Eccles observed that, despite the prominence given to the study of acoustics by Rayleigh, the new electroacoustics was not studied as eagerly in Britain as it was in the United States and Germany.[66]

Eccles was not the only one to identify a gap between his home nation's academic efforts in the new electrical communication sciences and those of the United States. In Germany, which Eccles saw as one of the leaders in the new acoustics research, Karl Willy Wagner, as president of the Telegraphentechnisches Reichsamt (Telegraph Technology Office), expressed similar concerns. Wagner left the Reichsamt in 1927 in order to initiate and direct the Heinrich Hertz Institut. The foundation of the institute was an explicit attempt by Wagner and others in the German research community to catch up and compete with American research, especially with the Bell Telephone Laboratories, which AT&T and Western Electric had formed in 1925 as a consolidated research laboratory.

Wagner did not attempt to copy the structures and practices of private American research laboratories. The patrons of the Heinrich Hertz Institut were the Studiengesellschaft für Schwingungsforschung (Association for the Study of Oscillation Research) and the Heinrich-Hertz-Gesellschaft zur Förderung des Funkwesens (Heinrich Hertz Association for the Advancement of Radio Communications), which was composed of representatives from the Reichspost (the Imperial Post, which administered the state monopoly of telephony, telegraphy, and radio broadcasting), the electrical industry, and the Technische Hochschule Berlin-Charlottenburg. While universities used to be the main arena for acoustics research before the Great War, the electrical industry, public research and testing laboratories, and Technische Hochschulen now took the lead in the new acoustics research in the interwar period. The Heinrich Hertz Institut brought all three actors together, forming an institution that was unique in the international context.

Just as its institutional composition was unique, so too was its research program.[67] The strong links among acoustics, communication technologies such as telegraphy and telephony, electric sound generation and amplification, and radio broadcasting led to the emergence of the research field of Schwingungsforschung, which brought together scientific and engineering disciplines.[68] The research program of Schwingungsforschung was not Wagner's invention, as we have already discussed, but grew out of Barkhausen's dissertation during their time at Göttingen. Wagner as the president of the Heinrich-Hertz-Gesellschaft acknowledged Barkhausen's contributions in the formation of the field of Schwingungsforschung by awarding him the first gold Heinrich Hertz Medal in 1928.[69]

In addition to Barkhausen and Wagner, there were others who formulated a research program for Schwingungsforschung in Germany. In an article in the *Zeitschrift für technische Physik* in 1922, Walter Hahnemann proposed problems arising from

[66] Ibid., 239.

[67] Friedrich-Wilhelm Hagemeyer, *Die Entstehung von Informationskonzepten in der Nachrichtentechnik: Eine Fallstudie zur Theoriebildung in der Technik in Industrie- und Kriegsforschung* (Berlin, 1979), 139–40, 267.

[68] Ibid., 326; Karl Willy Wagner, "Das Heinrich-Hertz Institut für Schwingungsforschung," *Elektrische Nachrichtentechnik* 7 (1930): 174–91; Wittje, *Acoustics, Atom Smashing and Amateur Radio* (cit. n. 12), 78.

[69] Wein, *Heinrich Barkhausen* (cit. n. 50), 1–2.

oscillation as the main frame for the new technical acoustics research.[70] The notion of electroacoustics gained momentum around the same time. The industrial physicist Walter Schottky, who spent most of his career at the Siemens Laboratories, wrote the chapter on electroacoustics in *Die wissenschaftlichen Grundlagen des Rundfunkempfangs* (The scientific basis of radio), a volume edited by Wagner in 1927 following a lecture series organized by the Heinrich-Hertz-Gesellschaft. In 1931, the *Zeitschrift für Hochfrequenztechnik,* a journal for electrical communication engineering, changed its name to *Zeitschrift für Hochfrequenztechnik und Elektroakustik,* reflecting the growing importance of electroacoustics within physics and electrical engineering.

Schottky, who worked on loudspeakers and microphones in the 1920s, is better known for his theoretical work on radio tubes and solid-state physics.[71] It was Ferdinand Trendelenburg who became the main acoustician at the Forschungslaboratorium (research laboratory) of Siemens & Halske and the Siemens-Schuckertwerke. Trendelenburg had obtained his doctorate at the University of Göttingen in 1922 with Max Reich, who had become Theodor Simon's successor as chair of applied electricity.[72] In 1932, Trendelenburg assembled a monograph on recent proceedings in acoustics, which he extended in a second edition in 1934.[73] In the section on hearing and speech, which was Trendelenburg's own research interest, he discussed a schematic representation of speech, which used an electric circuit of a self-exciting radio transmitter as an equivalent representation of the human voice (fig. 7). Trendelenburg referred to two scientists from the Bell Laboratories, Irving Crandall and Raymond Lester Wegel, the latter of whom had discussed representing the vibration of the vocal cords by an electric equivalent circuit.[74] John Q. Stewart, a research engineer in the Development and Research Department of AT&T, had already published a circuit diagram for producing simple speech sounds in 1922.[75]

Developing equivalent circuits for the human voice was certainly useful if one intended to build an analog electric human-speech synthesizer, an endeavor in which Bell Laboratories became a central actor. The electric-circuit analysis, however, could be applied to a whole range of acoustic systems that would never be built as real electric machines like the voder, the name under which the human-speech synthesizer came to be known. Noise abatement had become one of the most active

---

[70] Hahnemann, "Schwingungstechnische Probleme als Grundlage der technischen Akustik," *Zeitschrift für technische Physik* 3, no. 2 (1922): 44–6.

[71] Reinhard W. Serchinger, *Walter Schottky—Atomtheoretiker und Elektrotechniker* (Diepholz, 2008).

[72] Trendelenburg can thus be seen in the Göttingen tradition of electroacoustics research, established by Simon. In his thesis, Trendelenburg had worked on the thermophone as a sound transducer. For Trendelenburg, see Ernst Lübke, "Ferdinand Trendelenburg 60 Jahre," *Physikalische Blätter* 12 (1956): 270.

[73] Trendelenburg, *Die Fortschritte der physikalischen und technischen Akustik* (Leipzig, 1932); 2nd enlarged ed., 1934.

[74] Trendelenburg, *Fortschritte der physikalischen und technischen Akustik* 1934 (cit. n. 73), 77; see Hunt, *Electroacoustics* (cit. n. 1, on 66) for Wegel. Crandall introduced a range of electrical analogies in his *Theory of Vibrating Systems and Sound* (New York, 1926).

[75] Stewart, "An Electrical Analogue of the Vocal Organs," *Nature* 110 (1922): 311–2. Stewart, an astrophysicist, is better known for his engagement with social physics. During the Great War he served as chief instructor in sound ranging at the Army Engineering School, after which he worked for AT&T until 1921. See David H. DeVorkin, *Henry Norris Russell: Dean of American Astronomers* (Princeton, N.J., 2000), 208.

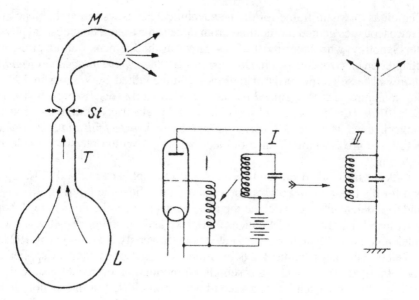

Figure 7. *The self-exciting radio-tube transmitter as an equivalent circuit for the human voice; Trendelenburg,* Fortschritte der physikalischen und technischen Akustik *1934 (cit. n. 73), 76.*

fields of acoustics research and development in the interwar period, owing its prominence partly to the battlefields of the Great War, partly to the pathogenic (though less deadly) sound levels of modern industrial and urban life.[76] Trendelenburg added a separate section on noise abatement in the second enlarged edition of his monograph in 1934.

The development of sound silencers was another field where equivalent-circuit diagrams became useful. Demand for sound silencers came, for example, from the growing automobile industry in Europe and North America, which needed to silence the noise produced by combustion engines of both cars and motorbikes. Most of the sound silencers were built on the principle of acoustic filters.[77] The theory of electric filters had advanced quite far. George W. Stewart, a physicist at Iowa State University, had designed an aircraft sound locator for the US National Research Council during the Great War.[78] In the early 1920s he worked on acoustic wave filters based on analogies with well-understood electric wave filters. Stewart sold some of his patent rights to Bell Laboratories but did not envision the sound silencer as an application.[79] This was done by Martin Kluge, a former student of Heinrich Barkhausen in Dresden. His habilitation thesis was on the silencing of the noise produced by

[76] Thompson, *Soundscape of Modernity* (cit. n. 1), 115–68.

[77] Trendelenburg, *Fortschritte der physikalischen und technischen Akustik* 1934 (cit. n. 73), 168.

[78] Stewart, "Location of Aircraft by Sound," *Physical Review* 14, no. 2 (1919): 166–7.

[79] Stewart, "An Acoustic Wave Filter," *Physical Review* 17, no. 3 (1921): 382–4; Stewart, "Acoustic Wave Filters," *Physical Review* 20, no. 6 (1922): 528–51; Harvey Fletcher, "George W. Stewart, 1876–1956," *Biographical Memoirs of the National Academy of Sciences* 32 (1958): 378–98. Stewart referred to George W. Pierce, professor of physics at Harvard University, who worked on wireless and electric circuits, with regard to electric filters. Karl Willy Wagner developed a theory of electric filters during the Great War as well.

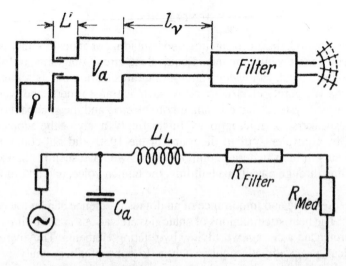

Figure 8. *Design and electric equivalent-circuit diagram of an automobile exhaust silencer with accumulation chamber by Martin Kluge; Trendelenburg,* Fortschritte der physikalischen und technischen Akustik *1934 (cit. n. 73), 169.*

combustion engines. As an electrical engineer, Kluge used electric equivalent-circuit diagrams in his design of an acoustic filter (fig. 8).

Architectural acoustics and the sound insulation of buildings were other prominent areas for the representation of acoustic systems by equivalent-circuit diagrams, as we have seen in figure 1. Not only Reno Berg and Johan Holtsmark in Trondheim, but also Erwin Meyer, the head of the Department of Acoustics at the Heinrich Hertz Institut in Berlin, used equivalent-circuit diagrams in studying how sound traveled through a wall.[80] In the 1930s Meyer had become Germany's most recognized acoustician and gave lectures in the United States, the Soviet Union, and Great Britain. The five lectures that Meyer gave at the Institution of Electrical Engineers in London in autumn 1937 were subsequently published in 1939 under the title *Electro-acoustics*, one among several monographs on electroacoustics at the time.[81]

By 1939 acoustics in Germany had experienced several years of remilitarization of its research agenda. The Berlin Institut für Schwingungsforschung was stripped of its eponym *Heinrich Hertz*, as Hertz, having been of partly Jewish ancestry, was unacceptable to the new National Socialist rulers. Within a few years Germany was at war with Britain, the Soviet Union, and the United States. Meyer headed research in underwater acoustics for submarine warfare. Under the auspices of Project Alberich he worked to make German submarines invisible to sonar. On the other side of the Atlantic, Harvard University's Frederick Vinton Hunt developed efficient sonar systems, which were supposed to detect and ultimately sink these very submarines. We must believe that the application of electrical analogies and equivalent circuits aided both Meyer and Hunt in their endeavors.[82]

[80] Meyer, *Electro-acoustics* (London, 1939), 112.

[81] Ibid. See also Philippe Le Corbeiller, *Electro-acoustique* (Paris, 1934); Agostino Gemelli and Giuseppina Pastori, *L'analisi elettroacustica del linguaggio* (Milan, 1934).

[82] Hunt made explicit reference to the advances that electroacoustics made in the United States during the Second World War; Hunt, *Electroacoustics* (cit. n. 1), vi.

CONCLUSION

In the interwar period, the electroanalog understanding of acoustics and its technologies and methods, especially the use of equivalent-circuit diagrams, unfolded into a new language of sound. Equivalent-circuit diagrams started out as a method to describe and analyze purely electric systems, such as transformers, and were then extended to coupled systems that contained both electric and mechanical components, such as loudspeakers and microphones. Interestingly, many of the acoustic systems represented by equivalent-circuit diagrams in the 1930s did not contain any electric components in their material composition, as we have seen in the examples of sound traveling through a wall in a building, the human voice, and automobile sound silencers.

The widespread use and importance of analogies in science can be traced throughout history.[83] The heuristic functions of analogies are various and multilayered. Analogies can propagate a certain worldview by relating elements. They can also serve more practical and pragmatic goals, such as to conceptualize, analyze, and solve scientific and technical problems. Analogies can allow the transfer of thought patterns, language, and mathematical formalism, and can be used to build models and machinery. They can point toward the likenesses of phenomena and supply arguments for identifying two apparently different phenomena as being physically identical. This was the case in Maxwell's and Hertz's identification of light as an electromagnetic phenomenon. But instead of claiming physical identity, analogies can also point toward structural identity and aid the visual and functional understanding of systems and processes by transferring from the well understood to the novel and unknown.[84]

How, then, did the electrical representation of sound in the 1930s differ from the Maxwellian representation of electromagnetic phenomena through the mechanics of the ether in the late nineteenth century? Victorian physicists like Maxwell, FitzGerald, and Lodge imagined electromagnetic phenomena as mechanical machinery in order to understand and explain their mode of operation. Even though the mechanical models were not thought of as true representations of the ether, the ether was believed to be of a truly mechanical nature. To conceive of a mechanical model meant to have understood the phenomenon. Despite the efforts of Polhem and Reuleaux, there was no mechanical alphabet or grammar available similar to the equivalent-circuit diagram. Some of these models were built as functional devices for lecture demonstration. These demonstration devices were not only meant to show the fundamentally mechanical nature of electromagnetism; through their analogy with well-understood mechanics, they also taught abstract and unfamiliar electromagnetic phenomena to scientists and engineers. Then, with the decline of a mechanical worldview and the increasing familiarity of scientists and engineers with electrical technology, the need to explain the mechanism of electromagnetic phenomena decreased.

The use of electrical analogies and equivalent-circuit diagrams in acoustics in the twentieth century was, in contrast, driven not by worldviews but by more pragmatic and technical considerations. With the rapid decline of the electromagnetic world-

---

[83] See Klaus Hentschel, ed., *Analogien in Naturwissenschaften, Medizin und Technik* (Stuttgart, 2010).
[84] Ibid., 32–3.

view no one claimed physical identity or argued for a fundamentally electrical nature of sound. Physical acoustics continued to be understood essentially as a section of the motions of elastic bodies within the field of mechanics as introduced by Helmholtz and Rayleigh. Why did these electrical representations of acoustic systems still become so powerful and widespread in the interwar period?

The answer to this question can be given by three arguments. First, measurement technology in acoustics research had become increasingly electrical. Before the electrification of acoustic measurement, a trained ear and an understanding of the system of European classical music were required. After electrification, the acoustician was not supposed to trust his own ear but had to develop an understanding of the design, behavior, and manipulation of electric circuits and transducers. Even if the acoustic system that was analyzed was purely mechanical, like a wall construction or a sound silencer, electrical technology and electrical thinking were already present. Electrical engineers and scientists proficient in electrical measurement technology therefore dominated acoustics research even in fields like civil engineering and automobile mechanics.

Second, electric oscillations were well understood and structurally analogous to acoustic vibrations. Acoustic systems could therefore be translated into electric systems. With the concepts of the voltage-source equivalent and the current-source equivalent, circuit diagrams had become a powerful tool to conceptualize, analyze, and solve all kinds of complex oscillation problems, whether they were electrical, electromechanical, or purely mechanical. The language of equivalent-circuit diagrams was a language of signs and relations that offered a reduced representation in which extraneous information was eliminated. Dealing with electric systems had become common practice for a large community of scientists and engineers who shared the language of the circuit diagram.

Third, equivalent-circuit diagrams were useful for designing, analyzing, and improving technology. This could be electroacoustic technology; for example, a loudspeaker, a microphone, or an audio amplifier. But purely mechanical technology like a wall construction or an automobile sound silencer could also be successfully analyzed and improved with the help of equivalent-circuit diagrams. As a structuralist language they created a direct link between sound and the industrial design and production of technology.

The electroanalog understanding of sound persists today as one of the foundations of acoustics in textbooks and manuals, and as an inherent component of the conceptual toolbox of the acoustician. Of course, since the 1960s a new digital concept of sound has emerged with the arrival of compact discs, computer music, and MP3 players, which has taken over the analog sound we had gotten so used to from radio, vinyl discs, and cassette tapes. Another translation has taken place from the analog to the digital, where electric signals are converted into bits and stored in binary form. Through digitization, our way to conceptualize, manipulate, and experience sound has again changed significantly. Together with computer algorithms, the concept of *information* has entered our aural world as a novel but fundamental entity. Again, a new phraseology of acoustics has become automatic "and is found to clear the mind and assist reasoning."[85] What has persisted into the twenty-first century is the essentially cultural as well as technological nature of our approach to sound.

[85] Eccles, "New Acoustics" (cit. n. 16), 233.

The Aesthetics of the Signal:
Noise Research in Long-Wave Radio Communications

by Daniel Gethmann*

ABSTRACT

This paper is based around two radio-reception experiments, which both originated as research on long-wave radio and serve to illustrate, from alternate perspectives, the constitutive role that noise plays in communication. The projects of Karl Guthe Jansky and Charles Francis Jenkins opened new research perspectives. The two experiments both ensued from acoustics research, but in different ways, turning disruptive noise into material for a new scientific research approach and even going so far as to investigate any information it might contain. Philosopher Michel Serres suggested that noise, forming the basis of all communication and signal or message transmission through space, might in fact help to reconceive communication and to foster a renewal of the concept of transmission. This article traces the history of Serres's suggestion.

The radio-reception experiments conducted in the early twentieth century by Karl Guthe Jansky and by Charles Francis Jenkins illustrate the constitutive role that noise plays in communication. Originating as research on long-wave radio, Jansky's and Jenkins's investigations turned disruptive noise into material for a new scientific research approach and even went so far as to explore any information it might contain. The philosopher and historian of science Michel Serres, in turn, fielded the suggestion that noise—which forms the basis of all communication and signal or message transmission through space—might in fact help to reconceive communication and to foster a renewal of the concept of transmission.[1]

In his book *Hermès II: L'interférence*, Serres describes the basic situation:

> What transmits of its own accord when no demons interfere? What can be heard in a world without people? The turbulence in its purest form, the flow of particles, the collision of individuals randomly dispersed in time, the streaming of the cloud due to the space-charge effect. Who converses inside this cloud? Strictly speaking, no one; though

---

* Institute of Architectural Theory, History of Art and Cultural Studies, Graz University of Technology, Technikerstr. 4, A-8010 Graz, Austria; daniel.gethmann@tugraz.at.
This article was translated by Dawn Michelle d'Atri.

[1] See Serres, *Hermès I: La communication* (Paris, 1968); *Hermès II: L'interférence* (Paris, 1972); *Hermès III: La traduction* (Paris, 1974); *Hermès IV: La distribution* (Paris, 1977); *Hermès V: Le passage du nord-ouest* (Paris, 1980); *The Parasite*, trans. Lawrence R. Schehr (1982; repr., Minneapolis, 2007).

© 2013 by The History of Science Society. All rights reserved. 0369-7827/11/2013-0004$10.00

OSIRIS 2013, 28 : 64–79

most certainly the object, the thing itself, the world. It speaks, as we say, yet here the third person retains the clear meaning that it had never lost: the accumulation of objects that form the universe; the interobjectivity as such.[2]

Research conducted in the 1920s and 1930s revealed that noise could grant insight into the function of communication devices regardless of whether or not it originated from someone communicating. The understanding of the signal-to-noise ratio estab-lished over the course of this research had two distinct features. The first was that the traditional linear concept of one-on-one communication was expanded to include a third party, which could also be a noise source. Second, the expansion of the com-munication model to include this third figure led to further questions: What does the new model contribute to the communication situation? Or, how does the listening-to-noise or listening-to-message status change based upon this new concept of com-munication?

STAR NOISE

Between 1928 and 1930, Karl Guthe Jansky, who worked in the Radio Research Division of Bell Telephone Laboratories, investigated interferences that occured dur-ing long-distance radio telephony. These interferences were remarkably strong, yet their source was unknown.[3] In 1927, a year before Jansky started his investigation, the American Telephone & Telegraph Company had begun offering a public radio-telephonic service between New York and London for the first time. The service op-erated on long-wave channels at a wavelength of 5,000 meters (60 kilohertz) that had previously been established for communicating across great distances. Working on the premises of a large farm in Holmdell, New Jersey, Jansky had, coincidentally, started recording long-wave static on 60 kilohertz. Already in 1929, the frequent oc-currence of static on this frequency had caused the wireless communication service to be switched to a short-wave frequency between 10 and 20 megahertz. This, in turn, motivated Jansky to pursue the construction of rotating antennae and also short-wave reception (20.5 megahertz) for troubleshooting the interference that he was now ex-periencing as well.

During his experiments with short-wave reception starting in the fall of 1930, Jan-sky encountered strong static that he attributed to thunderstorms both nearby and at a distance. This static, however, could be distinguished from an additional sort, "a very steady hiss type static the origin of which is not yet known."[4] Jansky de-scribed its acoustic performance thus: "It is, however, very steady, causing a hiss in the phones that can hardly be distinguished from the hiss caused by set noise."[5] After comparing his experimental data collected during the previous years, Jansky

[2] Serres, *Hermes II: Interferenz*, trans. Michael Bischoff (Berlin, 1992), 255.
[3] For more on Jansky, see K. Kellermann and B. Sheets, eds., *Serendipitous Discoveries in Radio Astronomy* (Green Bank, W.V., 1983); Woodruff Turner Sullivan III, "Karl Jansky and the Discovery of Extraterrestrial Radio Waves," in *The Early Years of Radio Astronomy: Reflections Fifty Years after Jansky's Discovery*, ed. Sullivan (Cambridge, 1984), 3–42; Sullivan, *Cosmic Noise: A History of Early Radio Astronomy* (Cambridge, 2009), 29–53.
[4] Jansky, "Directional Studies of Atmospherics at High Frequencies," *Proceedings of the Institute of Radio Engineers* 20, no. 12 (1932): 1920–32, on 1925.
[5] Ibid., 1930.

developed an astronomical explanation for this static in 1932: he localized the source of the radio-continuum noise outside of our solar system, but "the direction of the phenomenon remains fixed in space."[6] Jansky's hypothesis about the astronomical source of the noise established "cosmic static" as the basis for further astronomical research through the quest for additional "noiselike signals" from space. Jansky's colleagues did not immediately react to a lecture he gave titled "Electrical Disturbances of Extraterrestrial Origin" at the annual International Scientific Radio Union conference on 27 April 1933. It was not until Bell Labs issued a press release and the *New York Times* printed an article about Jansky's insights the following month that Jansky became famous overnight.[7] An invitation by the NBC Blue Radio Network followed. On 15 May 1933, Jansky facilitated a direct transmission of galactic noise from the station in Holmdell. According to a listener, it sounded "like steam escaping from a radiator."[8]

Experts in astronomy, by and large, initially ignored Jansky's experiments on radio reception—the astronomers in no way understood their practice to be a science of sound.[9] Or, put differently, astronomers and radio-communications engineers were not yet pursuing joint research interests. To illustrate: in August of 1933, Jansky's superior at Bell Labs was curtailing the research on "star noise" at the very same time that Jansky delivered an in-house progress report asserting that the Milky Way was the source of "interstellar interference."[10] Jansky's activity was not suspended due to his seemingly grandiose claim, nor because of the consequence of this claim—that knowledge of the universe might be obtained through the reception of electromagnetic waves, necessitating the founding of the field of radio astronomy.[11] Instead, Jansky's discovery of star noise had simply met the original research objective of improving radio reception.

So when Jansky gave a lecture titled "Hearing Radio from the Stars" in October 1933 at the American Museum of Natural History in New York, his superiors at Bell Labs were far from pleased that he had continued his research. As a result, Jansky figured that one option in this situation was to encourage commercial radio stations to support acoustic experiments in radio astronomy. While these efforts to obtain support proved unsuccessful, Jansky did find an enthusiastic amateur radio operator who was to be, for years to come, his only successor. Radio amateur Grote Reber for the most part pursued the idea of acoustic radio astronomy in his own backyard, where

[6] Jansky, "Electrical Disturbances Apparently of Extraterrestrial Origin," *Proceedings of the Institute of Radio Engineers* 21, no. 10 (1933): 1387–98, on 1388.
[7] See "New Radio Waves Traced to Centre of the Milky Way: Mysterious Static, Reported by K. G. Jansky, Held to Differ from Cosmic Ray," *New York Times*, 5 May 1933, 1.
[8] Quoted in Sullivan, "Karl Jansky and the Discovery of Extraterrestrial Radio Waves" (cit. n. 3), 18.
[9] See Jesse L. Greenstein, "Optical and Radio Astronomers in the Early Years," in Kellermann and Sheets, *Serendipitous Discoveries in Radio Astronomy*, 79–88; Greenstein, "Optical and Radio Astronomers in the Early Years," in Sullivan, *Early Years of Radio Astronomy*, 67–81 (Both cit. n. 3).
[10] Jansky mentioned "star noise" in a letter to his father, Cyril M. Jansky, dated 29 Mar. 1936; quoted in Sullivan, "Karl Jansky and the Discovery of Extraterrestrial Radio Waves," 23; see also A. C. Beck, "Personal Recollections of Karl Jansky," in Kellermann and Sheets, *Serendipitous Discoveries in Radio Astronomy*, 32–8, on 36 (Both cit. n. 3). In a short article published in *Nature* on 8 July 1933, Jansky had already written "that the direction of arrival of this disturbance remains fixed in space, that is to say, the source of this noise is located in some region that is stationary with respect to the stars." Jansky, "Radio Waves from Outside the Solar System," *Nature* 132 (1933): 66.
[11] See Sullivan, *Cosmic Noise* (cit. n. 3); Kristen Rohlfs and Thomas L. Wilson, *Tools of Radio Astronomy* (Berlin, 2003); Sullivan, *Early Years of Radio Astronomy*; Kellermann and Sheets, *Serendipitous Discoveries in Radio Astronomy* (Both cit. n. 3).

he erected an antenna and occupied himself with the reception of cosmic noise long past the Second World War.[12]

In 1937, in what was, for the time being, his last paper about the reception experiments, Jansky noted, "On the shorter wave lengths and in the absence of man-made interference, the usable signal strength is generally limited by noise of interstellar origin."[13] With the discovery of star noise, noise in general was introduced as a defining factor in communication just as its source was being determined. Because of its various (measurable) manifestations, noise could be calculated with significant precision, better enabling the long-distance transmission of messages. Jansky's discovery that star noise was an invariable, and basically external, source of sound that restricted radio communication established the practice of calculating noise as a constant in communication—and by no means only as a disruption thereof. Radio astronomy, therefore, can be understood to have originated from the sole aim of improving the quality of earth-bound conversation through the investigation of errors in wireless telephone connections at Bell Labs, an activity that was not of interest to the established astronomy community.

Jansky's experiments suggested that noise be integrated into the understanding of all communicative acts, both with and without technical media. According to the assumptions posited in his 1937 paper, this integration was inevitable. Jansky's experiments were complemented by a process of formalization in messaging-technology discourse on information theory initially articulated by Ralph Hartley in his 1927 paper "Transmission of Information." Hartley required that psychological factors be eliminated from the signs and symbols used for messaging and that, further, a measure of information in terms of purely physical quantities be established. He based this postulate on the visualization of a signal sequence in a "hand-operated submarine cable system in which an oscillographic recorder traces the received message on a photosensitive tape" (fig. 1).[14]

Hartley encountered difficulties, however, on the visual level of this signal sequence. When the signals of a deliberately chosen message were entered by a telegraph operator and "a similar sequence of arbitrarily chosen symbols might have been sent by an automatic mechanism which controlled the position of the key in accordance with the results of a series of chance operations such as a ball rolling into one of three pockets," then, at the sending end, even "those . . . not familiar with the Morse code" could recognize in both sequences the symbols selected.[15] Hartley continued, explaining that in this case a trained receiving operator "would say that the sequence sent out by the automatic device was not intelligible" and that this was due to the fact that "only a limited number of the possible sequences have been assigned meanings common to him and the sending operator."[16] Hartley concluded that it was psychological considerations that limited the number of symbols available and ultimately prevented a more effective signal transmission:

[12] See Reber, "Early Radio Astronomy at Wheaton, Illinois," *Proceedings of the Institute of Radio Engineers* 46, no. 1 (1958): 15–23.

[13] Jansky, "Minimum Noise Levels Obtained on Short-Wave Radio Receiving Systems" (summary), *Proceedings of the Institute of Radio Engineers* 25, no. 12 (1937): 1517–30, on 1517.

[14] Ralph Vinton Lyon Hartley, "Transmission of Information," *Bell System Technical Journal* 7, no. 3 (1928): 535–63, on 536. This paper was presented in Sept. 1927 at the International Congress of Telegraphy and Telephony, Lake Como.

[15] Ibid., 537.

[16] Ibid., 538.

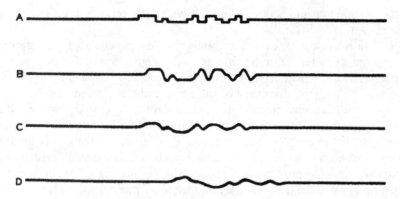

Figure 1. *"Fig. 1 shows at* A *the sequence of key positions, and at* B, C *and* D *the traces made by the recorder when receiving over an artificial cable of progressively increasing length. For the shortest cable* B *the reconstruction of the original sequence is a simple matter. For the intermediate length* C, *however, more care is needed to distinguish just which key position a particular part of the record represents. In* D *the symbols have become hopelessly indistinguishable. The capacity of a system to transmit a particular sequence of symbols depends upon the possibility of distinguishing at the receiving end between the results of the various selections made at the sending end." Hartley, "Transmission of Information" (cit. n. 14), 537.*

Hence in estimating the capacity of the physical system to transmit information we should ignore the question of interpretation, make each selection perfectly arbitrary, and base our result on the possibility of the receiver's distinguishing the result of selecting any one symbol from that of selecting any other. By this means the psychological factors and their variations are eliminated and it becomes possible to set up a definite quantitative measure of information based on physical considerations alone.[17]

As soon as the notion of information became perfectly arbitrary and was differentiated (as a random variable) from the system of meaning—that is, once the issue of interpreting a message could be ignored in the future—then it also became possible to consider the noise as part of the transmission.[18] It was still the linear concept of transmission between two points that was brought to bear here, although, after Jansky's experiments, a third noise source was essentially assumed in each transmission.

Technical devices conceived to modernize the realm of communication,[19] to el-

[17] Ibid. See also Friedrich-Wilhelm Hagemeyer, *Die Entstehung von Informationskonzepten in der Nachrichtentechnik: Eine Fallstudie zur Theoriebildung in der Technik in Industrie- und Kriegsforschung* (PhD diss., Freie Univ. Berlin, 1979), 207–57.

[18] "The fundamental problem of communication is that of reproducing at one point either exactly or approximately a message selected at another point. Frequently the messages have meaning; that is they refer to or are correlated according to some system with certain physical or conceptual entities. These semantic aspects of communication are irrelevant to the engineering problem. The significant aspect is that the actual message is one selected from a set of possible messages." Claude E. Shannon, "The Mathematical Theory of Communication," in *The Mathematical Theory of Communication*, ed. Shannon and Warren Weaver (Urbana, Ill., 1949), 29–125, on 31.

[19] "The word communication will be used here in a very broad sense to include all of the procedures by which one mind may affect another. This, of course, involves not only written and oral speech, but also music, the pictorial arts, the theatre, the ballet, and in fact all human behavior. In some connections it may be desirable to use a still broader definition of communication, namely, one which would include the procedures by means of which one mechanism (say automatic equipment to track an air-

evate communication as one of the "most remarkable theoretical inventions" of the twentieth century,[20] ended up revealing "background noise, jamming, static, cut-offs, hyteresis, various interruptions."[21] Serres, in his 1966 critique of the Platonic dialogue, inferred from these disturbances that communication as transmission entailed the materiality of the transmission channel being involved in the communication process. He emphasized the channel's noise when speaking of the transmission of signals and also expounded on the mediatic communications model of information theory, as had been advanced by Claude E. Shannon and Warren Weaver in the 1940s, to encompass all areas of communication: "If static is accidental, background noise is essential to communication."[22]

According to this assertion, which elevated noise from the status of disruption to an integral part of communication, the relation between transmitter and receiver had to be expanded to include a third element, that of noise arising in the channel of transmission:

> These interlocutors are in no way opposed, as in the traditional conception of the dialectic game; on the contrary, they are on the same side, tied together by a mutual interest: they battle together against noise. . . . To hold a dialogue is to suppose a third man and to seek to exclude him; a successful communication is the exclusion of the third man. . . . We might call this third man the demon, the prosopopoeia of noise.[23]

The existence of this third figure made plain that, as noted by Serres, "noise is part of communication."[24] In fact, Jansky's discovery of star noise had already expanded the transmitter-receiver model, even prior to Shannon and Weaver, to include this factor of noise and its localization. Jansky had spawned a new system with an additional, higher order of complexity than was previously known in linear communications. To put this in Serres's terms: "Noise is the fall into disorder and the beginning of an order."[25] Instead of following the engineering problem of "communication in the presence of noise" that initially motivated Jansky's research—implying, as would later be achieved by Shannon, an optimization of the connection within the linear transmitter-receiver route, despite all noise, by reducing it to the communications channel within which messages from then on would be "generable as selections or filtrations of noise"—here the "experimental relay of information and noise" in all its imperfection takes center stage and stands out as a sign against the noise, rather than being screened out.[26]

This means that in order to understand the complexity of the new system, we must

---

plane and to compute its probable future positions) affects another mechanism (say a guided missile chasing this airplane)." Warren Weaver, "Recent Contributions to the Mathematical Theory of Communication," in Shannon and Weaver, *Mathematical Theory of Communication* (cit. n. 18), 1–28, on 3.

[20] Dirk Baecker, "Kommunikation," in *Ästhetische Grundbegriffe: Historisches Wörterbuch in sieben Bänden*, ed. Karlheinz Barck et al. (Stuttgart, 2001), 3:384–426, on 384.

[21] Serres, "Platonic Dialogue," in *Hermes: Literature, Science, Philosophy*, ed. Josué V. Harari and David F. Bell, trans. Marilyn Sides (Baltimore, 1982), 65–70, on 66.

[22] Ibid.

[23] Ibid., 67.

[24] Serres, *Parasite* (cit. n. 1), 12.

[25] Ibid., 79.

[26] Claude E. Shannon, "Communication in the Presence of Noise" (1948), in *Collected Papers*, ed. Neil J. A. Sloane and Aaron D. Wyner (New York, 1993), 160–72; Friedrich Kittler, "Signal-Rausch-Abstand," in *Materialität der Kommunikation*, ed. Hans Ulrich Gumbrecht and K. Ludwig Pfeiffer (Frankfurt am Main, 1988), 342–59, on 345 and 347.

go back before its creation and consider the deviation, the disorder, and the noise as constitutive elements of the system on various levels. We must also question what it engenders, or what it is really conveying. This does not exhaust, in the functionalist aspect, the consequences of Jansky's findings, or of Shannon and Weaver's classical information theory regarding the optimization of information transfer through the model of its signal-to-noise ratio. Indeed, the linear relation between a message and noise (as the background against which the message is transmitted) is converted, between transmitter and receiver, to an interchangeable, oscillating model in which noise holds the independent position of a third figure in the communication.

The awareness that noise, in its position as third figure, is a part of all communication and, eventually, comes to be recognized as a problem for communication theory leads Serres to describe it as "a black thing, an obscure process, or a confused cloud of signals—what we shall soon call a problem."[27] This problem illustrates how all communication inheres in an order of elevated complexity as a linear bipolar arrangement. From this perspective, who or what presents this problem is certainly not clear. Here the noise can represent both message and channel, with something emerging in every possible position within this triadic constellation of communication. Thus each position may potentially be defined as a (disturbed) channel or (problematic) message. Serres elucidates this with an example of the conversational structure of noise during a banquet:

> At the feast everyone is talking. At the door of the room there is a ringing noise, the telephone. Communication cuts conversation, the noise interrupting the messages. As soon as I start to talk with this new interlocutor, the sounds of the banquet become noise for the new "us." The system has shifted. If I approach the table, the noise slowly becomes conversation. In the system, noise and message exchange roles according to the position of the observer and the action of the actor, but they are transformed into one another as well as a function of time and of the system. They make order or disorder. . . . The noise is the end of a system and the formation of a new one.[28]

This formation of a new system based upon a triode structure of communication facilitates an oscillation between the presence and absence of noise. Distributed within the triode structure, the signal as a whole oscillates between the poles of message and noise. For Serres, then, the figuration of the third therefore attains a pivotal status:

> A third exists before the second. A third exists before the other. As Zeno the Elder would say, I have to go through the middle before reaching the end. There is always a mediate, a middle, an intermediary. And in this three-handed game, the middle term can be any one of the three, depending.[29]

To Serres, the noise, the randomness, and also the misunderstanding inherent in all communication represent more than just a deviation against the background of conveyance of meaning that, if possible, must be excluded by the communication partners in order to ensure communication. To the question of what is the outcome of the new system that is based on the figure of the third, Serres explains: "Nothing is

---

[27] Serres, *Parasite* (cit. n. 1), 17.
[28] Ibid., 66–7.
[29] Ibid., 63.

conveyed by background noise. Or maybe sometimes. But that is another story. Precisely the story that we must write."[30]

To do so, we must reconsider the place of media in the history of science. We can start here with a concept of media-supported communication that takes into account the fact that media shape the content. Specific communicative content and forms are not preexistent but fostered through the operationality of the employed media.[31] If we assume the act of media communication to take place strictly through a fundament of signals that originate with the noise, thus first engendering meaning and logics of the sign, we obtain a new perspective on the consequences of Jansky's findings. If we allow noise, as per Jansky, to restrict not only the suitable signal strength for all communication, but also, as suggested by Serres, to be valid as a productive channel, then the objects it conveys become available under an entirely new logic of the sign. In addition to the traditional communications model, the reference to an entity (a thing or idea)—which, as the object of communication, is of course meant to be transmitted as precisely as possible by the signs—is thus employed in order to avoid producing semantic noise, thereby becoming a matter of the communication channel.[32] Significantly more can be ascertained from the noise now that the recording of the signals during the actual transmission either brings forth its signs and, in turn, the ideas as well, or concretizes them on its surface.

In order to illustrate this central point in an emblematic way, signals originating from background noise may be analyzed. The existence of these signals is therefore due entirely to their media-technological operations. They emerge through processes of ascription.[33] In these operations, specific manifestations of the difference between signals and signs are presented, as have been exhibited by technical media since the introduction of Lichtenberg figures. In 1777, Georg Christoph Lichtenberg detected figures in resin dust while conducting electrical experiments on his electrophorus (a capacitive generator). Through these experiments he substantiated processes of inscription in which something was happening that differed from all previous processes of visual representation: a physical carrier of a sign became inscribed therein as a new order of the signal. The fact that electricity was self-represented in the Lichtenberg figures provoked a fundamental break in the ontological mode of a sign.[34] Indeed, the formation of a new order of the signal anteceded the development of the

---

[30] Serres, *Hermes V: Die Nordwest-Passage*, trans. Michael Bischoff (Berlin, 1994), 45.

[31] See Friedrich A. Kittler, *Gramophone, Film, Typewriter* (Stanford, Calif., 1999); as relates to radio, see Daniel Gethmann, *Die Übertragung der Stimme: Vor- und Frühgeschichte des Sprechens im Radio* (Berlin, 2006).

[32] As John Locke noted in the year 1690: "To make Words serviceable to the end of Communication, it is necessary, (as has been said) that they excite, in the Hearer, exactly the same *Idea*, they stand for in the Mind of the Speaker. Without this, Men fill one another's Heads with noise and sounds; but convey not thereby their Thoughts, and lay not before one another their *Ideas*, which is the end of Discourse and Language." Locke, *An Essay Concerning Human Understanding*, bk. 3, ed. Peter H. Nidditch (New York, 1979), 478.

[33] According to Gerrit Verschuur, the transition to processes of ascription in radio astronomy was necessary to constitute meaning in the eye of the viewer: "Listening to the noise from space is of little practical value. The cosmic radio signals need to be translated into electrical currents which are then used to drive a moving pen over a paper chart or converted into numbers to be handled by a computer for later study . . . and displayed in a way which means something to the human eye." Verschuur, *The Invisible Universe Revealed: The Story of Radio Astronomy* (New York, 1987), 77.

[34] See Bernhard Siegert, *Passage des Digitalen: Zeichenpraktiken der neuzeitlichen Wissenschaften, 1500–1900* (Berlin, 2003).

transmission of signs in modern technical media, and this order has been consistently reappearing in constantly changing form ever since.

   The difference between signal and sign points to an origin of communication in these media, its message-related aspects coinciding with its semiotic ones. In 1949 Shannon and Weaver had determined the technological, signal-related problem apparent on the surface of the noise—"How accurately can the symbols of communication be transmitted?"—to be the central level of communication. This central level was said to influence not only the semantic problem of congruence between the signs used and the intended meaning but also the problem of the effectiveness of the message. The influence on the latter was so significant that an analysis of the technical problems "discloses that this level overlaps the other levels more than one could possible [sic] naively suspect." Therefore, an analysis of the technical problems was said to be, "at least to a significant degree, also a theory of [the other two] levels."[35] This is especially apparent when the signal-sign difference is ignored or overlooked and consequently produces semantic or ideographic noise. The break in the order of signifiers becomes obvious in phases when media-technological innovations upset conditions for communication and its channels so strongly that even fundamental differentiations are left open to question: whether the information received wirelessly is deducible in the form of images or sounds; whether it consists of random background noise, or rather of signals or even signifiers. Initially apparent during the reception of such new technologies is a materiality of the signal, which, as a precondition for media communication, determines its forms and content. The representation of this materiality thus also poses anew the question as to the reference of its signs, as will be elucidated in the following case study of another radio-reception experiment, conducted on National Radio Silence Day in the year 1924.

"LISTENING FOR MARS; HEARD ANYTHING?"

On 24 November 1894, American film pioneer Charles Francis Jenkins patented a projector, the Phantoscope, inspired by Thomas Edison's Kinetoscope. Among other inventions, he also developed a "kinetographic camera" with four rotating lenses synchronized with the transport of the film.[36] In 1916 he founded the Society of Motion Picture Engineers and was elected its first president. He was a prolific inventor who, during the course of his career, applied for over four hundred patents in the fields of cinematography, automotive engineering, and aircraft manufacturing. Jenkins is also considered to have pioneered mechanical scanning television in the United States, the development of which both he and his counterpart John Logie Baird from Great Britain, who was twenty-one years his junior, decisively furthered. Jenkins's first, utopian proposal for electromagnetic television dates back to 1913. At the time, Jenkins intended to transfer the principle of the selenium cell onto the surface of an entire photosensitive plate whose electrical impulses were then supposed to be transmitted to a further plate: "The plate would glow with varying intensity in different parts of the plate represented by the picture. This is somewhat analogous to the localized magnetic field which Prof. Poulsen employs to record and reproduce sound on a

[35] Weaver, "Recent Contributions to the Mathematical Theory of Communication" (cit. n. 19), 6.
[36] See Jenkins, "The Phantoscope," in *Animated Pictures* (1898; repr., New York, 1970), 25–42; Laurent Mannoni, *The Great Art of Light and Shadow: Archaeology of the Cinema* (Exeter, 2000), 429–32.

steel disc. . . . It will thus be seen that this thorium plate glows with a surface intensity corresponding to the picture at the distant station."[37]

By 3 October 1922, when Jenkins publicly introduced his procedure for sending photographs telegraphically (and later, wirelessly),[38] he had already been working with image transmission for many years.[39] His technique was based on two rotating, prismatic rings that originally replaced the sash shutter of a film projector, continually disrupting the shaft of light and transmitting to a photocell. On 2 March 1923, a special "radio photo" was sent,[40] for the first time, from the US Navy Radio Station at the Naval Operating Facility in Washington to the *Evening Bulletin* in Philadelphia and was printed by the newspaper.[41] The transmitted image of just several lines retained the logic of script; Jenkins had invented the Jenkins Picture-Strip Machine specifically for transmitting print. This machine made it possible for a long report to be transformed, according to the same principle as used by his wireless photo transmission, into an electrical signal sequence by means of a photocell and then sent via radio. For the reception of the signal sequence, either photographic paper or a filmstrip was pulled across a rotating cylinder in which the light source once again inscribed the report line for line.[42] The *New York Times* reported, "C. Francis Jenkins, who has been a prominent experimenter in the field of transmitting photographs by radio, set up a machine designed to receive on a moving film any message that might come smashing through the ether."[43]

Accordingly, Jenkins's apparatuses were among those selected for a radio-reception experiment conducted in August of 1924. David Todd, the former director of the astronomy institute at Amherst College and its observatory, convinced the Armed Forces to participate in the experiment. He brought together a fundamental astronomical understanding of the calculation of planetary orbits with the euphoric communications-related assumption that this experiment could succeed in determining the correct point in time to receive electromagnetic signals from Mars.

The *New York Times* explained that, in this experiment, over thirty high-performance

[37] Jenkins, "Motion Pictures by Wireless," *Motion Picture News* 8, no. 14 (1913): 17–8, on 18.

[38] See Albert Abramson, *The History of Television, 1880 to 1941* (Jefferson, N.C., 1987), 53; David E. Fisher and Marshall Jon Fisher, *Tube: The Invention of Television* (San Diego, Calif., 1997), 43–4; Russell W. Burns, *Television: An International History of the Formative Years* (London, 1998), 195–205.

[39] Jenkins made his first related proposal in the year 1894, for a "theoretical device" on the basis of selenium cells: numerous selenious wires, connected to the camera image as wired loops, were individually threaded to the image receptor, where incandescent filaments of varying intensity became illuminated. See Jenkins, "Transmitting Pictures by Electricity," *Electrical Engineer* 18, no. 325 (1894): 62–3.

[40] Jenkins used terminology to precisely differentiate between cable-bound media and the new wireless radio transmission technology: "In our laboratory we have found it convenient and informative to use the words radiogram, radiophone, and radio vision when we speak of radio-carried service; and to say telegram, telephone, or television when we speak of wire-carried service." Jenkins, "Radio Vision," *Proceedings of the Institute of Radio Engineers* 15 (Nov. 1927): 958.

[41] "Reproduced in the [*Evening*] *Bulletin*, and in the *Washington Star*, March 3, 1923." Jenkins, *Vision by Radio—Radio Photographs—Radio Photograms* (Washington, D.C., 1925), 119.

[42] "In the sending machine the rotating prisms sweep the image of the typewriter line across the light sensitive cell; and the strip is moved longitudinally by winding on a drum. In the receiving machine the strip is drawn along while it is curved around a rotating cylinder inside which the modulating light is located, turned off and on by radio. A corona glow lamp is preferably employed with the photographic paper." Ibid., 103.

[43] "Radio Hears Things as Mars Nears Us: A 24-Tube Set in England Picks Up Strong Signals Made in Harsh Dots," *New York Times*, 23 Aug. 1924, 9.

US Army and Navy radio stations were instructed to listen to space, together with American and European scientists as well as countless radio amateurs.[44] The *Washington Post* also referred to the experiment on 21 August 1924, in anticipation of the perihelion opposition of the planetary orbits of Earth and Mars to occur two days later, in which the planets would come closer than they had in nearly eighty years: "Army Radio Force to Listen for Signals from Martians."[45] The US armed forces promised "to 'pick up' any unusual radio phenomena . . . and to note strange signals in their logbooks."[46]

In the same article in the *Washington Post*, Todd announced that a recording device from Jenkins's laboratory would be implemented:

> Dr. Todd asserted [that] an automatic recorder, with slow-reeling tape, will be put in operation in the laboratory of the inventor, C. Francis Jenkins, at 1519 Connecticut avenue northwest. This recorder, he said, will be set to run 100 hours and will pick up any unusual radio signals.[47]

In order to ensure that the device in Jenkins's Washington lab operated appropriately, Todd campaigned for his idea of a National Radio Silence Day, as it was later loftily called: a five-minute-long break in the broadcast of all US radio stations nationwide once an hour for a day and a half on 22 and 23 August 1924, as the *New York Times* reported, while the planet Mars was most proximate to the Earth.[48] The scientist entered into related agreements with numerous embassies; with the head of US Signal Corps military stations, Charles M. Saltzman; with the US Navy; as well as with the respective presidents of the Radio Corporation of America and the National Association of Broadcasters.

On 22 August, initial news of "queer sounds in Vancouver" was already being reported from Canada:

> Mysterious signals picked up at Point Grey Wireless Station here during the last week culminated today in a strange group of sounds, causing wireless experts here to wonder if the planet Mars is trying to establish communication with the earth. Four distinct groups

[44] "Powerful radio stations of both the Army and Navy Departments will stand by from midnight tonight to 8 A. M. Monday to 'listen in' for possible signals from Mars. Admiral Eberlen, Chief of Naval Operations, issued orders tonight to naval stations, including those in Honolulu, Balboa, Canal Zone; San Juan, Sitka, Alaska and Cavite, Philippine Islands. A similar order was issued to army stations earlier in the day by the War Department." "Listening for Mars; Heard Anything? While Army and Navy 'Stand By' Vancouver's Radio Thinks It Caught Something," *New York Times*, 22 Aug. 1924, 13. Such reception experiments had up to that point been reserved for cinema: in one of the first feature-length 3D films—*Radio-Mania*, which premiered on 27 Dec. 1922 and also ran under the title *M.A.R.S.* in the United States—a scientist had already been seen communicating via radio waves with the planet Mars. See unsigned review of *M.A.R.S.*, directed by Roy William Neill, *New York Times*, 28 Dec. 1922.
[45] "Army Radio Force to Listen for Signals from Martians," *Washington Post*, 21 Aug. 1924, 9; the *New York Times* wrote, "Army operators would be instructed to 'listen-in' for any signals which the radio experts of Mars might attempt to make." "Asks Air Silence When Mars Is Near: Prof. Todd Obtains Official Aid in Washington Despite Doubts of Its Efficacy," *New York Times*, 21 Aug. 1924, 11.
[46] "Army Radio Force to Listen for Signals from Martians" (cit. n. 45).
[47] Ibid.
[48] "Professor David Todd, former head of the Astronomy Department at Amherst, obtained informal assurances from the Army and Navy departments today that his request for silence in the air would be observed as fas [*sic*] as possible during the period next Friday and Saturday when Mars will be nearest the earth. Professor Todd has announced that he intends to try to communicate with Mars. Willingness to cooperate was manifested by other Government departments." "Asks Air Silence When Mars Is Near" (cit. n. 45), 11.

of four dashes each came through the ether today, the operators stated. The signals were in no known code, starting on a low note and ending with a 'zipp,' and neither a spark nor a continuous wave is responsible for the sounds.[49]

As it turned out, Canadian radio engineers in Point Grey had already been receiving the signals for over a month: "The sounds had not been considered seriously by the operators until the last day or two,"[50] when a new contextualization became apparent due to the activity surrounding the Mars transit and related radio-reception experiment. The task of interpreting such signals was also pursued by British scientists and experts of the Marconi Company in a simultaneous reception experiment in London that aimed "to 'listen in' on Mars."[51] Incidentally, only three years later, Hartley suggested ignoring this interpretive endeavor in his aforementioned paper, "Transmission of Information." The *New York Times* reported on this experiment in the article "Radio Hears Things as Mars Nears Us":

> Tuning in started at 12:30 A. M., and at 1 A. M., on a 30,000-meter radius, sounds were heard which could not be identified as coming from any earthly station. The sounds were likened to harsh dots, but they could not be interpreted as any known code. The noises continued on and off for three minutes in groups of four and five dots.[52]

But appearing in recordings made by the Jenkins Picture-Strip Machine in Washington, alongside such regular signal sequences, was the real sensation of the reception experiment:

> The film, thirty feet long and six inches wide, discloses in black on white a fairly regular arrangement of dots and dashes along one side, but on the other side at almost evenly spaced intervals are curiously jumbled groups each taking the form of a crudely drawn face.[53]

Amid the ecstatic viewing of the image that ensued, the signal was taken to be a sign: the recordings made by the Jenkins device appeared to contain a "radio vision" that could also be viewed as a noisy signal from space.

In that the device recorded signals and reproduced them on a filmstrip, it was associating a variation of the Selenite concept—a utopian seventeenth-century idea that life forms existed in outer space; in the case of the Selenites, on the moon—with the scripted signal. As a result of this association, the Selenites were catapulted from twelfth place on a seventeenth-century list of "all kinds of wondrous people" directly into the age of technical media.[54] What is more, they were realized in a material form

---

[49] "Queer Sounds in Vancouver," *New York Times*, 22 Aug. 1924.

[50] "Radio Hears Things as Mars Nears Us" (cit. n. 43), 1.

[51] Ibid.

[52] Ibid.

[53] "The development of a photographic film record of radio signals during a period of about twenty-nine hours, while Mars was closest to the earth, has deepened the mystery of the dots and dashes reported heard at the same time by widely separated operators of powerful stations." "Seeks Sign from Mars in 30-Foot Radio Film: Dr. Todd Will Study Photograph of Mysterious Dots and Dashes Recently Recorded," *New York Times*, 28 Aug. 1924, 6.

[54] Johannes Praetorius, *Anthropodemus Plutonicus, das ist eine neue Welt-Beschreibung von allerley wunderbahren Menschen, als da seyn die 1. Alpmännergen, Schröteln, Nachtmähren. 2. Bergmännerlein, Wichtelin, Unter-Irrdische. 3. Chymische Menschen, Wettermännlein. 4. Drachenkinder, Elben. 5. Erbildete Menschen, Seulleute. 6. Feuermänner, Irrwische, Tückebolde. 7. Gestorbene Leute,*

and even instituted as a concrete communications partner who obviously possessed a certain degree of (radio-)technical training and intelligence while also, perhaps most significantly, embodying a presence.

We must turn again to Serres:

> Following scientific tradition, let us call noise the set of these phenomena of interference that become obstacles to communication. Thus, cacography is the noise of graphic form or, rather, the latter comprises an essential form and a noise that is either essential or occasional.[55]

Engendered through the cacography of the "Listening for Mars" experiment was an aesthetic of the signal, whereby the electromagnetic signal was interpreted as a sign and the sign as a symbol during the height of communications euphoria. This caused the signified, the Selenite idea, to become concretized in its media-induced existence during the radio-reception experiment, at the moment when its perceived communications value came to light. Serres explains that "the symbol is thus an abstract being that the graphs in question only evoke. This abstract being is recognized by the homeomorphism, if I dare say so, of the graphs."[56]

Consequently, because the play on differentiations between signal, sign, and symbol seemed to be realized through an interiority of ascription, where the signified itself was ultimately revived through its visualization, the perceived "signifying face" from Jenkins's signal film turned the "signified face" into a central media figure of the twentieth century.[57] While the semiosis of mediatic operations becomes visible on the basis of the signal, the fundamental condition of each and every sign within the signal is also deconstructing the concept of the sign itself, which has remained committed to ontology. It is now that the operation of the signal first reveals its sender, that it allows him to materialize as a conversation partner from a list of imaginary communicants and is already presenting his face in the signal recording of his supposed message. Considering the way in which the concept of the sign, in the new relational context of temporally bound signals of technical media, fosters deliberation, as can be ascertained from the cacography of their recording, it can be noted that, in the words of Jacques Derrida,

---

Wütendes Heer. 8. Haußmänner, Kobolde, Gütgen. 9. Indianische Abentheur. 10. Kielkröpfe, Wechselbälge. 11. Luftleute, Windmenschen. 12. Mondleute, Seleniten. 13. Nixen, Syrenen. 14. Oceänische oder Seemänner. 15. Pflantzleute, Alraunen. 16. Qual- oder Verdammte-Menschen. 17. Riesen, Hünen. 18. Steinmänner. 19. Thierleute, Bestialische, Weerwölfe. 20. Verwünschte Leute. 21. Waldmänner, Satyren. 22. Zwerge, Dymeken (Magdeburg, 1668–77). Astronomer Johannes Hevelius (1611–87) had already described the Selenites as moon beings in his *Selenographia sive Lunae Descriptio* from the year 1647.

[55] Serres, *Hermes: Literature, Science, Philosophy* (cit. n. 21), 66.

[56] Ibid., 68.

[57] Apparent in this operation is the discrepancy between mediatic, signal-supported semiotic practices and the classical linguistic concept of the sign: "And for modern linguistics, if the signifier is a trace, the signified is a meaning thinkable in principle within the full presence of an intuitive consciousness. The signified face, to the extent that it is still originarily distinguished from the signifying face, is not considered a trace; by rights, it has no need of the signifier to be what it is. . . . This reference to the meaning of a signified thinkable and possible outside of all signifiers remains dependent upon the onto-theo-teleology that I have just evoked." Jacques Derrida, *Of Grammatology*, trans. Gayatri Chakravorty Spivak (Baltimore, 1997), 73.

THE AESTHETICS OF THE SIGNAL

the trace affects the totality of the sign in both its faces. That the signified is originally and essentially (and not only for a finite and created spirit) trace, that it is *always already in the position of the signifier*, is the apparently innocent proposition within which the metaphysics of the logos, of presence and consciousness, must reflect upon writing as its death and its resource.[58]

CONCLUSION

The radio experiment of August 1924 predominantly used reception on wavelengths between 5,000 and 6,000 meters, meaning that it was geared to long-wave radio whose frequency range is propagated terrestrially as a ground wave following the curvature of the Earth. This information, however, was not divulged until later, during an attempt to explain how the experiment had been ill-fated to initially produce acoustic noise and, with Jenkins's device, also ideographic noise. The *New York Times* quoted Jenkins:

> I don't think the results have anything to do with Mars. . . . Quite likely the sounds recorded are the result of heterodyning or interference of radio signals. The film shows a repetition, at intervals of about a half hour, of what appears to be a man's face. It's a freak which we can't explain.[59]

Thus the fearful wonder as to what figure might be behind the background noise was promptly followed by disenchantment. And this after what had previously been disappointment about the planet's silence, as expressed by a headline that ran one day after National Radio Silence Day: "Mars Sails by Us without a Word."[60] The euphoric announcements of recent days were immediately renounced: "Strange Signals Not from Mars."[61] The signals received in London appeared to have resulted from a mixture of atmospheric electricity and static between individual stations,[62] while the signals received in Vancouver were identified as earthly radio beacons.

Incidentally, the reception experiment—in the act of its failure—ultimately verified not interplanetary communication but rather the famous hypothesis put forward by physicist Oliver Heaviside. In 1902 Heaviside formulated a concrete spatial-theoretical concept of a comprehensive, multidimensional media space designed to transmit electromagnetic waves on Earth.[63] Edward V. Appleton and his student

---

[58] Ibid.

[59] "Seeks Sign from Mars in 30-Foot Radio Film" (cit. n. 53), 6.

[60] "Mars Sails by Us without A Word: No Message Comes from Planet, Ruddily Glowing as It Nears the Earth," *New York Times*, 24 Aug. 1924, 30.

[61] "Strange Signals Not from Mars," *New York Times*, 24 Aug. 1924.

[62] "The object was to hear any signals that might be coming from Mars. As might have been expected, no such signals were received; but American broadcasting was heard on a small loop." *Scientific American* 131 (Nov. 1924): 336.

[63] "There may possibly be a sufficiently conducting layer in the upper air. If so, the waves will, so to speak, catch on to it more or less. Then the guidance will be by the sea on one side and the upper layer on the other." Heaviside, "Theory of Electric Telegraphy," in *Encyclopaedia Britannica*, 10th ed., vol. 33, s.v. "telegraphy." A similar hypothesis was formulated in 1902 by Arthur Edwin Kennelly. See Kennelly, "On the Elevation of the Electrically Conducting Strata of the Earth's Atmosphere," *Electrical World and Engineer* 39 (15 Mar. 1902): 473. After Guglielmo Marconi, stationed in St. John's, Newfoundland, received a radio signal from the Poldhu Wireless Station in Cornwall on 12 Dec. 1901—which had managed to cross the Atlantic despite the curvature of the Earth—Heaviside was quick to raise the question, in an unpublished letter to the editor of *Electrician* in the spring of

Miles A. F. Barnett corroborated this theory in England just a few months after National Radio Silence Day. This second 1924 radio-reception experiment used the diurnal variation of signals received at Oxford from the Bournemouth BBC station, over a distance of about 140 kilometers, though the investigations themselves were carried out during breaks in regular broadcasting after midnight into the early hours of dawn. Appleton convinced the BBC "slowly to change the operating wavelength of a broadcast transmitter and thus make for him a Lloyd's Mirror interferometer wherein the wave propagating directly along the ground could interfere with the wave propagating up and then back down from the ionized layer."[64]

The experiments showed a path difference that was "consistent with a reflecting layer at a height of about 85 kilometres."[65] Based upon their data, Appleton and Barnett determined that the received signal intensity made interference phenomena and the existence of a deflecting layer evident. The fact that the terrestrial media space for the transmission of electromagnetic waves was closed off through the restriction of their dispersion, and also through the reflection of the electromagnetic waves in the ionosphere, later referred to as the "Heaviside layer,"[66] engenders new configurations of the social within complex forms of electronic presences.[67]

If "the act of eliminating cacography, the attempt to eliminate noise, is at the same time the condition of the apprehension of the abstract form and the condition of the success of communication," then a presence became articulated in astronomer Todd's radio-reception experiment during the summer of 1924, a presence whose creation was first experienced following the recording of disruptions and static in electromagnetic waves.[68] In the fall of 1924, Appleton and Barnett determined the technical contingency of this presence in the media space spanning the Earth. This media space becomes concrete only when its boundless expansion is no longer conceived as infinite in space. Jansky had, as of 1930, inferred the fact that the available signal strength within this media space is generally limited by noise of interstellar origin. His experiments first targeted an analysis of the disruptions and static involved in terrestrial radio reception, but in the end he discovered a new method for acquiring astronomical knowledge through the reception of cosmic static.

This series of radio-reception experiments related to noise research has ultimately revealed that the order of electromagnetic signals goes beyond transforming methods of astronomical knowledge production. These experiments have also indicated the extent to which a theory of signs was subject to message-related aspects in the twen-

---

1902, as to whether "the recent success of Mr. Marconi in telegraphing from Cornwall to Newfoundland might not be due to the presence of a permanently conducting upper layer in the atmosphere." Quoted in W. H. Eccles, "Wireless Communication and Terrestrial Magnetism," *Nature* 119, no. 2987 (1927): 157.

[64] C. Stewart Gillmor, "Wilhelm Altar, Edward Appleton, and the Magneto-Ionic Theory," *Proceedings of the American Philosophical Society* 126, no. 5 (1982): 395–440, on 400.

[65] Edward V. Appleton and Miles A. F. Barnett, "Local Reflection of Wireless Waves from the Upper Atmosphere," *Nature* 115, no. 2888 (1925): 333–4, on 333.

[66] The name "Heaviside layer" was suggested by W. H. Eccles in 1912 in the context of his physical calculations of the spread of waves. See Eccles, "On the Diurnal Variations of the Electric Waves Occurring in Nature, and on the Propagation of Electric Waves round the Bend of the Earth," *Proceedings of the Royal Society A* 87 (1912): 79–99, on 94.

[67] See Daniel Gethmann, "Media Space: Networked Structures in Early Radio Communication," in *Re-inventing Radio: Aspects of Radio as Art*, ed. Heidi Grundmann et al. (Frankfurt am Main, 2008), 179–98.

[68] Serres, *Hermes: Literature, Science, Philosophy* (cit. n. 21), 68.

tieth century. For appearing behind the original signatum, which seems essential for the construction of meaning, is an entire economy of sign surfaces that foster meaning. The experiments have also clearly demonstrated how tying signals to a message-related concept of information ultimately also produced entanglements of engineers, scientists, and practitioners. An aesthetic of the signal, both the listening to noise and the ecstatic viewing of images, opens up a specific perspective on the fundaments of this communication. When considering the radio-reception experiment of August 1924, the technical conditions for effects of presence produced through media may be fleetingly recognized in the noise of the signal transcription. The fundamental difference between the transmission of signals and the sign surfaces, which is generally concealed by the media as that which makes their operations possible, is no longer blinded out. The disruption and its transcription instead reveal a figure that directly emerges from the noise. According to Serres, this forms the figure of the third within communication: "We might call this third man the demon, the prosopopoeia of noise."[69]

The discovery of his trace was in fact an accident, his outline having passed along the face of noise.

[69] Ibid., 67.

Listening to the Cold War:
The Nuclear Test Ban Negotiations, Seismology, and Psychoacoustics, 1958–1963

*by Axel Volmar**

ABSTRACT

This article shows how sound was used as an epistemic tool for seismological research during the Cold War, when the ability to distinguish underground nuclear explosions from natural earthquakes, the so-called detection or discrimination problem, became a critical issue of international arms control. In 1960, the psychoacoustician Sheridan Speeth created an "auditory display" for presenting seismographic records to the ear. Despite promising initial results, neither seismologists nor military officials seemed to welcome Speeth's work. I place his efforts within a Cold War framework and show how the sounds of science can be politicized and why it is important to take into account the disciplinary and sociopolitical contexts in which research enterprises are embedded.

INTRODUCTION

In June 1961, the *Journal of the Acoustical Society of America* published an article entitled "Seismometer Sounds."[1] In this paper, the psychoacoustician Sheridan Dauster Speeth (1937–95) presented the results of a series of experiments in which he had trained musicians to differentiate between the acoustic features of earthquakes and of underground nuclear explosions. Beforehand, Speeth had transposed the seismic signals into the audible range by means of digital signal processing. After three days of training, the listeners were able to make correct discriminations in 90 percent of the cases. But what was the goal of this procedure, and on what basis did Speeth develop an auditory method to estimate the differences between earthquakes and underground bomb explosions?

The 1950s were strongly influenced by the experience of the Cold War, especially

* Department of Media Studies, Siegen University, Adolf-Reichwein-Strasse 2, 57068 Siegen, Germany; volmar@medienwissenschaft.uni-siegen.de.

The article presented is closely related to a chapter of my PhD dissertation on the auditory culture of science. I am very grateful to Julia Kursell and Hans-Jörg Rheinberger from the Max Planck Institute for the History of Science (MPIWG), Berlin, for their continuous support, which greatly facilitated my research. I also wish to express my gratitude to Myles W. Jackson for kindly hosting me at New York University from Apr. to June 2011, and furthermore, to Lauren, Gillian, and Christopher E. Speeth for generously supporting my project. Finally, I wish to thank Milena Watzeck, Simone Turchetti, Sven Hannes, the participants in an informal workshop held at the MPIWG in Aug. 2011, and two anonymous reviewers for their critical remarks and valuable suggestions.

[1] Sheridan D. Speeth, "Seismometer Sounds," *Journal of the Acoustical Society of America* 33 (1961): 909–16.

© 2013 by The History of Science Society. All rights reserved. 0369-7827/11/2013-0005$10.00

the nuclear arms race and the associated fear of atomic destruction and radioactive contamination. In 1958, as diplomatic talks regarding a possible nuclear test ban between the United States, Great Britain, and the Soviet Union were established, the question of how to discriminate earthquakes from potential clandestine underground nuclear explosions became a serious scientific problem. This so-called detection or discrimination problem triggered a number of controversies among scientists as well as between scientists and policy makers. Speeth, who wished to help end the nuclear arms race and, eventually, the Cold War, regarded his psychoacoustic research as a direct scientific contribution to solving the political dilemma. For this purpose, he applied methods from psychoacoustics in order to solve a research question from the field of seismology.

Although Speeth obtained promising results at first, he did not find a favorable response among seismologists and the military establishment. For reasons that will be discussed in further detail below, Speeth's article went virtually unmentioned in academic discussions about the detection problem. Therefore, his auditory method soon fell into oblivion. Despite its fate, the seismometer sounds approach provides an excellent case study through which to investigate the use of sound and listening as a means for producing scientific knowledge. Taking Speeth's work as an example, I will show that in the history of the sciences, sound phenomena, sound technologies, and practices of trained listening not only represented *objects* of scientific research, but were also employed as *tools* to answer research questions and thus were able to serve as auditory "techniques of evidence,"[2] especially in the laboratory. Speeth developed his method in order to solve a specific problem from the field of seismology. Therefore, the history of seismometer sounds can be useful to question well-established associations of epistemological concepts such as *objectivity* or *observation* with visual practices in the sciences. A historiography of the auditory culture of science may therefore contribute to thinking about epistemic practices and scientific representations differently.

For this purpose, Speeth's method is particularly instructive with respect to its historical reception. Although his 1961 paper was barely noticed for nearly three decades, it has experienced a remarkable renaissance within the academic discourse on auditory display and scientific sonification since the early 1990s. Some forty papers presented at the International Conference for Auditory Display (ICAD) cite Speeth's auditory detection method,[3] most of which claim it to be an early example of a successfully implemented auditory display and an important forerunner of the so-called audification method—the direct conversion of real-world signals or data sets into audio signals.[4] Recently, the approach presented by Speeth has also prompted researchers to further explore the potential of an "auditory seismology."[5] Speeth's approach

[2] See Alexandra Hui, Julia Kursell, and Myles W. Jackson, "Music, Sound, and the Laboratory from 1750 to 1980," in this volume.

[3] It is interesting to note that most of the papers within the ICAD only reference Speeth's paper of 1961, without taking into account the circumstances surrounding the actual research process. Most of these papers are accessible online through http://www.icad.org.

[4] See Gregory Kramer, ed., *Auditory Display: Sonification, Audification, and Auditory Interfaces* (Reading, Mass., 1994), 37.

[5] Chris Hayward, "Listening to the Earth Sing," ibid., 369–404; Florian Dombois, "Using Audification in Planetary Seismology," in *Proceedings of the 7th International Conference on Auditory Display (ICAD 2001)* (Helsinki, 2001), 227–8; Dombois, "Auditory Seismology: On Free Oscillations, Focal Mechanisms, Explosions and Synthetic Seismograms," in *Proceedings of the 8th International*

has gained a prominent place and belated appreciation in the literature on auditory display long after his work was done.

Florian Dombois, one of the experts in the field, has claimed that auditory methods in seismology proved that "the ear is able to challenge the epistemological power of the eye."[6] If this is true, then why did geophysical experts ignore Speeth's auditory method? As historians of science have shown, it is not the senses as such that scientists rely upon in the course of research processes to generate stable facts, but rather the production of inscriptions and the use of technological transformations. Scientific objectivity does not depend on the use of a specific sense; thus, sight does not necessarily guarantee greater objectivity than audition. As will be shown below, in the field of seismology, particularly in finding a solution for the discrimination problem, scientific objectivity did not primarily depend on the "epistemological power of the eye," but on practices by which the obtained inscriptions could be processed and reduced into quantitative data. Based on Speeth's case, this article thus claims that the discussion about the auditory culture of science, and hence the epistemic status of sound in the laboratory, is not comprehensive enough if it is limited to a mere comparison of scientific senses. Instead, the outputs of scientific representations and transformations as well as the disciplinary and sociopolitical contexts of the research process should be taken into account.

This article consists of five sections. The first discusses sound and audition as representational practices in the history of science. In the second section, the emergence of the detection problem in the context of the test ban negotiations during the late 1950s will be examined. The third section provides some biographical facts on Speeth and reconstructs the genesis of his auditory method designed to distinguish recordings of earthquakes from those of bombs. The fourth section exposes the reserved reactions to Speeth's work and presents some reasons why his approach was generally neglected. The fifth section shows how Speeth became isolated as a person while other researchers were commissioned to continue his work. In the conclusion, I discuss how to assess the status of sound and listening as epistemic tools in the history of science.

SOUND AND AUDITION IN SCIENTIFIC PRACTICE

During the past three decades, the fundamental epistemological problem of representation has led to a stronger discussion of representational strategies in the history of science.[7] To a large degree, these studies have been concerned with the visual cul-

---

Conference on Auditory Display (ICAD 2002) (Kyoto, 2002), 27–30; Debi Kilb, Zhigang Peng, David Simpson, Andrew Michael, Megan Fisher, and Daniel Rohrlick, "Listen, Watch, Learn: SeisSound Video Products," *Seismological Research Letters* 83, no. 2 (2012): 281–6; Peng, Chastity Aiken, Kilb, David R. Shelly, and Bogdan Enescu, "Listening to the 2011 Magnitude 9.0 Tohoku-Oki, Japan, Earthquake," *Seismological Research Letters* 83, no. 2 (2012): 287–93.

[6] Dombois, "Using Audification in Planetary Seismology" (cit. n. 5), 229.

[7] See Hans-Jörg Rheinberger, *Toward a History of Epistemic Things: Synthesizing Proteins in the Test Tube* (Stanford, Calif., 1997), 104. The issue of representation has also been approached, e.g., by investigating scientific practices, the social construction of knowledge, the significance of inscriptions, and local contexts of knowledge production: see Ian Hacking, *Representing and Intervening: Introductory Topics in the Philosophy of Natural Science* (New York, 1983); Bruno Latour and Steve Woolgar, eds., *Laboratory Life: The Social Construction of Scientific Facts* (Beverly Hills, Calif., 1979); Latour, "Drawing Things Together," in *Representation in Scientific Practice*, ed. Woolgar and Michael Lynch (Cambridge, Mass., 1990), 19–68; Timothy Lenoir, ed., *Inscribing Science: Scientific*

ture of science and discussed the relevance of the scientific image and visualization practices. Bruno Latour, for example, has explicitly characterized the fundamental activities of scientific practice as "thinking with eyes and hands."[8] The sense of sight and techniques of visualization have doubtless played crucial roles in the history of producing and representing scientific knowledge. There are, however, many examples in which alternative modes of representation, such as practices of auditory display, came to the fore. Although scientists often approach their research objects and problems with all their senses, sound and listening as *means* or *tools* of scientific practice have only rarely been addressed by historians of science.[9] When it comes to our conception of scientific research practice, forms of auditory knowledge production literally do not "fit the picture."

Nevertheless, the existence of acoustic as well as auditory epistemic practices raises some interesting questions. Are sounds capable of producing evidence and answering scientific questions at all? Under what circumstances may sounds be useful to generate new knowledge in the laboratory? Why did auditory methods become accepted in some areas—for example, mediate auscultation in medicine, or the Geiger counter in physics—but not in others? And finally, what can we learn from the failures of auditory enterprises in the sciences? In order to answer these questions, I propose that we study sounds in the laboratory the same way as we would study images. In his work on scientific images and inscriptions, Latour conceptualized the process of scientific research as a constant transformation of things into inscriptions or, as he calls them, "immutable mobiles." While each transformation process induces the loss of something (matter, sometimes the studied objects themselves), at the same time something new is gained (order, symbolic form, representations, signs). These inscriptions allow scientists to take new perspectives on the scientific objects and, in doing so, to come to new insights.[10]

From this point of view, images and inscriptions develop their epistemological power not only through their mere visuality, but thanks to their ability to form elements in chains of representations and transformations. Therefore, it makes sense to also reconstruct the chains in which auditory methods are integrated, and to determine their specific place and purpose within these chains in order to estimate their epistemic value. Numerous studies from the field of actor-network theory and social construction of knowledge also show that the success of scientific research does not exclusively depend on factors intrinsic to science, but also on social and political circumstances. Therefore, the following study not only focuses on the conception of Speeth's auditory method, but also intends to reconstruct the context of its reception and diffusion. But first, the historical background that inspired Speeth to carry out his research on seismometer sounds will be illuminated.

Texts and the Materiality of Communication (Stanford, Calif., 1998); Rheinberger, Bettina Wahrig-Schmidt, and Michael Hagner, eds., *Räume des Wissens: Repräsentation, Codierung, Spur* (Berlin, 1997), 8.

[8] Latour, "Drawing Things Together" (cit. n. 7), 19.

[9] See Cyrus Mody, "The Sounds of Science: Listening to Laboratory Practice," *Science, Technology and Human Values* 30, no. 2 (2005): 175–98; Julia Kursell, ed., *Sounds of Science—Schall im Labor (1800–1930)* (Berlin, 2008); Trevor Pinch and Karin Bijsterveld, eds., *The Oxford Handbook of Sound Studies* (New York, 2012).

[10] Bruno Latour, "The 'Péofil' of Boa Vista: A Photo-Philosophical Montage," *Common Knowledge* 4, no. 1 (1995): 144–87.

THE *MISE EN SCÈNE*: THE NUCLEAR ARMS RACE,
THE TEST BAN NEGOTIATIONS, AND THE DETECTION PROBLEM

Between 1945 and 1958, more than 250 atomic bomb devices were set off.[11] By
1954, the growing fear of an impending nuclear disaster as well as the issue of increased radioactive fallout led officials to heed public outcries demanding an end to
the nuclear arms race.[12] In 1957, the protests reached their first climax. Among the
opponents of nuclear weapons were many scientists, including leading figures such
as Otto Hahn and Albert Schweitzer. In the United States, more than two thousand
scientists signed the petition of the Nobel Prize laureate Linus Pauling calling for a
cessation of nuclear testing.[13]

The Eisenhower administration and the US military in particular, however, had
little interest in abandoning their nuclear program. Positions began to change in the
fall of 1957 when new scientific advisers gained access to the president after the
Sputnik shock,[14] some of whom were in favor of abandoning the development of
nuclear arms.[15] In the spring of 1958, the Soviet Union carried out a large-scale test
series, which led to a significant increase of radioactivity in the atmosphere, creating
public agitation throughout the world. After its completion, the Soviet government
announced an indefinite suspension of its nuclear program and proposed beginning
negotiations toward a possible test ban treaty.[16] The Western powers agreed—on the
condition that an international control system would be established.

To this end, it was decided that scientists from the United States, Great Britain, and
the Soviet Union should first consent to the methods and technical requirements for
monitoring a test ban. During the so-called Conference of Experts (1 July to 21 August 1958), procedures for the identification of nuclear tests in four different environments were discussed: in the atmosphere, in outer space, in the oceans, and
underground. While it quickly became clear that the first three environments could
be monitored relatively well, the identification of secret underground explosions
proved to be problematic, for in many cases it was difficult to distinguish them from
earthquakes of similar magnitudes. As it turned out, the geosciences did not possess

---

[11] See Harold Karan Jacobson and Eric Stein, *Diplomats, Scientists, and Politicians: The United States and the Nuclear Test Ban Negotiations* (Ann Arbor, Mich., 1966), 100.

[12] *Fallout from Nuclear Weapons Tests: Hearings Before the Special Subcommittee on Radiation of the Joint Committee on Atomic Energy*, 86th Cong. (1959); Jacobson and Stein, *Diplomats, Scientists, and Politicians* (cit. n. 11), 19–21.

[13] In 1958 Pauling's petition was signed by more than nine thousand scientists worldwide. Jacobson and Stein, *Diplomats, Scientists, and Politicians* (cit. n. 11), 25, 34.

[14] As a direct response to the Sputnik shock, the Advanced Research Projects Agency (ARPA) was created in Feb. 1958. For the general impact of the Sputnik launch on the US sciences, see R. A. Di-vine, *The Sputnik Challenge: Eisenhower's Response to the Soviet Satellite* (Oxford, 1993); David Kaiser, "The Physics of Spin: Sputnik Politics and American Physicists in the 1950s," *Social Research* 73, no. 4 (2006): 1225–52.

[15] For the composition of the President's Science Advisory Committee, see Jacobson and Stein, *Diplomats, Scientists, and Politicians* (cit. n. 11), 47.

[16] For the history of test ban negotiations, see US Disarmament Administration, *Geneva Conference on the Discontinuance of Nuclear Weapon Tests: History and Analysis of Negotiations* (Washing-ton, D.C., 1961); US Arms Control and Disarmament Agency, *International Negotiations on End-ing Nuclear Weapon Tests, September 1961–September 1962* (Washington, D.C., 1962); Robert W. Lambert, *Review of International Negotiations on the Cessation of Nuclear Weapon Tests, September 1962–September 1965* (Washington, D.C., 1966); Jacobson and Stein, *Diplomats, Scientists, and Politicians* (cit. n. 11).

methods to solve this problem satisfactorily. The importance of the detection problem therefore led to the eventual formation of an entirely new branch of science:

> The new science of detection and identification of underground nuclear tests involves a challenging and exciting problem in pattern recognition. Out of a mess of seismic hash (called microseisms), picked up by any seismometer . . . , which to the nonspecialist may look little different from the recordings of brain waves, the explosion seismologist must detect that a significant event took place somewhere on the globe, and with records from other seismometers, he must determine as accurately as he can the location of the event. Then, by whatever means he has at his disposal, he must try to arrive at a decision as to whether or not the event was an explosion or an earthquake. Moreover, he must try to assess the yield or the energy released by the event.[17]

Nuclear explosions with a greater yield would be easier to detect, because there would not be many earthquakes with a comparable magnitude per year. However, the smaller the yield of a nuclear explosion, the more difficult and less likely its identification as such would be, because small earthquakes occur quite frequently, producing a great deal of low-level seismic noise. One of the major tasks therefore consisted in agreeing on a threshold down to which the future control system should be able to operate reliably.

The United States, which deeply mistrusted the closed political system of the Soviet Union, insisted that about 650 international control posts should be established, and that in the case of nonidentified seismic events, an annual number of onsite inspections should be allowed.[18] The Soviet Union, however, considered national control systems to be sufficient and accused the United States of deliberately using its position to foil a test ban in order to be able to continue its nuclear program.[19] The estimates obtained by seismologists as to whether a test ban could be reliably verified with the help of technical procedures were therefore of crucial importance for the formulation of political positions and influenced future diplomatic negotiations.[20]

In the final report of the Conference of Experts, the scientists reached a compromise by agreeing on a higher threshold, which eventually resulted in a significantly lower number of control posts (about 170) and onsite inspections.[21] The report stated that the construction of a comprehensive control system was considered feasible and thus formed the basis for the subsequent diplomatic negotiations. For this reason, the Conference of Experts was regarded as a success and positively influenced the political climate.[22] However, the meeting of the scientists also showed that many questions regarding the identification of underground tests were still unanswered. Up

[17] Nilo Lindgren, "Earthquake or Explosion? The Science of Nuclear Test Detection," *IEEE Spectrum* 3 (1966): 66–75, on 68.

[18] The calculation of the number of control posts was based on the assumption that the system should be able to detect and identify with 90 percent accuracy underground explosions with a yield of approximately 1 kiloton. Jacobson and Stein, *Diplomats, Scientists, and Politicians* (cit. n. 11), 76.

[19] Lindgren, "Earthquake or Explosion?" (cit. n. 17), 66.

[20] See US Congress, Senate, Committee on Foreign Relations, Subcommittee on Disarmament, *Control and Reduction of Armaments: Detection of and Inspection for Underground Nuclear Explosions, Replies from Seismologists to Subcommittee Questionnaire*, Staff Study no. 10 (Washington, D.C., 1958).

[21] In this calculation, the system was intended to be able to identify 90 percent of continental earthquakes whose signals were equivalent to 5 kilotons, and only a small percentage of continental earthquakes equivalent to 1 kiloton. Jacobson and Stein, *Diplomats, Scientists, and Politicians* (cit. n. 11), 77.

[22] Jacobson and Stein, *Diplomats, Scientists, and Politicians* (cit. n. 11), 80.

to that point, the discipline of seismology had received comparatively little financial support. When politicians requested a solution to the discrimination problem, geophysicists were initially unprepared.[23] In addition, the United States had only carried out a single underground test—the so-called Rainier shot on 10 September 1957.[24] Therefore, in the summer of 1958, there was a lack not only of appropriate methods and instrumentation but also of available data. Since all parties had agreed on a moratorium on nuclear tests during the period of diplomatic negotiations, the United States and Great Britain scheduled as many tests as possible before the beginning of the talks. As part of the Hardtack II test series, several underground tests were carried out, five of which delivered new seismic data that allowed for the evaluation both of the recommendations made by the Conference of Experts and of new seismological methods.

The diplomatic negotiations, which started on 31 October 1958, in the form of the Geneva Conference on the Discontinuance of Nuclear Weapon Tests, were soon to be overshadowed by a discussion of the new data initiated by test ban antagonists. In early 1959, a couple of US scientists, including the air force seismologist Carl F. Romney, had claimed in public hearings that, on the basis of the new data, the identification of explosions in the low kiloton range would be even more problematic than assumed by the Conference of Experts. Hence, these scientists considered the reliable monitoring of a test ban improbable.[25] In collaboration with Edward Teller and others,[26] Albert Latter of the RAND Corporation also presented the so-called decoupling theory. The core of the theory was the assumption that the strength of seismic waves could be damped by a factor of 10 or more, and thus, that the degree of detectability could be reduced should tests be performed in large underground cavities.[27] Although such a science fiction–like scenario seemed rather unlikely, the theory did not fail to achieve its intended effect: its mere plausibility highlighted the fact that not only could the technical means for monitoring a test ban advance but "the art of clandestine underground explosions" could as well.[28] The immediate result of the debate about the interpretation of the new data was a growing skepticism among US policy makers toward a comprehensive test ban.[29]

The detection problem also had a direct impact on science. In December 1958, the president's special assistant for science and technology, James R. Killian, convened the Panel on Seismic Improvement under the chairmanship of Lloyd V. Berkner. In

[23] See Lindgren, "Earthquake or Explosion?" (cit. n. 17), 68.

[24] See Jacobson and Stein, *Diplomats, Scientists, and Politicians* (cit. n. 11), 58.

[25] See ibid., 150; Kai-Henrik Barth, "Science and Politics in Early Nuclear Test Ban Negotiations," *Physics Today* 51 (1998): 34–9, on 36.

[26] The Hungarian-American nuclear physicist Edward Teller (1908–2003) was one of the key figures in nuclear arms research in the United States. He is generally considered the "father of the hydrogen bomb" and remained an active promoter of a vigorous US nuclear testing program throughout his life. Due to his strong advocacy of nuclear weapons Teller became a well-known and controversial public figure in the 1950s and is said to be one of the inspirations for the character of Dr. Strangelove in Stanley Kubrick's eponymous 1964 movie. See Peter Goodchild, *Edward Teller: The Real Dr. Strangelove* (London, 2004); István Hargittai, *Judging Edward Teller: A Closer Look at One of the Most Influential Scientists of the Twentieth Century* (Amherst, N.Y., 2010).

[27] Latter, R. E. LeLevier, E. A. Martinelli, and W. G. McMillan, *A Method of Concealing Underground Nuclear Explosions*, RAND Corporation report RM-2347-AFT (Santa Monica, Calif., 1959); Latter, Martinelli, and Teller, "Seismic Scaling Law for Underground Explosions," *Physics of Fluids* 280 (1959): 280–2.

[28] Lindgren, "Earthquake or Explosion?" (cit. n. 17), 69.

[29] Jacobson and Stein, *Diplomats, Scientists, and Politicians* (cit. n. 11), 140.

March 1959, the Berkner Panel recommended substantial investment in further research in seismology in order to achieve the intended capacity of the proposed control system, despite the additional challenges. To meet these expectations, ARPA launched an extensive seismological research and development program called Project Vela Uniform on 2 September 1959. The program was to shape the earth sciences well into the 1970s; thus, thanks to the detection problem, seismology became Big Science during the Cold War.[30]

At the Geneva Conference, the differing assessments of the ability to detect underground nuclear tests and to distinguish them from natural seismic events led to a continuing dispute between the Western powers and the Soviet Union. Even a second scientists' meeting, termed Technical Working Group II, which took place in November and December 1959, offered no solution to the diplomatic deadlock. According to historian Kai-Henrik Barth,

> US and Soviet experts did not find a common language to solve the problems of the diplomats, because their own language of seismology, physics and statistics allowed different readings with significantly different consequences for international relations. The two delegations disagreed about the capabilities of instruments, the value of theories, how to handle raw data and how to define a widely used scientific concept such as seismic magnitude.[31]

The range of interpretation allowed for the use of scientific arguments to influence the course of diplomatic negotiations. "While the US scientists consistently emphasized the limitations of a monitoring system, the Soviet scientists used the same data to defend their government's claim that underground nuclear explosions could be adequately monitored."[32]

Again, the Soviet Union accused the United States of undermining the goal of the conference in order to create a loophole for future tests in the underground environment. And, in fact, this was exactly the goal that the majority of US military officials and scientists were trying to pursue.[33] According to John M. Swomley, the military establishment, as well as many scientists involved in defense research, opposed a test ban because of the substantial budget cuts that would have been a natural consequence of a comprehensive treaty.[34] Scientific arguments discrediting the control system seemed to be a promising strategy to circumvent a test ban agreement, or at least to delay it.[35] Considering the fact that a partial test ban would have been more

---

[30] Between 1959 and 1971, ARPA spent about $245 million on Vela Uniform. In addition, federal funding for seismological research at universities and research institutions increased significantly. One of the major outcomes of the project was the establishment of the World-wide Standardized Seismographic Network (WWSSN) in 1961. See Kai-Henrik Barth, "The Politics of Seismology: Nuclear Testing, Arms Control, and the Transformation of a Discipline," *Social Studies of Science* 33 (1 Oct. 2003): 743–81, esp. 754; Carl-Henry Geschwind, *California Earthquakes: Science, Risk, and the Politics of Hazard Mitigation* (Baltimore, 2001), 130; Jacobson and Stein, *Diplomats, Scientists, and Politicians* (cit. n. 11), 151, 178–9. An overview is also provided by *Developments in the Field of Detection and Identification of Nuclear Explosions (Project Vela) and Relationship to Test Ban Negotiations: Hearings Before the Joint Committee on Atomic Energy*, 87th Cong. (1961).

[31] Barth, "Science and Politics in Early Nuclear Test Ban Negotiations" (cit. n. 25), 39.

[32] Ibid., 34.

[33] Jacobson and Stein, *Diplomats, Scientists, and Politicians* (cit. n. 11), 229.

[34] See Swomley, *The Military Establishment* (Boston, 1964), 139–76.

[35] As a matter of fact, most of those who were skeptical of the reliability of a control system were also vehement proponents of nuclear weapons. For example, Romney was a seismologist in the air force, and Teller was deeply involved in the development of nuclear weapons.

beneficial to the Western powers than to the Soviet Union, Swomley's argument seems entirely plausible:

> Although it is impossible to weigh accurately the advantages and costs which would accrue to each side, it is probably true that the USSR would not benefit as much from a partial test ban as it would from a total cessation of nuclear weapons tests. It was the United States, not the Soviet Union, which expressed great interest in the development of tactical nuclear weapons and in testing such devices underground. Putting it in a different way, American military leaders viewed tactical nuclear weapons as a way of compensating for the USSR's superior strength in conventional forces. Because of this superiority the USSR showed little interest in tactical nuclear weapons.[36]

In some quarters of public opinion, especially among peace activists, the continuing insistence of the US negotiators on unrealistic security standards as a condition for an agreement led to the impression that reaching an agreement was not a high priority of the US government. In light of the Eisenhower administration's lack of a clear political will to achieve a test ban, the detection problem seemed to pose a crucial obstacle to diplomatic negotiations, for it served as a satisfactory explanation for the diplomatic deadlock. Yet an early resolution of the scientific and technical difficulties appeared unlikely, as the Vela Uniform program, which had just been launched, had shown only little progress. In late 1959, the atmosphere between the negotiators had significantly soured, and the prospects of achieving an agreement in the near future seemed to have moved well beyond reach.

SEISMOMETER SOUNDS: AN AUDITORY APPROACH TO THE DETECTION PROBLEM

One of the peace activists questioning the goals of the US negotiators was the psychology student Sheridan Speeth. Speeth was not formally involved in test ban negotiations, in either political or scientific terms. He followed the progress of the Geneva Conference as an outsider but with great interest, as he was politically very active.

Speeth, born in Cleveland, came from a very left-wing family. His father, Henry William Speeth, was a well-known lawyer and had been part of the local political scene.[37] As a confirmed Roosevelt Democrat and a supporter of Harry Hopkins, he harbored strong sympathies for socialist ideas. The philosopher Gerald Heard and the writer Aldous Huxley were regular visitors at the Speeth house. This background strongly shaped the political views of the young Sheridan Speeth, who, growing up under the influence of the emerging Cold War, became a resolute pacifist. While studying psychology at Harvard, where he worked for the behavioral psychologist B. F. Skinner as a graduate assistant, Speeth became deeply involved with the peace movement. As a result, he joined several liberal organizations, including the Committee for a Sane Nuclear Policy.[38]

After completing his masters degree in 1959, Speeth was granted a research position in the Visual and Acoustics Research Laboratory at the Bell Telephone Laboratories in Murray Hill, New Jersey, where he became particularly interested in the

[36] Jacobson and Stein, *Diplomats, Scientists, and Politicians* (cit. n. 11), 128.
[37] Henry Speeth served as councilman for the Fifth Ward in Cleveland, as president of the city council, and as one of the three commissioners of Cuyahoga County. Christopher Eric Speeth (Sheridan Speeth's brother), interview by author, 22 May 2011.
[38] Philip G. Schrag, "Scientists and the Test Ban," *Yale Law Journal* 75 (1966): 1340–63, on 1356.

psychology of human hearing. Bell Labs was one of the leading research institutes for psychoacoustics in the country. This allowed Speeth to work side-by-side with key figures such as John R. Pierce, Max V. Mathews, and Manfred Schroeder while preparing his doctoral thesis in experimental psychology at Columbia University. In his first experiments, Speeth tested the ability of the human ear to make out signals in very noisy environments.[39]

Speeth was frustrated by the slow pace of the Geneva Conference. In the course of 1959, he came to the conclusion that his political commitment to the peace movement would not have the least impact on the course of the negotiations. Diplomatic progress in the field of arms control seemed to depend entirely on scientific and technical developments. Therefore, Speeth regarded the detection problem as the major obstacle on the path to agreement between the nuclear powers and decided to pursue his political goals through scientific means.[40] In 1960, Speeth intensively studied seismology in order to understand the difficulties encountered in seismological detection of underground nuclear testing. The solution Speeth would eventually develop was rather unique and differed substantially from other geophysical approaches. This was primarily due to the fact that Speeth did not look at the problem through the eyes of a seismologist but listened to it with the ears of a psychoacoustician. Or, to put it differently: when Speeth began to address the detection problem scientifically, he perceived it on the basis of his everyday psychoacoustic research practice. As a result, he regarded the task of distinguishing between the seismograms produced by earthquakes and explosions as a general problem of signal detection. He therefore translated it into the auditory domain and, specifically, into the experimental system of psychoacoustics. In doing so, he aimed to transform the seismological problem into an acoustic one to which acoustic methods could be applied:

> Distinguishing an explosion from an earthquake may in many ways be similar to (and as complicated as) deciding which one of two of your friends is speaking on the telephone. Let us press this analogy by considering the input and response of both the seismometer and the telephone microphone. If your friend is in a normal room with plaster walls, then there will be multiple arrival times for each of his vocal pressure waves, a parallel to the seismologists' P, pP, PP, PKP, and other waves. The band limiting performed by telephone transmission corresponds to the narrow bandpass of most seismometers. If the friend's room were to contain machinery or other sources of noise, you would have to perform a task not unlike distinguishing a seismic signal from the noise of microseisms. If the voice decision were then made on the basis of vowel pronunciation, you would have demonstrated the ear's ability to use the information contained in the temporal dynamics of the short-time audio spectrum. The analogy could be indefinitely extended, but by now the experimental question should be obvious: Would any benefits for the seismologist accrue from having his seismometer output presented as an auditory display?[41]

Thus, Speeth described the discrimination task in terms of an acoustic model and suggested delegating the task to the ear and its "ability to use the information" conveyed by an audio signal. Like human voices, he assumed, the different causes of seismic events should be discernible by their acoustic patterns.

[39] Sheridan D. Speeth and Max V. Mathews, "Sequential Effects in the Signal Detection Situation," *Journal of the Acoustical Society of America* 33 (1961): 1046–54.
[40] Schrag, "Scientists and the Test Ban" (cit. n. 38), 1356.
[41] Speeth, "Seismometer Sounds" (cit. n. 1), 909.

Since Speeth was free to choose his research subjects at Bell Labs, he was able to fit his experiments on the discrimination problem into his own psychoacoustic research plan. Nevertheless, the representation of seismograms in the form of an auditory display in order to distinguish characteristic signal properties from one another does not necessarily appear to have been an obvious solution. Rather, Speeth's personal background seems to have been the key trigger for this transfer—especially the fact that the psychologist had turned to seismological issues primarily due to his political motivations.

In order to create acoustic representations of seismic events, the seismological records had to be transposed into the audible range by technical means. Seismic signals are located in a spectrum of about 0.3 millihertz to 20 hertz and are therefore inaudible if directly presented to the ear, since the audible range for humans is generally considered to be 20 hertz to 20 kilohertz. Around 1960, shifting this low-frequency spectrum into the audible range could be achieved by playing back the recorded seismograms, stored on a magnetic tape or an optical film, at a higher speed. As early as 1951, Beno Gutenberg, Hugo Benioff, and Charles Richter of the Seismological Laboratory of the California Institute of Technology had reported on new possibilities provided by new instruments such as the "seismic tape recorder which enables Lab workers to actually *listen in* on the movements of the earth."[42] Using this method, the record producer Emery Cook had been able to release audible earthquake seismograms on one of his sound-effects records.[43] Yet Speeth aimed to retain full control over the process of signal conversion, for as a psychoacoustician he knew that the sensitivity of the human ear is dependent on frequency and hence so is any pattern-recognition task. Therefore, he decided to use a digital signal-processing technique performed on an IBM 7090 computer that was available at Bell Labs. In his first experiment, Speeth determined the optimum acceleration factor:

> An auditory display was created by using time compression to shift recorded seismogram frequencies into the audible range. . . . A pair of seismograms, one of an explosion, the other of an earthquake, . . . had been digitalized at a sampling rate of ten samples/sec, and were available on punched cards. To equalize intensities, the two sets of cards were fed into an IBM 7090 where every sample of one set was multiplied by a constant to produce equal rms [root mean square] amplitudes. Both were then read to a magnetic tape through a digital-analog converter at sampling rates of 1000, 2000, 4000, and 8000 samples/sec. This provided time compression factors of 100, 200, 400, and 800. The resulting analog tape was played through an AR-1 loudspeaker, and a clearly discriminable difference between the two seismograms could be heard.[44]

As a comparison of these first seismometer sounds showed, the difference could be heard most clearly at a compression factor of either 200 or 400. Armed with this insight, Speeth prepared his actual test series. By means of the IBM computer and a digital-analog converter, he created two magnetic tapes, each containing paired sequences of earthquakes and underground explosions in a random order. For the explosions, the seismic data at hand stemmed from the Hardtack test series of autumn 1958:

[42] See Caltech's Seismological Laboratory, "Earthquakes—Recorded on Tape," *Engineering and Science* 15 (1951): 7–11, on 7.
[43] Hugo Benioff, "Earthquakes around the World," on *Out of this World*, Cook Laboratories/Road Recordings, LP 5012, 1956, 33⅓ rpm.
[44] Speeth, "Seismometer Sounds" (cit. n. 1), 909.

Recordings of the 19-kt *Blanca* explosion were made at the distances, 600, 900, 1000, 1200, 3000, and 4000 km from the test site, and of the 5-kt *Logan* explosion at the distances 500, 700, and 4000 km. The earthquake recordings were made with a similar seismometer system at the telemetered seismometry laboratory of the Bell Telephone Laboratories. All records were digitalized originally at a sampling rate of 10 samples/sec and converted to audio tapes at 3000 samples/sec.[45]

On the first tape, Speeth recorded eighty trials of audified seismograms consisting of two orders of forty pairs (five bombs by eight quakes). All test subjects were trained with this material. The second tape contained sixty-four additional trials (two orders of four bombs by eight quakes) to be used for a final test of the previously trained subjects.

From psychoacoustic studies it was known that training, familiarization, and experience often played an important role in the performance of specific listening tasks. Therefore, Speeth tested the effects of different training periods as well as the use of special training methods.[46] In addition, Speeth looked into the performance of listeners showing differing levels or backgrounds of listening experience. Speeth, who had played the violin since his childhood days, suspected that habituation to a certain frequency range might have an effect on the discrimination task. For this reason, he conducted listening tests with double bassists and cellists, among others, and found that the cellists achieved the most reliable results. Interestingly enough, there is no reference in Speeth's later publication to the choice of musicians as test subjects. It is possible that Speeth did not want to give the impression that musical training or knowledge of some sort might be a prerequisite for a successful application of his method.[47] In the test series, four high school students and a fellow worker at Bell Labs completed a three-day training program of approximately one hour per day with the eighty pairs of the first tape (240 trials in total). Four other employees received only a brief training, which covered only 30 to 60 trials, on the same day as the final test.[48] After each trial, all subjects were told the correct answer. The results of the final test showed that the listeners who had completed the extended training series were able to correctly discriminate the seismometer sounds in over 90 percent of the trials.[49]

Encouraged by the positive results of the test series, Speeth intended to perform a comprehensive study with larger and better data sets. The collection of suitable data had turned out to be troublesome and unsatisfactory, particularly because most underground nuclear tests were considered to be classified information, and therefore, seismic records of them were difficult to obtain. Since Speeth had received no official order for his research, he had only been able to use the sparse data from the Hardtack series, which were publicly available. But since these seismograms were made with portable seismometers, which had been immediately removed after the

[45] Ibid., 910–1.

[46] "Speeth had studied with Professor B. F. Skinner at Harvard and employed Skinner's techniques of positive reinforcement in training his subjects." Schrag, "Scientists and the Test Ban" (cit. n. 38), 1362.

[47] Unfortunately, I was not able to verify exactly to what extent musicians were used. Cecil H. Coker, the only notably mentioned listener of the test series, passed away on 12 Apr. 2011. I rely on an oral account from Speeth's brother, with whom Speeth frequently spoke about the progress of his work in the initial phase of the experiments. C. E. Speeth, interview (cit. n. 37).

[48] Speeth, "Seismometer Sounds" (cit. n. 1), 910.

[49] Ibid., 913.

test series, no recordings of earthquakes with similar magnitudes were made with the same equipment and at the same place:

> The Air Force . . . had not left them in place long enough to record any West Coast earthquakes. Nor had the Air Force ever taken seismometers back to record such earthquakes. Thus although Speeth used bombs and earthquakes recorded over the same distances and of comparable magnitudes, he did not have "matched pairs" which were recorded over precisely the same path of travel through the ground.[50]

The lack of data pairs in which the circumstances of the recording of the seismic events—the path of the propagation wave, the recording location, and the seismic equipment—were identical raised the question of whether the excellent discrimination performance had actually been achieved thanks to characteristic auditory features in the signals, or whether it had been boosted by artifacts. Speeth was convinced that the possible presence of artifacts had not influenced the test results, but only new experiments would bring certainty. Therefore, Speeth suggested that another test series with better data sets be performed.[51]

Nevertheless, the auditory method had shown impressive results and seemed to be far superior to the previous detection methods. Its inventor was convinced that he had discovered an important approach to solving the discrimination problem, and thus had made a considerable contribution to the advancement of nuclear disarmament. However, as it turned out, other researchers would not take up Speeth's study. Both in the documentation of the Geneva Conference as well as in the two most comprehensive historical accounts of the role of seismology in the area of nuclear arms control,[52] Speeth's name is not mentioned once. What had happened? Why was Speeth being ignored? In order to understand this development, we need to take a closer look at the environment that surrounded Speeth and his work. In the next section, I will proffer some reasons as to why Speeth's research was barely noticed or appreciated.

REJECTION AND ISOLATION: THE LONESOME QUEST OF DR. SPEETH

After the completion of his study, Speeth had hoped to raise strong interest in his method. But instead, he was to face some serious difficulties.[53] In the summer of 1960, Speeth summarized his findings in a technical report and requested a quick release for publication. He also asked his department head for permission to discuss his results with Hans Bethe.[54] Bethe was known to the public as a proponent

---

[50] Schrag, "Scientists and the Test Ban" (cit. n. 38), 1357.

[51] "Even more convincing would be the duplication of the reported results using the recordings of a seismometer system with which both explosions and earthquakes have been recorded. Such systems exist at most major universities." Speeth, "Seismometer Sounds" (cit. n. 1), 910.

[52] Bruce Alan Bolt, *Nuclear Explosions and Earthquakes: The Parted Veil* (San Francisco, 1976); Carl Romney, *Detecting the Bomb: The Role of Seismology in the Cold War* (Washington, D.C., 2009).

[53] This section relies considerably on the historical account given by Schrag ("Scientists and the Test Ban" [cit. n. 38], 1355–63). In 1965, Schrag interviewed Speeth about the response to his work.

[54] Like Edward Teller, the German-American nuclear physicist Hans Albrecht Bethe (1906–2005) was a key figure in US nuclear weapons development. Bethe served as head of the Theoretical Division in the Manhattan Project and also made substantial contributions to the development of the hydrogen bomb during the early 1950s. Although Bethe was deeply involved with the development of nuclear weapons as a scientist, he voiced a critical attitude toward the usefulness of these weapons after the outbreak of the Korean War. See Silvan S. Schweber, *In the Shadow of the Bomb: Bethe, Op-*

of the test ban, and it appeared to Speeth, especially after the public confrontation between Bethe and Teller in the spring of 1960 (the so-called Bethe-Teller debate),[55] that Bethe would be his best bet for promoting his work and possibly securing him a Vela contract for further research. But not only was his request denied, he was also informed that his work had been classified; therefore, he was forbidden to discuss it with anyone. Consequently, an exchange with other scientists and even the prospect of prompt publication were rendered impossible. Speeth subsequently filed an official request for the release of his research, but was left in the dark about how long the process would take. Nevertheless, Speeth began to write up his paper in order to have it ready for publication—a scientific routine that brought him, perhaps unwittingly, right back into the political arena:

> Eventually, Speeth was called into the office of Bell's Vice President, John Tukey (a member of the Berkner Panel) who suggested to Speeth that he rewrite his article, replacing the word "bomb" wherever it occurred with the word "explosion," and omitting an entire section entitled "Suggestion for a detection system" which discussed the implications of the research for a test-ban monitoring system. Tukey revealed that Carl Romney (the Air Force's seismologist on the Berkner Panel) had been called in to review the paper, though Speeth was not told if it was Romney who insisted on the alterations.[56]

Speeth's revisions were rejected several times. When his secretary accidentally sent out several copies of his paper to the *Journal of the Acoustical Society of America* and various liberal American scientists, such as Linus Pauling and Harrison Brown, in December 1960, it was intimated to Speeth that his work was "politically loaded."[57] The paper was not cleared for publication until January 1961. The editors of the *Journal of the Acoustical Society of America* were apparently convinced of the importance of the work and thus gave it publication priority.

Eventually, "Seismometer Sounds" appeared in the July issue of the journal. It did not attract a great deal of attention, and responses to the paper were scarce. Why was the publication of Speeth's work repressed for such a long time? After all, it was a proposal for a method possibly suitable for international disarmament measures. Suggestions to solve the discrimination problem therefore should have been a matter of public interest. Apparently, it was not only the content of Speeth's paper, but also the character of the author that had raised suspicion among his superiors, army personnel, and Vela seismologists. Three main issues rendered the relationship between Speeth and his environment a problematic one.

First of all, Speeth's political activities did not go unnoticed by the authorities: according to Philip G. Schrag,[58] Speeth had donated money to the activist Committee for Non-violent Action and had attended a seminar on Marxism given by Marxist historian Dr. Herbert Aptheker. Speeth had even been briefly arrested after protesting against a "take-cover" air raid drill. In 1960, he went to postrevolutionary Cuba

---

penheimer, and the Moral Responsibility of the Scientist (Princeton, N.J., 2000); Schweber, *Nuclear Forces: The Making of the Physicist Hans Bethe* (Cambridge, Mass., 2012).

[55] See Jacobson and Stein, *Diplomats, Scientists, and Politicians* (cit. n. 11), 251–2; see also *Technical Aspects of Detection and Inspection Controls of a Nuclear Weapons Test Ban: Hearings Before the Special Subcommittee on Radiation and the Subcommittee on Research and Development of the Joint Committee on Atomic Energy*, 86th Cong. (1960).

[56] Schrag, "Scientists and the Test Ban" (cit. n. 38), 1357.

[57] Ibid., 1358.

[58] Ibid., 1356.

with his wife for about a month, and he had also been a temporary member of the Fair Play for Cuba Committee—an organization that advocated a more moderate policy toward Cuba and that is now primarily known for the fact that Lee Harvey Oswald was one of its local members. It is not a great surprise that a peacenik like Speeth was considered to be a political opponent—if not a left-wing extremist—by the air force administration that headed the Vela project.

Second, the fact that Speeth was not a seismologist but a psychologist caused problems with his colleagues from the field of geophysics, even though seismologists were also interested in developing methods to distinguish earthquakes from bombs. The seismologists were members of a discipline that, thanks to interest spurred by the Cold War and especially the Vela Uniform program, had transmogrified from a rather small, notoriously underfinanced scientific field to a large-scale research branch with extraordinary financial resources at its disposal.[59] Consequently, the seismologists pursued goals that hardly coincided with Speeth's. As Barth has shown, the seismologists did not completely respond to the desires of their supporters, but were equally concerned about pushing fundamental research agendas in their field.[60] Romney, for instance, was part of a circle of comparatively young seismologists who had a great interest in shaping their field. Being an air force seismologist, Romney not only represented his discipline but also the institutional views of his employer. Therefore, Romney proved to be an unfavorable reviewer of Speeth's approach. While Barth has argued that virtually all research projects were approved and funded in the initial phase of the Vela Uniform program,[61] Schrag points out that there were in fact certain exclusion mechanisms present that were not solely based on scientific criteria, but rather respected the interests of the air force. Schrag blames the then-common "research by contract" system that offered more opportunities to scientific networks than to "'lone wolf' scientists who are not in the club, who have been engaging in some 'unfashionable' line of research, or who have not previously shown an interest in government programs."[62] Schrag even explicitly adduces Speeth as one of his exemplary case studies for this practice. By converting seismograms into acoustic events and by choosing auditory patterns as criteria for distinction among events, Speeth had also left the territory of seismology's "symbolic order," which was essentially based on the measurement of individual oscillations, especially the first motion of seismic events. Moreover, the seismologists' task did not end with the distinction between natural and manmade events. Once a possible underground shot was detected, the location of the event, its magnitude, and the kiloton range of the device had to be assessed. While the auditory method might have opened up a promising way to determine the cause of a seismic event, it hardly allowed for calculating these additional parameters. In the eyes of a Vela seismologist, the auditory display presumably represented a dead end in the chain of transfor-

[59] In his history of seismology in California, the historian Carl-Henry Geschwind explicitly highlights the growing monetary dependency of the geosciences after World War II: "In the years after 1945, . . . first earthquake engineering and then seismology came to depend more and more on funding from military agencies and the National Science Foundation, as federal support for earthquake research increased a hundredfold." Geschwind, *California Earthquakes* (cit. n. 30), 121.
[60] See Barth, "Politics of Seismology" (cit. n. 30), 764.
[61] Ibid., 744.
[62] Schrag, "Scientists and the Test Ban" (cit. n. 38), 1355. Schrag's attention seems to have been attracted to Speeth's case by Swomley, *Military Establishment* (cit. n. 34), 170–2.

mations of scientific data, for the acoustic representation did not seem to offer a way to obtain any quantifiable results nor any further use within seismological research.

The third hurdle to Speeth's acceptance was his insensitivity to the realities of the defense-research bureaucracy.[63] Speeth was convinced that procedures associated with disarmament measures should be accessible to all negotiating parties. Therefore, in his article, unlike many Vela scientists, he intended to point out clearly the application context of his research. He emphatically insisted that this context should remain visible even in the final version of his revised paper:

> Because of the obvious relevance of this work to the area of international nuclear controls, it is particularly important that all possible sources of artifacts, which could influence the results, be carefully studied. Despite these caveats, auditory methods seem to be a promising addition to other means for interpreting seismograms. . . . Exploration of listening methods should be continued with vigor.[64]

From the administration's point of view, it seemed obvious that Speeth lacked political integrity and caution.

THE METHOD, NOT THE MAN: THE SECOND TEAM

In spite of the denial of support from ARPA and the geophysical community, Speeth's method was not entirely ignored. For instance, an article published by the *Washington Post* on 25 July 1961, which reported on Speeth's recently published study, caught the attention of the public. During the public hearings held by the congressional Joint Committee on Atomic Energy in July 1961, the RAND Corporation physicist Richard Latter, and the chief of Vela Uniform, ARPA's Charles Bates, were asked to evaluate Speeth's results. Although Latter did not conceal his doubts regarding the method's potential, he nevertheless stated, "It is a method which must be looked at. If we are to have any chance of a breakthrough, we have to take advantage of every possible opportunity in our research program."[65] Bates pointed out the success of other auditory methods, such as in sonar technology, and also supported pursuing the research. Bates even announced that a group of scientists at the University of Michigan had also taken up research regarding the auditory method about two months earlier, in May 1961.[66]

Speeth was greatly surprised and concerned to learn from Bates's testimony that his research on auditory discrimination had been handed over to another team, and that he had not been given the opportunity to work with these other scientists. Shortly after the hearings, Speeth finally was awarded an official Vela contract and thus gained some formal recognition for his work. However, the contract would not result in any closer collaboration with other Vela seismologists.

Speeth now began to prepare a follow-up study to reassess his first test series. For this purpose, he turned to the Department of Defense and asked for data sets better than the ones that had previously been available to him. While visiting Bell Labs in the fall of 1961, a military officer affirmed that Speeth would be provided with such

---

[63] Obviously, Speeth was not only a radical idealist, but also someone with his "head in the clouds." Gillian Speeth (Sheridan Speeth's sister-in-law), interview by author, 22 May 2011.

[64] Speeth, "Seismometer Sounds" (cit. n. 1), 913.

[65] *Developments in the Field of Detection and Identification of Nuclear Explosions* (cit. n. 30), 25.

[66] Ibid., 69–74.

data in the near future.[67] However, Speeth would never receive any of the promised records. Quite the contrary: "One day in the spring of 1962, he was called into
the office of his superior and told that he would not be able to get even the lowest
grade of security clearance, and therefore he would not be able to see the improved
recordings."[68] Speeth could have foreseen that the Pentagon would not trust any
security-related data to a man who had—among other things—visited postrevolutionary Cuba. Nevertheless, he felt increasingly harassed and bitter after this setback:
his work, he once told his family, was considered to be so important that he wasn't
even allowed to read what he himself had written.[69]

Meanwhile, Speeth had managed to get hold of at least a couple of matched data
pairs from the then almost outdated 1958 Hardtack series, and he used these in his
follow-up study conducted in the winter of 1961/2. The best result reached by listeners in this test series was an average discrimination rate of only 74 percent. However,
the results of the new experiments enabled Speeth to describe the characteristic auditory features in the signals that he considered to be essential to the discrimination by
means of a mathematical formula. Based on this formula, Speeth outlined an algorithm "that would replace the human ear."[70] In February 1962, he summed up his new
results and filed another research report.[71] Speeth seems to have received no official
feedback on this report. And it got even worse. Since it was felt that, under the circumstances, Speeth would hardly be able to make any further contribution, he was
excluded from the Vela Uniform program without further ado. Speeth had no other
choice but to devote himself to less controversial studies in psychoacoustics, including research to obtain his PhD in experimental psychology.

Through his work, Speeth had acquired significant skills in the then-young field of
digital audio processing. For this reason his colleagues Max V. Mathews and John R.
Pierce invited him to appear on a phonographic record featuring computer-generated
music created at Bell Labs. For the long-playing version of *Music from Mathematics*,
which was released in 1962, Speeth contributed a musical piece called "Theme and
Variations," which closes with an audified seismogram of a nuclear explosion.[72] In
doing so, Speeth succeeded in smuggling one of his explosion sounds out of the laboratory and into the world. Thus, while musicians contributed to solving the discrimination problem in the psychoacoustic laboratory, the seismograms that were created
for acoustic representation were eventually turned into aesthetic objects in the form

[67] Schrag, "Scientists and the Test Ban" (cit. n. 38), 1359. Speeth was optimistic because on 1 Sept.
1961, as a result of the continuing diplomatic deadlock, the Soviet Union announced that it would
resume its nuclear program. An extensive test series started on the same day and included the largest
nuclear explosion to date. Due to the revocation of the moratorium, the Western powers resumed testing as well. From these new tests, Speeth hoped to obtain new data.
[68] Ibid., 1360.
[69] C. E. Speeth, interview (cit. n. 37). Lauren Speeth, Sheridan Speeth's daughter, offers the following explanation: "The way I heard the story is this: at one point a secretary was typing up one of his
papers, and used 'top secret' paper (either by mistake, or because his research was classified over his
head). Regardless of the cause, the result was above the threshold of what was allowed for him to read,
and he couldn't read his own work." E-mail message to author, 23 Sept. 2012.
[70] Schrag, "Scientists and the Test Ban" (cit. n. 38), 1359.
[71] Speeth, *A Sound Detection Technique*: *Technical Report on Contract NObsr 85206* (Murray Hill,
N.J., Feb. 1962). This report may be lost. Schrag gives some information about its contents; see "Scientists and the Test Ban" (cit. n. 38), 1360.
[72] *Music from Mathematics: Played by IBM 7090 Computer and Digital to Sound Transducer*,
Decca, DL 9103, 1962, 33⅓ rpm.

of computer music.[73] As this episode shows, the entwined histories of Western science and music remain closely linked,[74] even in the era of computer-generated music and practices of so-called time axis manipulation.[75]

Meanwhile, Speeth continued to work on the detection problem in his free time. Further publications were either denied or constrained in almost absurd ways. For instance, it took Speeth over nine months—from November 1962 to August 1963—to publish a very short article about the current state of his research requested by a scientific journal, because clearance was not granted by the Directorate of Security Review for over five months. This note consisted of just five paragraphs and contained no quantitative results.[76] From the documents available, it cannot be exactly determined whether or not the authorities purposefully tried to prevent the publication of further results. However, testimony given at the hearings of the Joint Committee on Atomic Energy in 1963, in which the auditory method was discussed once again, indirectly hints that this was indeed the case. While Latter and Bates, who had testified in the 1961 hearings, appeared to be rather hopeful, Jack Ruina, director of ARPA, considerably minimized the potential of the method in 1963:

> REPRESENTATIVE PRICE. Doctor, in 1961 hearings I asked the question concerning an experiment at the Bell Telephone Laboratories that had to do with audio tapes. Has there been progress made in that field since that time?
> DR. RUINA. We have done more work in the field. It still doesn't look like it would be a very promising technique in itself. It was an extremely interesting idea of whether the human acting as a computer can recognize a pattern. . . . In some early experiments which were not done as carefully as we would have liked to see them done, the results looked very good. I believe that Dr. Speeth may have reported results as high as 85 to 90 percent of the records were correctly identified by a trained observer. When the experiments were done under slightly more careful conditions, the result was between 60 and 70 percent correct identification.[77]

Only a few days earlier Speeth had been summoned to Washington to report on the state of his research to ARPA officers. Speeth had used this opportunity to explain his approach in detail once again.[78] It is little wonder that he regarded Ruina's summary to be a deliberate distortion and misrepresentation of his work. Ruina's statements were obviously based on the results of the second team that had been created behind Speeth's back. Ruina further declared:

> A 60 percent number is hardly an impressive number. However, there seems to be enough to this that we are not quite ready to drop the work and we want to explore it as far as we can, using other techniques, electronic techniques for shifting signals. A variety

---

[73] In 1957, the Greek composer Iannis Xenakis already had incorporated seismic recordings into his first electroacoustic piece, *Diamorphoses*, composed at the famous recording studio of the Groupe de recherches musicales in Paris. Thus, Speeth was not the first to use accelerated seismograms for composition purposes, but he certainly was the first to introduce subterranean nuclear blasts into experimental music.

[74] See John Tresch and Emily Dolan, "Toward a New Organology: Instruments of Music and Science," in this volume.

[75] Friedrich A. Kittler, *Grammophone, Film, Typewriter* (Stanford, Calif., 1999), 34–6, 109, 127.

[76] Speeth, "Test Detection," *International Science and Technology* 2 (1963): 20.

[77] *Developments in Technical Capabilities for Detecting and Identifying Nuclear Weapons Tests: Hearings Before the Joint Committee on Atomic Energy*, 88th Cong. 82 (1963).

[78] Schrag, "Scientists and the Test Ban" (cit. n. 38), 1361.

of techniques have been suggested, and we can see if indeed this would be a helpful technique for identification.

REPRESENTATIVE PRICE. Is it an extensive experiment? Has much effort gone into it?

DR. RUINA. No. It is not the sort of thing that is capable of very extensive work. There are a few researchers working at the University of Michigan and a few working at the Bell Telephone Laboratories independently. They are now going to get a lot more data to work with, a lot more records, and try to handle the records in an identical way so there is no built-in bias in the experiment and then try it again.

REPRESENTATIVE PRICE. It is not one of the several techniques that you stated in your prepared statement that have emerged as a new method toward distinguishing an earthquake from an explosion?

DR. RUINA. No, it is not. We have not included that technique.[79]

Most likely Ruina said "we are not quite ready to drop the work" only in respect to the team at the University of Michigan, because contrary to his announcement that "they" were "going to get a lot more data to work with," Speeth did not receive any data at all that would have allowed him to resume his work. Apparently, ARPA was ready to support research on the auditory method, but not its original inventor. While Ruina's statement suggested that, in addition to the Michigan team, "a few working at the Bell Telephone Laboratories" were studying the method independently, it only disguised the fact that Speeth had been completely isolated and silenced.

John DeNoyer of the Institute for Defense Analysis also took a firm stand against the practice of "auditory recognition" in testimony at the hearings:

> Since the Joint Committee has expressed previous interest in the subject, I will discuss briefly the topic of auditory recognition. I do not propose this as one of the methods that might be of immediate application in a test ban monitoring system.
>
> Attempts to recognize the difference between earthquakes and underground nuclear explosions by listening to speeded up magnetic tape recordings have not been very successful. The most recent work by this method indicates that events can be identified correctly about 64 percent of the time. This is better than half correct answers and as such it should be encouraging.
>
> The recordings on which this percentage was based were not equally divided into explosions and earthquakes. This method does contain a hidden danger. Seismograms from explosion sources were called earthquakes about as often as seismograms from earthquakes were called explosions. Any diagnostic aid must be viewed with caution that classifies an explosion as an earthquake.[80]

DeNoyer did not even mention Speeth's work. Instead, he referred to "the most recent work by this method," done by the team at Michigan. The members of this team were Gordon E. Frantti and Leo A. Levereault, two geophysicists working at the Acoustics and Seismics Laboratory, which was part of the University of Michigan's Institute of Science and Technology in Ann Arbor. In May 1961, ARPA commissioned Frantti and Levereault to reproduce the experiments on auditory discrimination within a period of two years.[81] For this purpose, the material submitted by Speeth was sent to

---

[79] *Developments in Technical Capabilities for Detecting and Identifying Nuclear Weapons Tests* (cit. n. 77), 82–3.

[80] Ibid., 220–1.

[81] Frantti and Levereault's work was sponsored by the Air Force Office of Scientific Research under contract AF 49(638)-1079 as part of the Vela Uniform program. See Thomas W. Caless, *Report of Vesiac, Compendium of Contract Information in the Vela Uniform Program*, report 4410-8-T (Ann Arbor, Mich., 1961).

them.[82] As Vela researchers and members of the Acoustics and Seismics Laboratory, they enjoyed ARPA's full confidence and also had access to the seismographic database of their institution, an extensive library of earthquake and bomb seismograms just created within the framework of the Long Range Seismic Measurement program.[83] In their study, Frantti and Levereault converted about 200 short-period seismic signals into "seismic sounds" by means of time compression and filtering. The approach was based on Speeth's original test series: 19 subjects, divided in groups of 9 to 13 participants, were asked to take approximately 1,500 auditory trials, and averaged around 65 to 68 percent correct decisions. In their publication, which did not appear until 1965, a rather understated conclusion was given: "The results of the experiments suggest that a trained listener can identify approximately two-thirds of the seismic sounds presented, where one half corresponds to chance performance."[84]

Frantti and Levereault published the results of their study in the *Bulletin of the Seismological Society of America*, being far more reserved and vague regarding the purpose of their research than Speeth. Since they showed little enthusiasm about the auditory method in their article and did not suggest further research, seismologists gained the impression that the auditory method was of little value. While shortly after the hearings in June 1963, for instance, an article published in the *Bulletin of Atomic Scientists* had listed the "auditory discrimination between earthquakes and explosions using speeded-up tapes of the seismic signals" along with other diagnostic aids,[85] Nilo Lindgren only casually mentioned the auditory method in his 1966 overview on nuclear test detection in the *IEEE Spectrum*, and without referring to any sources:

> Also, interesting experiments have been conducted with transforming seismic records into audio signals on the assumption that the unique pattern-recognition properties of the human ear would distinguish the sounds peculiar to explosions and earthquakes, but these studies have not proved remarkably fruitful.[86]

Frantti and Levereault had been able to access a large inventory of high-quality data, and they certainly exercised reasonable scientific care in carrying out their experiments. However, the question arises whether or not Speeth would have obtained other, if not more favorable, results. For the geologists, the psychoacoustic experiments represented an excursion into a field where they lacked experience and interest. It seems worth mentioning that Frantti and Levereault did not discuss auditory features that might enhance the detection of earthquakes and explosions. Instead of actually training their subjects for optimal performance, they only published results regarding a general ability to carry out the discrimination task.

[82] It must be assumed that Speeth's work was received well by at least some Vela researchers, because, interestingly enough, the only two US libraries (apart from that of Columbia University) that hold copies of Speeth's doctoral thesis are precisely those located at the University of Michigan, Ann Arbor, and the University of Nevada, Las Vegas, which is the closest university to the Nevada test site.

[83] This library consists of high-quality seismograms from earthquakes, nuclear explosions, and chemical explosions, stored on magnetic tape. See David E. Willis and James T. Wilson, *Seismic Studies and Experimental Evaluations*, final report, contract AF 19(604)-6642 (Ann Arbor, Mich., 1963).

[84] G. E. Frantti and L. A. Levereault, "Auditory Discrimination of Seismic Signals from Earthquakes and Explosions," *Bulletin of the Seismological Society of America* 55, no. 1 (1965): 1–25, on 1.

[85] Richard Preston, "Test Ban: Optimism on Project Vela," *Bulletin of the Atomic Scientists* 19, no. 6 (June 1963): 33–7, on 34.

[86] Lindgren, "Earthquake or Explosion?" (cit. n. 17), 71.

The fact that Frantti and Levereault, unlike Speeth, were not driven by an absolute will to succeed, might also have added to the rather poor results. Apparently, they did not make the same "fanatic effort" as the politically motivated psychologist.[87] Quite the contrary; they probably considered their study on the auditory method to be contract work and were only partly convinced of its merit, for they described it as a "subjective" technique that was only capable of providing "auditory results" rather than "analytical results."[88] On the other hand, one could argue as well that Speeth's political enthusiasm may have biased his own judgment in favor of the auditory method and that the noncommitted geophysicists had the necessary scientific distance to evaluate the method correctly. Whatever the case may be, the low recognition rate that resulted from the new experiments provided ARPA with a suitable justification to ultimately drop further research regarding auditory discrimination.

The detection problem itself was finally resolved in two ways. Within the test ban negotiations, the difficulty of detection eventually led to the exclusion of the underground environment from the treaty.[89] The Moscow treaty, consequently known as the Limited or Partial Test Ban Treaty, was signed on 5 August 1963 and came into effect on 10 October of that year. It banned nuclear arms tests in the atmosphere, in outer space, and underwater. The radioactive contamination of the atmosphere was thereby reduced, and the emergence of new nuclear powers was hampered. It also enabled the test ban antagonists to continue their work without further interruption: of all 1,030 nuclear tests conducted by the United States between July 1945 and September 1992, a total of 815 were performed underground. Thus, after the signing of the test ban treaty, the detection problem ultimately became a problem of intelligence and reconnaissance within the national control system of the United States.

To tackle the detection problem scientifically, Vela scientists had turned to the development of seismic arrays—large groups of seismic units within a certain area connected to a central data-processing station.[90] After preliminary studies, engineers from MIT's Lincoln Laboratory completed the Large Aperture Seismic Array (LASA) in the winter of 1964/5. The system consisted of 525 seismic units that were divided into 21 subgroups and spread over an area of 125 square kilometers.[91] LASA significantly improved the signal-to-noise ratio of seismic measurements and therefore represented a "substantial improvement in seismic discrimination capabilities."[92] Needless to say, extravagant projects like LASA, which allowed for obtaining high-quality seismic data, were far more attractive to seismologists than the rather simple and straightforward method proposed by Speeth. LASA was not only capable of detecting extremely weak underground explosions, it also constituted a powerful system that produced seismographic inscriptions of the highest quality. It could profitably be used by seismologists, for example, for the study of very small earthquakes. The auditory method

[87] Schrag, "Scientists and the Test Ban" (cit. n. 38), 1362.
[88] Frantti and Levereault, "Auditory Discrimination of Seismic Signals" (cit. n. 84), 1, 7.
[89] In the summer of 1962, by which time the continuing deadlock in diplomatic negotiations had lasted for several years, the Western powers decided to propose an agreement that consisted only of national control systems and did not require any further onsite inspections, provided that underground tests would not be covered by a test ban. About a year later, in July 1963, the Soviet Union withdrew its demand for a comprehensive test ban and accepted an agreement based on the Western draft treaties.
[90] Lindgren, "Earthquake or Explosion?" (cit. n. 17), 71–4.
[91] P. E. Green, R. A. Frosch, and C. F. Romney, "Principles of an Experimental Large Aperture Seismic Array (LASA)," *Proceedings of the IEEE* 53 (1965): 1821–33.
[92] Lindgren, "Earthquake or Explosion?" (cit. n. 17), 72.

did not provide such additional value for the discipline of seismology, and consequently, it sank into oblivion.

CONCLUSION: AUDIFICATION AND COGNITION

Today, Speeth's work on auditory discrimination is known primarily from references in papers from ICAD in which it was rediscovered. As shown above, these authors regard Speeth's 1961 paper "Seismometer Sounds" as a successful ancestor of their own research agenda on auditory display. At the same time it is assumed that Speeth's method was not able to reach wide acceptance among seismologists because scientists in general are too accustomed to working with only their "eyes and hands." In contrast, this article shows that Speeth's failure to popularize his auditory method was due less to the prevailing "epistemological power of the eye" in the geosciences than to both the actual limitations of the auditory display as well as Speeth's difficult political, institutional, and disciplinary relations—as a peace activist and a psychologist—with ARPA and the seismological community.

After its publication, seismologists showed little interest in the auditory method of discriminating earthquakes from underground explosions. This lack of interest cannot be reduced to the auditory nature of the method. Instead, it can be explained much better by the fact that it failed to answer questions seismologists considered to be of critical importance. Whereas Speeth was searching for auditory patterns in the entire signal in order to distinguish earthquakes from underground explosions, seismologists primarily focused on single oscillations of seismic events and different types of seismic waves present in the signal. In addition, discrimination of the seismic events simply was not a sufficient result for seismologists, as they were also expected to calculate the geographic origin and the magnitude of seismic events as well as the possible yield in case of an underground explosion. These quantitative parameters could not be assessed by the auditory method. Therefore, if they noticed it at all, seismologists seem to have considered the auditory representation of seismic data to be an oddity that did not offer connections to the theories, concepts, and issues of their field.

Of course, it could be argued that Frantti and Levereault's discouraging results as well as the success of other discrimination methods, especially the seismic arrays, indicate that the performance of the auditory method was simply more limited than its inventor had wished it to be. But even if this was the case, why then was Speeth denied the opportunity to at least try his best? Speeth was an "ingenious tinkerer" and an idealist who valued the social and humanitarian benefits of his work more highly than a successful scientific career.[93] Within the framework of a larger project, such as an interdisciplinary collaboration of experienced seismologists and psychoacousticians, new possibilities might have emerged for the auditory representation and assessment of seismological data in general, and for the detection problem in particular. Instead, considerable efforts were made on different levels to suppress Speeth's work: by his superiors, seismologists, and ARPA officials, many of whom obviously did not favor a cessation of nuclear testing. Therefore, the general rejection of Speeth's work was not entirely for scientific reasons, but rather due to a predominant climate among members of the defense sciences, most of whom viewed a comprehensive test ban not as a desirable step toward global disarmament but merely as

[93] G. Speeth, interview (cit. n. 63).

an inhibition to their work.[94] Even John Tukey, vice president of Bell Labs at the time and a member of the Berkner Panel, did not support the innovative research that came out of his own institution. In this respect, the story of the seismometer sounds can also be read as a lesson in the marginalization of science. Almost nobody recognized Speeth's effort to resolve the detection problem as an attempt to contribute to global peacekeeping. Instead, Speeth was seen as both a disciplinary outsider and a political opponent—if not a loose cannon—who was to be controlled and silenced. The classification of Speeth's work functionally resulted in censorship and eventually led to his isolation as a scientist. Further research was rendered impossible by the denial of security clearance and cooperation. Neglect and deliberate misrepresentations by officials and scientists in public hearings helped to discredit the auditory method and its inconvenient inventor.

Yet this article is not intended to tell the story of a fallen hero. Rather, the case of the seismometer sounds demonstrates that sound and listening were being used to produce scientific facts and to assess scientific questions in the laboratory. This application of sound and listening as techniques of evidence reveals the desideratum of a history of auditory practices as epistemic tools, or, more generally, a history of the auditory culture of science and scientific practice. In writing such a narrative, I suggest that we not focus entirely on the scientific senses that are harnessed in the lab, but on the transformations and representations of inscriptions, with special attention to the question of how "sound facts"—quantifiable results, but also qualitative data—are produced by auditory methods. In doing so, we may escape the trap of explaining the dominance of visual practices in the sciences by ontological preconceptions of the senses. Speeth may have failed in his endeavor to promote his method among geophysicists, but this was certainly not due to the prevalence of a general hegemony of vision within Western culture and the sciences. Instead Speeth faced concrete scientific, disciplinary, and sociopolitical problems. Historians of sound should therefore be careful not to hold a general visual primacy responsible when auditory methods for creating scientific knowledge fail to gain popularity. In evaluating the question of why the use of sound in laboratories proves to be successful in some cases but not in others, we should try to determine the place auditory practices occupy within the chain of inscriptions and the network of the research process as a whole. Regardless of how promising, or not, Speeth's method of auditory detection and discrimination actually was compared to other seismological approaches, the major difference between Speeth and the rest of the Vela seismologists was that they were asking different questions and following different goals. If we wish to determine the status of auditory knowledge production in the sciences, we therefore should not only emphasize the scientific answers that can be provided by such methods, but also consider which of them are actually desired in the course of a specific research process, and to what extent these answers are shaped by reasoning based on technological transformations and representational practices.

---

[94] "So the argument *against* the test ban, people will raise the issue of whether it can be monitored, which is what you're talking about when you talk about decoupling. But that's not the *main* reason for wanting the test ban to go away. It's because they feel that it's inhibiting. . . . The people who are against it—this monitoring issue is *much* argued but it *wasn't* what motivated people. What *motivated* people was that they either *did* or *didn't* want the limitation itself." Herbert F. York, interview by Mary Palevsky, 22 July 2004, La Jolla, Calif., available on the website of the Nevada Test Site Oral History Project, University of Nevada, Las Vegas, http://digital.library.unlv.edu/objects/nts/1307, 12.

PHENOMENOTECHNIQUE OF SOUND

Mineral Sound or
Missing Fundamental:
Cultural History as Signal Analysis

*by Bernhard Siegert**

ABSTRACT

The cultural and the scientific history of the bell sound are linked together, and thereby the methodological premises are created to analyze cultural phenomena in terms of signal analysis. Because of the strange phenomenon that the tone after which a bell is named, the so-called strike note, cannot be found within the spectrum of the bell, the sound of a church bell can be described as the deconstruction of Western acoustic culture. In its sound the secondary precedes the primary. The primary—the fundamental tone—is projected first of all by the secondary, its harmonics. This article, which introduces (to a modest extent) media-theoretical and Lacanian terminology into the history of science discourse, argues that bells represent and signal a state of emergency in the symbolic order because they *are* this state of emergency in the acoustical real.

POETRY GENERATION BY NOISE FILTERING

The day before Christmas, 1814, Johann Wolfgang von Goethe wrote a poem, which he entitled "Boldness" ("Dreistigkeit"):

> Worauf kommt es überall an,
> Daß der Mensch gesundet?
> Jeder höret gern den Schall an
> Der zum Ton sich rundet.
>
> Alles weg! was deinen Lauf stört!
> Nur kein düster Streben!
> Eh er singt und eh er aufhört,
> Muß der Dichter leben.
>
> Und so mag des Lebens Erzklang
> Durch die Seele dröhnen!
> Fühlt der Dichter sich das Herz bang
> Wird sich selbst versöhnen.[1]

---

* Bauhaus-Universität Weimar, Internationales Kolleg für Kulturtechnikforschung und Medienphilosophie, 99421 Weimar, Germany; bernhard.siegert@uni-weimar.de.
  All translations, unless otherwise credited, are by the author.
[1] Goethe, "Dreistigkeit," in *Werke*, ed. Erich Trunz, Hamburg ed., vol. 2, 14th ed. (Munich, 1989), 16.

© 2013 by The History of Science Society. All rights reserved. 0369-7827/11/2013-0006$10.00

The English translation by John Whaley does not translate *Erzklang* ("mineral sound"), which is essential for an understanding of Goethe's precise politics of sound:

> What's the universal measure
> For man's health propounded?
> All men hear a sound with pleasure
> When as tone it's rounded
>
> Clear away whatever hinders!
> No more striving dully!
> 'Fore he's sung himself to cinders
> Poet must live fully.
>
> Let life resonate and thunder
> Through the poet's soul!
> Though he fear his heart may sunder
> Soon again he's whole.[2]

In the year that Goethe wrote his poem, Napoleon was forced to abdicate after Blücher's victory at La Rothière. Allegedly, there was still fighting in December. But, while it makes an acoustic difference whether *Erzklang* referred to the thunder of guns or to the thunder of bells, in Goethe's acoustic world they both belonged to the same Napoleonic paradigm, as one can deduce from a letter that Goethe wrote to Charlotte von Stein in 1812 from Carlsbad, in which he signed off in the following way:

> But what would you say, if I had no other choice than to date this letter as:
> Carlsbad
> the 15th of August
> at the Napoleon feast
> amidst the heaviest tolling of bells and the thunder of cannons truly dedicated
> 1812.   Goethe.[3]

Bells and cannons mingle not only within the perception of Napoleonic feasts but also in reality. All of the artillery used by the revolutionary armies and the armies of General Bonaparte had formerly been church bells. Since 1793 a large proportion of the church bells of France had been cast into cannons. It was estimated that 100,000 bells from 60,000 church steeples were melted down during the time of the revolution.[4] But metamorphosis into cannons used to be the destiny of church bells in the German states, too. The confiscation of bells in times of war was a long-standing tradition in Europe. In early modern Europe the chief of the artillery had a right to the bells of a conquered city.[5]

The question with which the poem begins subordinates the lyrics directly to the discourse of medicine: "What's the universal measure / For man's health propounded?"

---

[2] Goethe, "Boldness," in *West-Eastern Divan—West-östlicher Divan: Bi-lingual Edition of the Complete Poems*, verse translation by John Whaley (New York, 1998), 37.

[3] Goethe, *Goethes Werke*, ed. on behalf of Grand Duchess Sophie von Sachsen, pt. 4, vol. 23 (Weimar, 1905), 73.

[4] See Alain Corbin, *Die Sprache der Glocken: Ländliche Gefühlskultur und symbolische Ordnung im Frankreich des 19. Jahrhunderts* (Frankfurt am Main, 1995), 44.

[5] Ibid., 28.

The arrogant phonocentrism of the answer negates the technical fact that all lyrics begin with the letters of the alphabet, and instead insinuates—via the image of the minstrel—that lyrics originate in an acoustic operation: "All men hear a sound with pleasure / When as tone it's rounded."

Poetry is a system that must generate the boundary separating it from its environment, by filtering noise—or, more precisely, filtering out all those elements of a sound that do not contribute to the roundness of a tone; so, in terms of acoustics, filtering out all nonperiodic parts of the acoustic spectrum.[6] Idiophones of ore, stone, or wood produce more noise than other musical instruments because their harmonics are not integral multiples of the fundamental and entertain among each other irrational and therefore chaotic proportions. The elements of poetry according to Goethe's poem are initially generated by filtering noise instead of being given by the alphabet. But the poem does not merely speak of this operation of rounding. It performs in a self-referential way precisely that about which it speaks. Half of the vocalic rhymes end with a spondee, which replaces here the trochee (a spondee has two beats [– –]; a trochee, beat and offbeat [– ∪]). While the spondee simulates inhuman ore sounds ("bang!"), the trochee rounds the sound (as in "rundet"). Alternating from A-rhymes to B-rhymes performs a reconciliation, a sublation of the mere sound to the interiorized tone.

Poetry is therefore a system that also generates the subject—since poetry belongs to the discourse of medicine. To filter out all that is chaotic from the "mineral sound of life" is to constitute the subject first, the poet especially. In other words, the bell is the Other of culture and the Other of the subject, inasmuch as culture can be seen as being equal to the medium of the alphabet. This subject according to Goethe is an effect of the feedback loops of poetry. In the process of self-perception, in which the author is the subject and object of poetry at the same time, the human is constituted by way of negating the unwritable and incalculable.

CULTURAL SEMIOTICS OF BELL SOUNDS

The cultural and scientific history of the bell sound provides an example of how one can write cultural history in terms of signal analysis. What sorts of history can the sound of a bell have? I think there are two sorts: The first results from the interpretation of the bell sound as a sign. In this case one writes a history of cultures of emotions (this is exactly what Alain Corbin has done[7]). The second results from the analysis of the bell sound as a signal in the sense of physics. In this case one would stick to the definition of the sound object as given by signal analysis. This presupposes the technical possibility of recording and processing those acoustical data, which resist the code of musical notation.

A media-historical approach, as it was developed in Germany over the last twenty-five years, will not try to do either the one thing or the other, but will combine both to derive a concept of cultural semiotics from signal analysis. The use of media-theoretical concepts is still uncommon in large parts of the history of science. Historians of science prefer to speak of "instruments,"[8] but usually shy away from studying

[6] See Friedrich A. Kittler, "Ein Subjekt der Dichtung," in *Das Subjekt der Dichtung: Festschrift für Gerhard Kaiser*, ed. Kittler, Gerhard Buhr, and Horst Turk (Würzburg, 1990), 399–410.

[7] See Corbin, *Sprache der Glocken* (cit. n. 4), 28.

[8] See Albert Van Helden and Thomas L. Hankins, eds., *Instruments*, vol. 9 of *Osiris*, 2nd ser. (1993).

the question of how instruments can turn into media.[9] They further shy away from even discussing the question of how instruments can be properly distinguished from media, even though media are quite often rooted in histories of scientific instruments and vice versa; for instance, in the case of the electric telegraph or wireless telegraphy.[10] The problem of the discernibility of the history of scientific instruments and media history has not led to a broader reception of media theory within most parts of the history of science, which is mostly due to the fact that media theory—at least German media theory—has strong roots in French poststructuralist thinking. This approach implies the use of philosophical and psychoanalytical concepts, which are uncommon or even unwanted among many historians of science who see their discipline as more closely connected to the methods of the natural or engineering sciences than to the methods of the humanities. But it is exactly the distinction between the sciences and the humanities that is questioned by media. In contrast to the aloofness of historians of science, historians of media make wide use of the publications of historians of science and technology; for instance, Bruce J. Hunt's article on Faraday and cable telegraphy or Peter Galison's book *Einstein's Clocks.*[11] Thus, a bidirectional transfer of concepts and methods between historians of science and historians of media has yet to be achieved. It is precisely because media subvert the all-too-comfortable distinction between the sciences and the humanities that I am suggesting here not the analysis of signals *instead* of the interpretation of signs, but the interpretation of signs *as* or *via* signal analysis.

The Western system of harmonics was stabilized over the centuries by the system of musical notation based on the phonetic alphabet. The parameters, which were available to composers and musicians as variables that could be manipulated, were pitch (i.e., frequency), duration (time), and volume (amplitude), the latter of which could not be represented by the technical code of notation but rather through a linguistic code (from *fortissimo* or *fff* to *pianissimo* or *ppp*). What could not be written down within the notational system was the timbre or acoustic color. Acoustic color is the parameter of sound that refers to the materiality of the musical instruments. It is due to acoustic color that we identify a fluid as a fluid, a violin as a violin, and a clarinet as a clarinet. The majority of this information is contained in the transient phenomenon of the signal and in its formants.

What is a bell sound in terms of cultural semiotics? Originally, the sacred function of bell sounds consisted of their signaling character: the tolling of the church bells is a sign that announces a sign. The language of the bells that Martin Heidegger knew so well when, as a boy, he was a sacristan's assistant and his favorite playground was

[9] This has been done in an exemplary essay by Joseph Vogl: "Becoming-Media: Galileo's Telescope," *Grey Room*, no. 29 (Fall 2007): 14–25.

[10] See, e.g., Timothy Lenoir, "Farbensehen, Tonempfindung und der Telegraph: Helmholtz und die Materialität der Kommunikation," in *Die Experimentalisierung des Lebens: Experimentalsysteme in den biologischen Wissenschaften, 1850/1950,* ed. Hans-Jörg Rheinberger and Michael Hagner (Berlin, 1993), 50–73. Most interestingly, in the English version of Lenoir's article the medium—the telegraph—vanished from the title: "Helmholtz and the Materialities of Communication," in Van Helden and Hankins, *Instruments* (cit. n. 8), 183–207.

[11] See Hunt, "Michael Faraday, Cable Telegraphy and the Rise of Field Theory," *History of Technology* 13 (1991): 1–19; Galison, *Einstein's Clocks and Poincaré's Maps: The Empire of Time* (New York, 2003).

the church steeple,[12] is a call that gathers together what it calls: "The gathering call is the tolling."[13] To put it in the words of Pink Floyd on *Dark Side of the Moon*:

> Far away, across the field
> The tolling of the iron bell
> Calls the faithful to their knees
> To hear the softly spoken magic spell.[14]

Friedrich Kittler has analyzed the church as a medium in a language that combines concepts of Lacanian psychoanalysis with the terms of media technology:

> Even if in sermons and the liturgy the church is relying essentially on the word of the Great Other, which it has to pronounce and to amplify, no human voice is able to carry this word far enough beyond the naves. Although the naves were constructed as sound amplifiers, only noises reach beyond the naves. Therefore discursive powers form alliances with superior powers, and creations with a *tohubohu*—even if they are threatened by the possibility of being overpowered and drowned.[15]

Heidegger in his later years tried to define the language of bells as the essence of language itself: "Language speaks as the tolling of silence."[16] In terms of communications technology, if we replace "silence" with "background noise" (which is perfectly correct in the physiological, thermodynamic, and communications-technological sense of silence) and "tolling" with "signal," this means that language speaks as the signal-to-noise ratio. The essence of language is not found in its being a means of communication, but in its being a signal. The church bell has always already articulated this essence of language. From this call to gather in prayer the more secular functions of the bell as an alarm mechanism have been derived. If not to signal that God is calling the faithful to the "magic spell," bells toll when an enemy is approaching or a fire has broken out.

Corbin has shown that the right to make some noise or to deafen the ears became the object of numerous local conflicts in nineteenth-century postrevolutionary France. The new regime in Paris demonstrated the sovereign power of the nation to local authorities by prohibiting the ringing of the bells or by giving the order to take down the bells (until the revolution of 1830). He who rules the signals rules the nation.

In Dorothy Sayers's novel *The Nine Tailors*, the nine giant bells of a country parish church turn into murderers. In the graveyard of Fenchurch St. Paul a body with horribly distorted facial features is found shortly after Christmas Eve. Lord Peter Wimsey ruminates until the end of the novel on what might have caused the mysterious death until by chance he finds himself locked in the bell chamber of the church steeple during the ringing of the bells on a holiday. After being rescued at the very last second, Wimsey knows who the murderers were: the mineral sound of the bells first drove the victim to madness and then killed him.[17] The substratum of a "culture

[12] See Heidegger, "Das Geheimnis des Glockenturms," *Gesamtausgabe*, pt. 1 (Frankfurt am Main, 1983), 13:113–6.
[13] Heidegger, "Die Sprache," in *Gesamtausgabe* (cit. n. 12), 12:7–30, on 27.
[14] Pink Floyd, "Time," *Dark Side of the Moon*, EMI 3C 064-05249, 1973, 33⅓ rpm.
[15] Kittler, "Ein Subjekt der Dichtung" (cit. n. 6), 402.
[16] Heidegger, "Sprache" (cit. n. 13), 27.
[17] See Sayers, *The Nine Tailors* (London, 1934).

of emotion" that produces a communitarian identity through the tolling of bells turns out to be, if experienced up close, something dreadfully violent.[18]

Bells mark off the border between sanity and madness, between sobriety and frenzy, between the state of normalcy and the state of emergency (in the case of war or natural catastrophes), between God and man, between life and death. I now wish to combine this very short outline of the church bell's cultural semiotics with an argument derived from signal analysis. It will be my thesis here that bells represent and signal a state of emergency in the symbolic order because they *are* this state of emergency in the acoustical real.[19] This is the reason why bells are media, because they place that which is mediated through them under the conditions that they produce and that they themselves are.[20]

The bell sound established in the heart of rural and small-town Europe a network of distributed local centers in which culture constituted itself by an alliance with chaos, which on the level of Christian symbols and ideas signified either the *tohubohu* or the devil, and on the level of media (codes and instruments) signified something that could not, on principle, be written.

BELL SOUNDS AND THE HISTORY OF MATHEMATICAL ACOUSTICS

After that very abridged outline of the cultural semiotics of the bell sound now follows an equally abridged outline of the role the bell played in the history of scientific acoustics. Textbooks on the physics of musical instruments still today begin with Jean le Rond d'Alembert's wave equation, and with good reason.[21] The wave equation provided the initial basis for a mathematical acoustics, which was deconstructed soon after by Leonhard Euler and Jean Baptiste Joseph Fourier. On the one hand, one cannot tell the story of experimental acoustics without mentioning Simon Stevin, Marin Mersenne, and Robert Hooke, who made experimental contributions to a modern mathematical concept of pitch and of the interval based on absolute frequencies. Stevin was bold enough to maintain in 1595 that the octave necessarily consisted of twelve equal halftones and thus calculated the intervals on the monochord for the equally tempered scale on the basis of $\sqrt[12]{1/2}$.[22] Most theoreticians of music during

[18] See Corbin, *Sprache der Glocken* (cit. n. 4), 15.

[19] Here as well as in the following I am making use of Jacques Lacan's conceptual triad of the symbolic, the imaginary, and the real. Whereas the symbolic signifies the structural network of the digital signifier, the operations of which are for the most part unconscious, the imaginary, whose original scene is the "mirror stage," is responsible for the constitution of an I and of our conviction of an objective reality. The real (which must not be confused with reality) underwent various changes of definition in the course of Lacan's writings from the 1930s to the 1970s. In this article the term is used in the sense of that which escapes all kinds of symbolization. The real is that which cannot be written, or the impossible, in the sense of Bataille. Encounters with the real, which is the leftover from a primordial split from which the subject originated, usually have traumatic or psychotic consequences. Kittler's media theory identified the symbolic with typographic media, the imaginary with cinema, and the real with acoustic media. In the digital age, which led to the convergence of analog media, all three are processed within the computer.

[20] See Lorenz Engell and Joseph Vogl, eds., editorial, *Archiv für Mediengeschichte*, vol. 1, *Mediale Historiographien* (2001): 5–8, on 6.

[21] See, e.g., Neville H. Fletcher and Thomas D. Rossing, *The Physics of Musical Instruments* (New York, 1994).

[22] See Mark Lindley, "Stimmung und Temperatur," in *Hören, Messen und Rechnen in der frühen Neuzeit*, vol. 6 of *Geschichte der Musiktheorie*, ed. Frieder Zaminer (Darmstadt, 1987), 180.

the Renaissance could not accept the idea that the sweet sound of the fifth was based on such an "unspeakable, dishonest, irrational number" as 1 $\sqrt[12]{1/128}$,[23] but for Stevin musical harmony no longer had to be represented by rational proportions. Mersenne lengthened the cord of a monochord to a record-breaking twelve cubits.[24] Although he could not hear the oscillations any longer, he could see them with his eyes, and thus was able to extrapolate the frequencies of audible tones by shortening the cord. Hooke pursued flies and other buzzing insects through the streets of London. He reproduced their hum by singing it until he returned home and was able to identify its pitch on his piano, and thus managed to calculate the frequency of the small whirring wings of the insects.[25]

On the other hand, it was d'Alembert's wave equation that for the first time provided a closed mathematical formula for all oscillatory events of a certain type.[26] The wave equation in its classical form applies to all periodic acoustic events that are produced by one-dimensional vibrating bodies—that is, by cords or air columns in wind instruments. The situation becomes much more difficult if one moves from point sources of sound, which are mainly responsible for the sound of a standard Western European orchestra, to surface sources like membranes or idiophones, which make up the instrumentation of most non-European cultures. Around 1800 Ernst Florens Friedrich Chladni explicitly declared that his namesake figures of sound, which made visible the knot lines of vibrating copper plates, should compensate for the impossibility of representing the sound of two-dimensional vibrating bodies in an analytical way,[27] which would have required analytically representing all arbitrary functions that were solutions to the two-dimensional wave equation. The case of three-dimensional idiophones like bells, which simultaneously execute two kinds of oscillations, cross-sectional and longitudinal, was of course even worse, since it would have required the solution of partial complex differential equations of the third order.[28] Additionally, the frequencies of the overtones (the "harmonics") of huge bells (and I am talking here only about church bells and not about so-called carillons, the small bells of chimes) are not multiples of the basic frequency, as is the case in cord or wind instruments. Because its harmonics do not deserve their name, the sound of a bell, the form of oscillation of which is not purely periodic, is called "anharmonic."

Fourier's treatise on the distribution of heat, which he presented to assembled mathematicians at the Institut de France in 1807, bestowed upon mathematical

[23] Stevin, *Vande Spiegeling der Singconst*, in *The Principal Works of Simon Stevin*, ed. Ernst Crone et al. (Amsterdam, 1966), 5:440.

[24] See Mersenne, *Harmonie universelle: Contenant la théorie et la pratique de la musique*, facsimile of the 1636 edition (Paris, 1975), 3:150. Mersenne proudly remarked that not even Galileo managed to count the vibrations of cords or airwaves. See Mersenne, *Les nouvelles pensées de Galilée, mathématicien et ingénieur du duc de Florence*, ed. Pierre Constabel and Michel-Pierre Lerner (Paris, 1973), 1:93.

[25] See *The Diary of Samuel Pepys: A New and Complete Transcription*, ed. Robert Latham and William Matthews (London, 1974), 7:239 (8 Aug. 1666). See also Sigalia Dostrovsky and John T. Cannon, *Entstehung der musikalischen Akustik (1600–1750)*, in Zaminer, *Hören, Messen und Rechnen* (cit. n. 22), 32.

[26] See Jean le Rond d'Alembert, "Recherches sur la courbe que forme une corde tenduë mise en vibration," *Histoire de l'Académie royale des sciences et belles lettres de Berlin* 3 (1747): 214–9, on 216.

[27] See Chladni, *Die Akustik* (Leipzig, 1802), 70.

[28] See Andreas Weissenbäck and Josef Pfundner, *Tönendes Erz: Die abendländische Glocke als Toninstrument und die historischen Glocken in Österreich* (Graz, 1961), 7.

acoustics the analytical tool that still remains essential to its existence today.[29] Without Fourier analysis we would know nothing about the real foundations of our acoustic culture. By means of Fourier analysis the material world, inasmuch as it consists of elastic-body, sound, heat, or ether vibrations, can be analyzed and written down in terms of infinite sums of sines and cosines. Fourier determined that every periodic function that possesses a well-defined graph can be represented by an equation of this type:[30]

$$f(t) = \frac{a_0}{2} + \sum_{k=1}^{\infty} (a_k \cos k\omega t + b_k \sin k\omega t)$$

This had been already suggested by Daniel Bernoulli fifty years earlier as the most general solution of d'Alembert's wave equation. Bernoulli's method was based on the assumption that "all sonorous bodies potentially include an infinity of tones and an infinity of corresponding ways to execute their periodic oscillations."[31] Euler added: be they continuous or discontinuous oscillations.[32] In fact, the process Bernoulli suggested and Fourier performed is mathematically identical to the decomposition of a musical tone into its fundamental and its harmonics. In music such a series converges very rapidly into the identity of acoustic colors.

Until the twentieth century, however, in mathematical acoustics, because its form demanded nearly unsolvable partial complex differential equations of the third order, the bell represented the great beyond of the computable.[33] As long as it was impossible to determine the harmonics relevant for the sound of a bell, even though it was known that this sound was contained in a Fourier integral, the Fourier analysis of bells remained an experimental science. But experimental acoustics experienced a big surprise when, in 1890, John William Strutt, third Lord Rayleigh, applied experimental Fourier analysis to mineral sound for the first time. Rayleigh must have realized that it was impossible to determine with tuning-fork resonators the idiophonic fundamental responsible for the pitch of a bell. The bell designates the state of emergency of physical acoustics.

BELL SOUNDS AND THE HISTORY OF PHYSICAL ACOUSTICS

The acoustic composition of a bell sound consists of two components: first, the physically real oscillations of the hum tones and the higher overtones, which can be de-

[29] See Ivor Grattan-Guinness, *Convolutions in French Mathematics, 1800–1840* (Boston, 1990), 2:585, 595.

[30] Jean Baptiste Joseph Fourier, "Théorie de la propagation de la chaleur," in Ivor Grattan-Guinness, *Joseph Fourier, 1768–1830: A Survey of His Life and Work, Based on a Critical Edition of His Monograph on the Propagation of Heat, Presented to the Institut de France in 1807* (Cambridge, Mass., 1972), 193.

[31] Bernoulli, "Réflexions et éclaircissemens sur les nouvelles vibrations des cordes exposées dans les mémoires de l'Académie de 1747 et 1748," *Histoire de l'Académie royale des sciences et belles lettres de Berlin* 9 (1753): 151.

[32] See Leonhard Euler, "Remarques sur les mémoires précédens de M. Bernoulli," in *Opera Omnia* (Leipzig, 1947), ser. 2, 10:244. This is discussed in great detail in my book *Passage des Digitalen: Zeichenpraktiken der neuzeitlichen Wissenschaften, 1500–1900* (Berlin, 2003), 211–20 and 240–50.

[33] See Fletcher and Rossing, *Physics of Musical Instruments* (cit. n. 21). The "general bell formula" that Kittler mentions in this context has nothing to do with this problem of mathematical acoustics, but with a bell founder's formula. It is also related only to carillons, which represent a much less difficult case. See André Lehr, "A General Bell Formula," *Acustica* 2, no. 1 (1952): 35–8.

tected by measuring apparatuses, and second, the strike note. The strike note dominates the impression of the pitch of the strike sound that is produced by the tolling of a bell. The systematic position of the bell with respect to pitch within a peal is defined according to this strike note. Yet this very note remained, until the 1960s, an inexplicable phantom. While it was possible to detect the single harmonics of a bell by using Hermann von Helmholtz's method of resonators and to record them oscillographically, the one tone that is most prominent and determines the pitch of the bell could not be determined among the measurable frequencies (nor among the difference tones). "With church bells the strange phenomenon appears that the tone after which a bell is named, the so-called *strike note*, cannot be found in general within the spectrum of the bell."[34] Rayleigh therefore had to content himself with an empirical rule, according to which the strike note was located one octave below the fifth eigenfrequency.[35] Even in 1961 the campanological magnum opus about the bells in Austria had to admit "that it has not been possible yet to determine the pitch of the strike note with technical means, but only with the help of one's ears."[36]

In 1930, Franklin G. Tyzzer succeeded in classifying the partial tones of a bell for the first time with the help of knot circles and meridian circles. In the thirties the technical media became available to disclose the secret of the bell with means other than tuning-fork resonators. Ferdinand Braun in Strassburg had already used the cathode-ray tube as a medium for the visualization of oscillations—that is, as a cathode-ray oscilloscope—in 1896. In 1921 Carl Stumpf in his Berlin Psychological Institute made use of the telephone to confirm his theory of the formants of consonants.[37] From now on microphones, oscilloscopes, and choking chains (lattice networks consisting of coils and condensators that act as low-pass filters), which made experimental Fourier analyses by electromagnetic means possible, were the standard equipment of experimental acoustics, which was also carried out in the laboratories of big telephone companies like AEG, Siemens, and AT&T.[38] The Dutch company Philips, which manufactured radio tubes and other electronic equipment, had also founded a research lab in 1914. It was therefore not by chance that in 1940 Jan Schouten published his residuum theory of the strike-note pitch in *Philips' technische Rundschau*. Only after an algorithm was implemented that could perform Fourier transformations did it become possible to subject the acoustic signal of church bells to an objective pitch analysis. Since 1982 one obtains the frequencies of the partial tones and their amplitudes from a digital frequency spectrum computed via the Fast Fourier Transform algorithm.[39]

As mentioned, in 1940 the Dutchman Jan Schouten published his residuum theory. Schouten was a fervent fan of the idea, already promoted by Helmholtz, that the

---

[34] Jan F. Schouten, "Die Tonhöhenempfindung," *Philips' technische Rundschau* 5, no. 10 (1940): 294–302, on 301.

[35] Rayleigh, "On Bells," *London, Edinburgh, and Dublin Philosophical Magazine and Journal of Science*, 5th ser., 29, no. 176 (1890), 1–17.

[36] Weissenbäck and Pfundner, *Tönendes Erz* (cit. n. 28), 13.

[37] See Stumpf, "Über die Tonlage der Konsonanten und die für das Sprachverständnis entscheidende Gegend des Tonreiches," *Sitzungsberichte der Preußischen Akademie der Wissenschaften zu Berlin*, 1921, 2nd half vol., 636–40, on 639.

[38] See E. Meyer and J. Klaes, "Über den Schlagton von Glocken," *Die Naturwissenschaften* 21 (1933): 697–701.

[39] See Ernst Terhardt and Manfred Seewann, "Auditive und objektive Bestimmung der Schlagtonhöhe von historischen Kirchenglocken," *Acustica* 54, no. 3 (1984): 129–44, on 131.

basilar membrane in the cochlea "was able . . . to perform Fourier analysis."[40] But the ear, when it perceives a sound that consists of various harmonics, does not make use of its analytical ability but only transmits the impression of a sound of a certain pitch and color. The physiological production of an identity of acoustic color over-rules the ear's ability to decompose the tone into its overtones. Since it became pos-sible to decompose a sound into its overtones using technology, it is known that we always ascribe to a tone the pitch of its fundamental even if that fundamental does not exist within the spectrum.[41] It was the telephone in particular, which cuts off the whole spectrum of frequencies below 300 hertz, that taught acousticians about this phenomenon. Although the fundamentals of the male and, partially, of the female voice lie within the range of frequencies that is cut off by the telephone, the pitch of a speaker's voice is not altered.

In Schouten's experiments with periodic impulses, which represented a series of harmonics without a fundamental oscillation, a "strong, sharp tone" with the fre-quency of the absent fundamental oscillation was nevertheless perceived. Even when more of the low harmonics were removed from the spectrum, this tone remained au-dible. It disappeared only after "the highest harmonics had been removed from the sound."[42] This sound component Schouten called "the residuum." It was located in a range in which the harmonics lie so close together that the resolution capacity of the ear is not sufficient to distinguish them from each other. For instance, the frequen-cies of 2,000, 2,200, 2,400 hertz, and so forth, simultaneously cause a sensation of 200 hertz in the basilar membrane. The only question is why they form a component of such a low pitch. Schouten's explanation was that it is not the frequency of the har-monics, nor the space between them, that constitutes the imaginary fundamental, but the frequency of their envelope (see figs. 1 and 2).[43] From the perspective of physical acoustics it is a matter of chance that, in an anharmonic spectrum of harmonics such as that of the bell, a number of frequencies occur that are multiples of a fundamental oscillation that does not actually exist in the spectrum but becomes audible as the re-siduum of these frequencies.[44] Today physicists speak of the "virtual pitch" or "miss-ing fundamental."[45]

DECONSTRUCTING WESTERN ACOUSTIC CULTURE

Thus, a bell sound consists primarily of acoustic color. The acoustic being of a bell is not determined by pitch but by a mixture of harmonics that are responsible for its acoustic color. The fundamental of a bell is a product of its acoustic color. It may seem like an exaggeration to call the bell the deconstruction of Western acoustic culture, but I think it might be admissible to say that a bell possesses overtone oscil-lations without a fundamental oscillation (which renders the concept of the "over-tone" rather meaningless). In its sound the secondary precedes the primary. The primary, the fundamental, is initially projected by the secondary, the overtones, its

[40] Schouten, "Tonhöhenempfindung" (cit. n. 34), 295.
[41] See ibid., 296.
[42] See ibid., 298.
[43] The curve that describes the contour of the variation of amplitude is called the "envelope" of the wave.
[44] See Schouten, "Tonhöhenempfindung" (cit. n. 34), 302.
[45] Bill Hibbert, "The Strike Note of Bells—an Old Mystery Solved," *Ringing World* 20 (2003): 586.

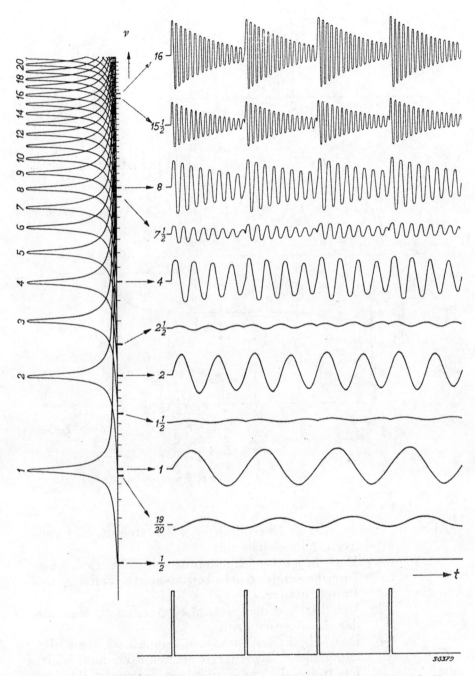

Abb. 9. Links über einer logarithmischen Frequenzskala Reizkurven, die für jede Harmonische der auf das Modell wirkenden Kraft (des periodischen Impulses) angeben, welche Resonatoren (Eigenfrequenz ν) in Bewegung gebracht werden und in welchem Maß. Die entstehende Bewegung ist hier für eine Anzahl Resonatoren (relative Eigenfrequenz $1/_2$, 19/20, 1, $1^1/_2$, 2, $2^1/_2$, 4, $7^1/_2$, 8, $15^1/_2$, 16) gezeichnet. Man sieht, daß im Gebiete der höheren Harmonischen, wo die Reizkurven einander stark überlappen, die Schwingung aller Resonatoren deutlich die Periodizität der Grundfrequenz zeigt.

Figure 1. Representation of the "residuum"; Schouten, "Tonhöhenempfindung" (cit. n. 34), 300.

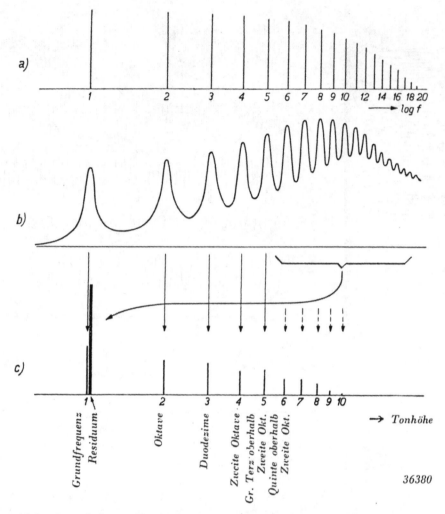

Abb. 10. Schematische Darstellung der Entstehung des sub-
jektiven Klangspektrums.

    a) Das objektive Spektrum eines periodischen
Impulses mit einer Impulsbreite von $^1/_{20}$ der
Periodenlänge.

    b) Die durch diesen Impuls entstehende Reizung
der Basilarmembran.

    c) Das subjektive Spektrum, gemäß welchem das
Ohr den in *b*) wiedergegebenen Reiz analysiert.
Die Reize, die von denjenigen Teilen der Basilarmembran herrühren, welche lokal den höheren
Harmonischen entsprechen, werden zusammen
als eine Komponente von n i e d r i g e r Tonhöhe,
das Residuum, wahrgenommen.

Figure 2. Schematic representation of the creation of the subjective sound spectrum;
Schouten, "Tonhöhenempfindung" (cit. n. 34), 301.

partial tones. The partial tones are part of something that they themselves produce in the first place.

Signal analysis and cultural history are connected by the Lacanian triad of the real, the symbolic, and the imaginary, which perfectly determine the nature of the bell. Although the use of these basic Lacanian concepts may still be rather uncommon among historians of science, they are quite frequently used in media theory.[46] As they are connected to the Freudian theory of the unconscious and the role of the subject of the unconscious, they provide a conceptual framework that no longer allows the history of media to be subsumed to a narrative of more or less ingenious human actors and their achievements. Instead, what become effective within the history of objects like the bell are the inherent laws of the symbolic, the imaginary, and the real, and their technical implementations, forming to a certain degree the conditions of what within a certain epistemic constellation appears for the human subject to be "scientific truth" or "objective reality." In this way the epistemic objects in the history of scientific knowledge are also constituted by desire, narcissism, and suppression.

The symbolic is the order of integral proportions according to which bells are named and described, in which they are classified by virtue of their pitch and the profile of the bell (octavo bell, seventh bell, sixth bell, ninth bell).[47] In the case of the bell the symbolic is an effect of the imaginary, in that the strike note or fundamental of the bell is first of all produced by the imaginary. In fact it literally is the imaginary, the sound Goethe described as rounded as a tone, of which physicists speak as an "imaginary fundamental" or "virtual pitch." Electroacousticians at the end of the twentieth century also spoke of "spectral pattern recognition."[48] But the symbolic is also the signal, which is the strike of the church bell, the call that summons the people within and beyond the borders of the parish to gather to hear the word of God, to put out the fire, to fight the enemy. The real of the bell, finally, is that which cannot be written down, and which only technical media—magnetic tape recorders and harmonic analyzers—are able to record and to process.[49] The real of frequencies, which in the case of the bell is characterized by a missing fundamental, in psychoanalytic terms is nothing but a fundamental missing.[50] Insofar as the sound spectrum of a bell is characterized by this fundamental lack, it consists only of overtones. But insofar as these overtones contain mainly the information of the acoustic color, the mineral sound is pure timbre, amplitude, and reverberation time. Hence the bell is a sound generator that generates a pitch by means of acoustic color. The fragmented body of sound not only precedes the imaginary wholeness, it produces the latter in the first place as if the imaginary were already realized on the physiological level, as if the ear by the constitution of a residuum wished to protect the subject from the horror of a fragmented body. If the residuum is cut off by a low-pass filter, the madness of the real is established in the ear.

The terminology of the imaginary and the real also covers the methodological difference between musicology, which favors the auditive method as a *via regia* to

[46] See n. 19 above.

[47] See Weissenbäck and Pfundner, *Tönendes Erz* (cit. n. 28), 7.

[48] See Terhardt and Seewann, "Auditive und objektive Bestimmung" (cit. n. 39).

[49] Hear audio 1 (1,760KB; WAVE) in the electronic version of this article for the sound of the church bells of the village of Legnaro, Italy. Digitally recorded by Tom Gebhardt in 2009 and reproduced with his kind permission.

[50] See Hibbert, "Strike Note of Bells" (cit. n. 45), 586.

campanological knowledge of pitch, and experimental acoustics, which uses technical media from the choking chain to the digital harmonic analyzer. Whereas campa-
nologists, who have a musicological education and are proud of their sense of hearing, insist that the strike note can be heard by "every musical man" and has to be
"judged absolutely as a real musical fact,"[51] for acousticians like Johannes Biele the
strike note is nothing but an "imaginary figure of tone [*imaginäres Tongebilde*]."[52]
The musicologists are convinced that the physicists only negate the real existence of
the strike-note pitch because physicists in general "do not have a good musical sense
of hearing."[53] At the end of the day the musicological discourse turns out to be based
on the prejudice of the superiority of a bourgeois "high culture."

The surface sound sources of blues and jazz and the electric instruments of rock
music did not just blow up the alphabetic order of Western music. Surface sound
sources, electric distortions, feedback effects, and the Moog synthesizer combined
with other sound cultures to take possession of our ears, too—which could be recognized in the sixties by the fact that various bands, like the Beatles and Pink Floyd,
made use of not only bells but also Asian instruments like the sitar or gong. Moreover, it is significant in this context that Claude Lévi-Strauss placed at the beginning
of his great book on the myths of the South American Indians a long reflection on
the musique concrète of Pierre Schaeffer and the serial music of Pierre Boulez—as
if gaining access to the thinking of the other was only made possible by, on the one
hand, cut-up and distorted magnetic-tape recordings of sounds and, on the other, a
compositional method "in which there is no preconceived scale anymore."[54]

But church bells had always already established an alliance with this other of European sound culture. The word formed an alliance with the real, the *logos* with the
alogon, to propagate its message across and beyond the borders of the parish. Its
anharmonic acoustic-color sounds, which called the Western man to the borders of
reason, of war, and of his earthly finiteness, had always already opened up an abyss
within the European culture of sound.

[51] Weissenbäck and Pfundner, *Tönendes Erz* (cit. n. 28), 11.
[52] Biehle, "Die Analyse des Glockenklanges," *Archiv für Musikwissenschaft* 1, no. 2 (1919): 289–
312.
[53] Weissenbäck and Pfundner, *Tönendes Erz* (cit. n. 28), 13.
[54] See Lévi-Strauss, *Mythologica I: Das Rohe und das Gekochte* (Frankfurt am Main, 1976), 40–2;
the quoted sentence is on 42.

Changeable Ears:
Ernst Mach's and Max Planck's Studies of Accommodation in Hearing

by Alexandra Hui*

ABSTRACT

This article offers an examination of the psychophysical studies of accommodation in hearing by Ernst Mach and Max Planck, natural scientists better known for their accomplishments in physics and philosophy. Early in his career, Mach sought to experimentally locate the possible mechanism of accommodation in hearing, the phenomenon in which individuals can alter their experience of sound by changing their attention. Planck, employing a microtonal harmonium, studied the role of attention in vocalists' abilities to hear tempered intervals—what *he* termed accommodation in hearing. Both mobilized music as a means of argument and experiment. This article shows how each physicist's conception of accommodation in hearing drew on music and, in turn, informed his ideas about the historicity of hearing, the universality of the nineteenth-century Western musical aesthetic, and the nature of knowledge itself.

INTRODUCTION

When presenting his work on accommodation in hearing, the young physicist Ernst Mach would employ the following demonstration of the phenomenon. He urged his reader to play the chord E + G$^{\#}$ + B + e′ followed by the chord a + A + c$^{\#}$′ + e′ (see fig. 1 for a visual aid) on a guitar or keyboard instrument.[1] Then, the reader was to play the chord sequence again but this time was instructed to listen carefully to the highest tone in each chord (that high E). The reader, according to Mach, would have the

* Department of History, Mississippi State University, Box H, Mississippi State, MS 39762; ahui@history.msstate.edu.

I would like to thank Erwin and Elfrieda Hiebert, Myles Jackson, Paul Josephson, Julia Kursell, Barton Moffatt, Peter Pesic, Norton Wise, and the two anonymous referees for their helpful feedback on previous drafts of this article. Earlier versions were presented at the History of Science and Medicine Colloquium at the University of California at Los Angeles, the Long Eighteenth Century Reading Group at the University of Mississippi, and the preparatory workshops for this volume, hosted by Mississippi State University, the Dibner Family Chair of the History and Philosophy of Science and Technology of the Polytechnic Institute of New York University, and the Max Planck Institute for the History of Science in Berlin; the discussions afterward were extremely stimulating. Research for this article was funded by a grant from the German Academic Exchange Service and the Department of History at Mississippi State University. Portions of this article appear in my monograph, *The Psychophysical Ear: Musical Experiments, Experimental Sounds* (Cambridge, Mass., 2012).

[1] *Do* try this. A recording can be heard in audio 1 (300 KB; MP3) in the electronic version of this article. Performed by the author on an M-Audio Keystation 49e USB MIDI keyboard in Los Angeles, Calif., 2006. Special thanks to Daniel Crosby.

© 2013 by The History of Science Society. All rights reserved. 0369-7827/11/2013-0007$10.00

Figure 1. Chord example used to demonstrate the phenomenon of accommodation in hearing. Mach, "Bemerkungen über die Accommodation des Ohres"(cit. n. 2), 344. Mach used this same example in his 1865 lecture "Erklärung der Harmonie" (cit. n. 40).

impression that the tones remained the same and only the tone quality had changed between chords. Playing the chord sequence one last time, the reader was asked to instead focus on the lowest tones (that transition from the low E to the low A). In this case she would hear a clear step down in pitch, as if the entire chord had dropped down significantly. Mach explained that because the tones of the piano were even, and because the ear was not fatigued by such constant tones, there was no explanation for the reader's changed sound sensation other than that she had changed her attention.[2]

I offer this example of Mach's to highlight his use of music to examine a psychophysical phenomenon. For Mach, music was the best way to demonstrate the phenomenon of accommodation in hearing: the individual's changed experience of sound in relation to the individual's changed attention to sound. Deliberately directed attention altered an individual's aural experience. Accommodation explained the listener's ability to hear, for example, the cello part in a symphonic performance once the listener chose to direct his or her attention to the cellos. The role of the individual in both creating and experiencing accommodation made it a particularly difficult phenomenon to study in others. Accommodation in hearing was also a difficult phenomenon to explain convincingly; better to mobilize readers' subjective experience by walking them through a demonstration.

Did you hear it? If you have had musical training then you likely did. Mach's readers would have. Their *Bildungsbürger* upbringing would have guaranteed them facility with at least one musical instrument as well as music-reading ability. His readers seeing this example in print would know of the phenomenon that he was referring to. For Mach, a discussion of one of the more curious features of sound sensation was also, necessarily, a discussion of music.

The bulk of this article focuses on the sound-sensation work of Mach and examines the motivations and means by which he mobilized music toward his goal of locating the mechanism of accommodation in hearing. Mach's use of music had implications for his historicism and, later, his epistemological thinking. I then turn to Mach's great critic, the physicist Max Planck, to discuss his use of a microtonal harmonium to understand the role of accommodation in the hearing of tempered rather than pure intervals, and I attempt to contextualize Planck's experiments in relation to his own epistemological thinking. The article ends with a short discussion of the work of the twentieth-century American composer Harry Partch, presented as an example of the continuing tension between Western musical aesthetics, the dominance of equal temperament in particular, and the individual's experience of sound. Ultimately Partch was able to accomplish his compositional goals by abandoning known sounds. Partch's hard-won freedom to forsake tradition and develop his own tuning systems,

---

[2] Mach, "Bemerkungen über die Accommodation des Ohres," in *Sitzungsberichte der kaiserlichen Akademie der Wissenschaften* 51 (1865): 343–6, on 344.

complete with innovative new musical instruments, was only possible upon his dual realization that (as Mach believed) hearing was historical and that (as Planck argued, though not in quite such negative terms) creators of music were "both the exponents and victims of system, philosophy, and attitude determined for them by their milieu."[3]

More broadly, this article shows how psychophysical studies of sound sensation were bound up with practitioners' relationships with music and music culture, both material and immaterial. These relationships allowed Mach and Planck to comfortably employ music as a means of both argument and experiment. They applied their respective ideas about accommodation in hearing to theories on the origins and development of musical systems. At stake in these theories were such issues as the historicity of sound sensation, the universality and supremacy of the nineteenth-century German musical aesthetic, and, potentially, the nature of knowledge itself.

The world of sound in the second half of the nineteenth century was highly unstable. New tuning systems, new tones, new music, and the fledgling discipline of musicology all jostled to establish position. The transition from earlier forms of tuning to equal temperament meant that the pitches themselves were not fixed, standardized neither between instruments nor within individual ones. Later, a growing interest in non-Western music introduced entirely new sounds.

The development of new acoustic instruments, the tuning fork tonometer being perhaps the most important, had brought great promise of standardization and equivalence both within individual instruments and between them. Equal temperament, with its associated freedoms of composition and performance (its use meant that concert programs or even single works could modulate between a greater variety of keys without requiring the instruments to be retuned), was practicable, no longer mere theory.[4] It was the latest of several attempts at a solution to the problems presented by the Western tuning system's adherence to a scale in which intervals were based on pure ratios that were repeatable over many octaves.[5] The twelve acoustically pure fifths were almost, but not quite, equivalent to seven acoustically pure octaves. A twelve-tone scale in which all tones were related by pure fifths could not be enclosed within an octave. Before the developments in the late nineteenth century, the Western tuning system had prioritized keeping certain intervals pure—the most commonly played ones—by sacrificing others. The Pythagorean tuning system, used through the Renaissance, made fourths and fifths pure. In order to do so, the Pythagorean "comma," or the microtonal discrepancy between twelve perfect-fifth ratios and seven octaves, was placed where it would not be noticed. So to listeners at the time, thirds and sixths in this system sounded quite jarring and out of tune. Alternative systems, those of just intonation and mean tone temperament, instead prioritized

[3] Partch, *Genesis of a Music: An Account of a Creative Work, Its Roots, and Its Fulfillments*, 2nd ed. (New York, 1974), xvii.

[4] Aristoxenus, a pupil of Aristotle, first proposed the theory of equidistant smallest intervals for Western scales. It was embraced by some lutenists as early as the fifteenth century, mostly motivated by convenience. Equal semitones allowed the same fret to mark off a diatonic semitone on one string (a B♭ on an open A string, for example) and a chromatic semitone on another (an F♯ on an open F string), commonly required in Renaissance lute music. Mark Lindley and Ronald Turner-Smith, *Mathematical Models of Musical Scales* (Bonn, 1993), 44–6.

[5] A harmonic interval in music is the distance between two pitches heard simultaneously. The number of steps between the pitches in a scale traditionally determines the name of the interval. From the pitch C up to E or down to A is a third. Further up to G or down to F is a fifth. From C to the C above or below is a perfect octave, an eighth.

triads (and therefore thirds) as well as fourths and fifths as pure, distributing the Pythagorean comma among other intervals. Equal temperament sought to distribute the
comma evenly—equally—among all the twelve tones of the scale.

It was, however, inconsistently applied. This was in part due to the fact that while
the means of measuring equal temperament did exist, the techniques for equal-
tempering a piano or organ of eighty-eight or more keys with precision did not. For
some, that was just fine. Many believed that the ease of transposition achieved with
equal temperament did not outweigh aesthetic sacrifices of coloration and tonality.
With previous tuning systems, while limiting the intervals and therefore keys (of
diatonic scales) on the instrument that could be performed without a complete re-
tuning, those keys were said to have distinct coloration and character of sound, a
quality coveted by composers and listeners alike.[6] Many lamented the abandonment
of pure intervals and the associated loss of key coloration with the tempering of the
Western scale. Equal temperament's treatment of the Pythagorean comma could, for
example, be cynically understood as spreading the error everywhere. Helmholtz, for
one, went so far as to document the beats—his criterion for dissonance—of the tempered triad, implying that equal temperament was unnatural and unmusical.[7] Partch
would describe Helmholtz's impatience with equal temperament as "a salutary and
long-overdue influence."[8]

In addition to shifting tones within the Western tuning system, sounds altogether
new to Europe were introduced toward the end of the nineteenth century. Increasingly, non-Western music ensembles visited Europe to perform. The introduction
of the phonograph to field studies in the late 1870s granted music further ability to
travel; field ethnomusicologists returned to Europe with wax cylinders containing
never-before-heard music. This non-Western music, some of which was based on
highly complex scale systems, undermined European beliefs in the inherent superiority of Western intonation and fueled the development of new questions and theories
about Western musical aesthetics.

In this same period, certainly related to the instability of sound in the music world,
there was a growing interest in the role of attention in hearing, and the phenomenon
of accommodation in particular. The phenomenon had already been established and
physiologically explained for vision. Helmholtz's sign theory of vision had showed
(very much building on Hermann Lotze's model) that the contraction of the muscles
of the eye allowed an individual to spatially locate the object of his or her focused
observation.[9] Many believed that this model for the role of attention in vision could
be extended to the sensation of sound.

[6] When questioned as to whether an individual key had absolute character or only relative character
in comparison to another, Hermann von Helmholtz raised the possibility that the distinct character of
keys was due, in part, to a particularity of the human ear. But, at least for pianos and bowed and wind
instruments, the more likely cause of the different characters of keys, according to Helmholtz, was
the way in which a particular key was played on the instrument. Piano keys, for instance, were struck
differently depending on whether they were the black or white keys. For bowed and wind instruments,
the different lengths of the strings or wind chamber as a particular tone was sounded contributed to the
supposed character of the key. Helmholtz, *Die Lehre von den Tonempfindungen als Physiologische
Grundlage für die Theorie der Musik*, 3rd ed. (Brunswick, 1870), 501–4.
[7] He complained: "I do not know that it was so necessary to sacrifice correctness of intonation to the
convenience of musical instruments." Ibid., 529.
[8] Partch, *Genesis of a Music* (cit. n. 3), 389.
[9] Helmholtz had presented a broadly sketched "theory of signs" in his 1855 lecture celebrating the
centennial of Kant's inaugural lecture at Königsberg, "Über das Sehen des Menschen," based on the

Historian Michael Hagner sees mid-nineteenth-century psychophysical studies of attention as an indicator of the extent to which attention was redefined from late eighteenth-century conceptions.[10] He explains that attention had previously been a virtue, making individuals masters of themselves and the exploration of their world. The early work of the psychophysicist Gustav Fechner, however, showed that attention was actually quite difficult to control and maintain and threw into relief the instability of the human perceptual condition. Hagner points to Mach specifically and his belief that attention was a purely motor-based phenomenon.[11] He argues that this was part of a growing acceptance among psychophysicists that conscious control and self-discipline influenced perception as motor skills. If Fechner had determined that perception changed in spite of, possibly even because of, focused attention, Mach's interest in accommodation was to determine why this happened.

To study the means by which an individual's deliberately altered attention to sound affected his or her aural experience would be difficult enough under any circumstances. Doing so in a period of dramatic sonic upheaval both confounded and further advanced Mach's and Planck's studies. It follows, then, that their respective explorations of accommodation in hearing must be understood in relation to the shifting music world. If the sounds and harmonies of music were changeable and if hearing itself was changeable, then the individual, subjective experience of sound was potentially valid, perhaps even more valid than the theories and aesthetics advanced by music critics. For Mach, this validity buttressed his phenomenological view of the world. For Planck, it supported his antipositivist stance. An awareness of the connections between Mach's and Planck's differing conceptions of sound sensation, and of accommodation in particular, allows for a new approach—and hopefully new insight—into the two giants' later epistemological clash. The world appears different when it includes sound.

MACH'S ACCOMMODATION EXPERIMENTS

Much of the historiographical work on Mach has focused on his physics, in particular his work on the shock waves of supersonic projectile motion, and his philosophy, the fields for which he was so well respected. His phenomenology, which Mach insisted could be reconciled with experimental science, drew the ire of Planck, among several others. Historians of science often point to Mach's 1872 treatise *Die Geschichte und die Wurzel des Satzes von der Erhaltung der Arbeit* (History and root of the principle of the conservation of energy) as the first full articulation of his position on

---

idea of "local signs" proposed by Lotze and others. He believed that vision operated according to an optimization principle. The contraction of the eye muscles as the eye arced from a specific peripheral spot to a spot of sharpest vision (moving the eye so that the object of observation was most clearly visible) corresponded to a series of changing feelings of position. This series of feelings was stored in the memory and recalled whenever that specific peripheral spot was stimulated. The local sign consisted of the physical and physiological actions required to orient each spot on the retina to the visual axis. Gary Hatfield, *The Natural and the Normative: Theories of Spatial Perception from Kant to Helmholtz* (Cambridge, Mass., 1990); Timothy Lenoir, "The Eye as a Mathematician: Clinical Practice, Instrumentation, and Helmholtz's Construction of an Empiricist Theory of Vision," in *Hermann von Helmholtz and the Foundations of Nineteenth-Century Science*, ed. David Cahan (Berkeley, Calif., 1993), 109–53.

[10] Hagner, "Toward a History of Attention in Culture and Science," *Modern Language Notes* 118, no. 3 (2003): 670–87.

[11] Ibid., 680, 681.

the historicist nature of ideas; that is, that ideas were specific to time and place. This historicism directly informed the logical positivist movement that developed in the twentieth century.

The musical aspect of Mach's psychophysical work has never been the focus of serious study by historians of science, and it is interesting. But perhaps even more exciting is that Mach's engagement with the music world and mobilization of music to scientific ends contributed to the maturation of his historicist thinking. Mach's use of music in his psychophysical experiments on accommodation—at a time when established musical aesthetics were being called into question—in turn informed his eventual belief that hearing was historically contingent. Mach came to believe that the sensation of sound was not just psychophysical, and certainly not just a physiological mechanism, but also cultivated and cultured. This ultimately eliminated the need to locate a mechanism of accommodation, as it would be changing constantly anyway, which explains why Mach was no longer discussing the accommodation mechanism in 1885. The following examination of Mach's use of music in his study of accommodation in hearing suggests that he was thinking in a historicist way, at least about sensory perception of sound, much earlier than credited by historians of science. As early as 1863 Mach believed that hearing—how one heard, what one heard, what one focused one's attention on—was bound to culture and therefore specific to time and place. Hearing itself was historical.

Mach began his career interested in acoustic phenomena. His earliest work was an examination of the controversy between Christian Doppler and Joseph Petzval over the relation of motion to changes in color or tone.[12] Recall that the Doppler effect is the phenomenon in which the tone or color of a wave changes as an observer moves in relation to the source of the wave. Now, Mach's study dealt mostly with the physics and mathematics of the Doppler-Petzval controversy, but it should be noted that from the very beginning he was dealing with problems that involved the specificity of the observer's experience; the explanation of the Doppler effect was bound to the location of the observer relative to the wave source.

In the following year, Mach began to study the observer's experience more directly, undertaking his initial work on the phenomenon of accommodation in hearing. He first developed a model of the mechanism of accommodation in hearing analogous to Helmholtz's sign theory of vision. Just as the eye muscles allowed the individual to spatially orient, Mach posited that the individual differentiated tone pitch through the contraction of various muscles in the ear in response to changed attention. He believed that the phenomenon of accommodation in hearing was psychophysical but the mechanism was physiological.

Mach suspected that the accommodation mechanism was rooted in the tensor tympani and possibly also the stapedius muscles, which would contract in response to altered attention, changing the transmission of sound waves to the cochlea. In his 1863 article "Zur Theorie des Gehörorgans," Mach sought to reconcile physiological theory with investigative technique through a kymographic theory of the ear.[13] Like the kymograph, which recorded blood pressure through a stylus on a rotating band

[12] Mach, "Ueber die Kontroverse zwischen Doppler und Petzval, bezüglich der Aenderung des Tones und der Farbe durch Bewegung," *Zeitschrift für Mathematik und Physik* 6 (1861): 120–6.

[13] Mach, "Zur Theorie des Gehörorgans," *Sitzungsberichte der kaiserlichen Akademie der Wissenschaften* 48, no. 2 (1863): 283–300.

of paper, Mach believed that the ear drew (*zeichnen*) sound waves in the labyrinthian fluid of the inner ear. These sound waves were then absorbed by the auditory nerve. The entire ear—the eardrum, the middle ear muscles and bones, and the labyrinthian fluid—functioned to transcribe the sound waves from the medium of air molecules to the medium of labyrinthian fluid. Though the sound wave was modified (regulated and damped), the system did not perform an analysis of it. Mach demonstrated all of this through a series of experiments and mathematical derivations.[14]

One of the implications of this kymographic theory that was of particular interest to Mach was the simultaneous reflection of the sound waves transmitted by the eardrum, an analogue to Gustav Kirchhoff's theorem of the equal absorption and emission of light waves. Mach performed a series of experiments mobilizing this effect toward direct observation of the accommodation phenomenon in another person.[15] Placing an assistant with a long rubber tube in his ear in another room, Mach very softly sang a constant tone while moving the other end of the tube back and forth, relative to his own ear. The tone was loudest, according to the assistant, when Mach's end of the rubber tube was nearest Mach's ear, when the sung tone was amplified by the reflection of its sound waves in Mach's ear. In another experiment, Mach softly sang a tone with one end of the rubber tube in each of his own ears. When he pinched off the tube in the middle he noticed a decrease in the volume of the sung tone, presumably because his pinching had eliminated the reflection of the sound waves back and forth between his ears through the tube.[16]

Thus, Mach believed he had both theoretically derived and experimentally demonstrated the mechanics of his kymographic model of the ear, the transmission of sound waves through the ossicles to be transcribed in the labyrinthian fluid. But the model did not necessarily explain the ability of the listener to actively distinguish a single tone from other tones sounding simultaneously. It could not explain accommodation in hearing. The musical examples clearly demonstrated the phenomenon of accommodation, and yet Mach could not locate and directly observe the mechanism. Still, he maintained his belief that attention was a bodily function and that, therefore, the phenomenon of accommodation had its foundation in the mechanisms of the body.[17] So he continued with his search.

Elaborating on these early investigations, Mach performed a series of experiments in the summer of 1863 with Joseph Popper and students of the Vienna Physical

[14] Mach's physical proof was a mathematical demonstration that the eardrum, ossicles, and labyrinthian fluid all vibrate to regulate the transmission of sound waves in two significant ways: even absorption and quick dissipation of the initial state of the sound waves. Mach showed that the restoring force of the vibrating ear bones and the viscosity of the labyrinthian fluid both equalize the absorption of sound waves of varying frequency (tones of different pitch) and also damp resonance (harmonic overtones of the original sound wave), allowing for the transmission of a quick succession of tones. Mach further posited the possibility that different types of transmission occur in the ear depending on the wavelength of the sound waves. He suggested that the eardrum, ossicles, and labyrinthian fluid all vibrate together to transmit lower tones but vibrate separately to transmit higher tones. Although this hypothesis did not provide a mechanism of accommodation, it was at least a physical explanation in which different pitches were treated differently in the ear. Ibid., 285–7.

[15] Ibid., 289.

[16] Mach later revisited this experiment and concluded that it was more likely that the pinching created reflected waves in the tube and that the interference of these waves with the original ones caused the weakened volume of the sung tone in his ears. Mach, "Über einige der physiologischen Akustik angehörige Erscheinungen," *Sitzungsberichte der kaiserlichen Akademie der Wissenschaften* 50 (1864): 342–62.

[17] Mach, "Zur Theorie des Gehörorgans" (cit. n. 13), 297.

Institute. For these experiments Mach placed a vibrating tuning fork in his teeth and one end of a rubber tube in one of his ears.[18] The other end of the rubber tube was placed in an assistant's ear. As the tuning fork sounded Mach slowly changed his attention from the fundamental or ground tone to various harmonic overtones. Mach could hear these overtones as strong and distinct from the ground tone as he moved his attention from one overtone to the next. But the assistant could not. Although other experimental work electrically stimulating the tensor tympani had established the muscle's ability to change the tension on a prepared (nonliving) eardrum, and although Mach had mathematically demonstrated that changed tension on the eardrum would result in higher tones appearing louder, Mach's tuning-fork experiment could not confirm that changed attention correlated with changed eardrum tension, which in turn correlated with changed sound sensation (hearing the overtones more strongly). He was left to conclude that while his kymographic theory held promise, further experimental proof was required to show that it explained the phenomenon of accommodation.[19]

In the 1870s Mach began a series of collaborative projects in relation to sound sensation with Johann Kessel. Developing earlier animal-based research by physiologists Charles-Édouard Brown-Séquard, Jean Pierre Marie Flourens, and Friedrich Goltz, Kessel was investigating the role of the semicircular canals and labyrinth in individuals' ability to balance. He believed this function of the hearing organ could be employed to better understand inner-ear diseases and injuries.[20] Mach would, in the next few years, publish two articles on further physiological experiments on the sense of balance in humans, as well as a lengthier piece on the sense of acceleration.[21]

In relation to this work on balance, Mach was also examining the mechanics of the middle ear, the rotation points and axes of movement, as well as making experiments on the eustachian tube that consisted of observing the function of hearing through rapidly changing air pressure.[22] This study of the topography of the middle ear was developed more extensively with Kessel in a series of measurements and experiments on the ligature and musculature of middle ears removed from cadavers.[23] Incorporating a stroboscopic apparatus and technique for determining pitch developed by Mach, the two scientists then made a series of observations on the middle ear system in motion.[24] One set of these cadaver experiments focused on the influence of the middle ear muscles on the movement and vibration of the eardrum, a relationship that could provide Mach insight in his search for the accommodation mechanism.[25]

The experimental preparation for Mach and Kessel's accommodation experiments

---

[18] Mach, "Bemerkungen über die Accommodation des Ohres" (cit. n. 2), 345.

[19] Mach, "Zur Theorie des Gehörorgans" (cit. n. 13), 299–300.

[20] Mach, "Vereinsangelegenheiten," *Lotos* 21 (1871): 196–8.

[21] Mach, "Physikalische Versuche über den Gleichgewichtssinn des Menschen," *Sitzungsberichte der kaiserlichen Akademie der Wissenschaften* 68 (1873): 124–40; Mach, "Über den Gleichgewichtssinn," *Sitzungsberichte der kaiserlichen Akademie der Wissenschaften* 69 (1874): 44–51; Mach, *Bewegungsempfindungen* (Leipzig, 1875).

[22] Mach, "Vereinsangelegenheiten" (cit. n. 20).

[23] Mach and Kessel, "Beiträge zur Topographie und Mechanik des Mittelohres," *Sitzungsberichte der kaiserlichen Akademie der Wissenschaften* 69 (1874): 221–43.

[24] Mach, "Über die stroboskopische Bestimmung der Tonhöhe," *Sitzungsberichte der kaiserlichen Akademie der Wissenschaften* 66 (1872): 67–74.

[25] Kessel, "Ueber den Einfluss der Binnenmuskeln der Paukenhöhle auf die Bewegung und Schwingungen des Trommelfels am todten Ohre," *Archiv für Ohrenheilkunde* 2 (1874): 80–92. Kessel dated this article July 1873.

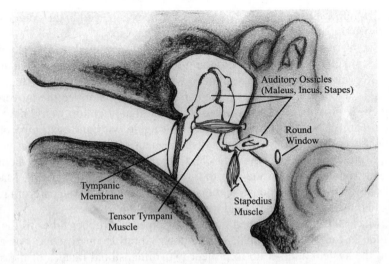

Figure 2. The middle ear. Hui, Psychophysical Ear *(cit. n. *), 99.*

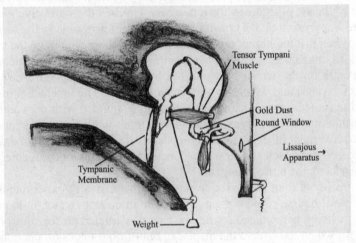

Figure 3. Mach and Kessel's prepared ear. Hui, Psychophysical Ear *(cit. n. *), 100.*

on middle ears was as follows (see figs. 2 and 3). First they carefully removed the tensor tympani in order to attach a thread to it. Once it was reinserted into the middle ear cavity, weight on the order of a few grams could be added to the other end of the thread in order to produce tension in the muscle. The lower, bony portion of the middle ear cavity was cut out to allow space for the thread to pass over a pulley and hang vertically. Another thread was attached to the head of the stirrup bone. This thread was passed through the canal alongside the stapedius muscle (the connection of this muscle to the stirrup bone was left intact). Most of the bony portion of the inner ear—everything past the oval and round windows—was cut away to provide space both for the stirrup thread to pass over a pulley and hang vertically and for a microscope and ocular micrometer to measure the displacement of the stirrup thread and view the movement of the ossicles, which were dusted with gold flakes, through the round window. Lastly, one end of a rubber tube was placed in the outer ear canal with its other end attached to the opening of an organ pipe.

In the first series of experiments Mach and Kessel measured the displacement of the stirrup thread, which correlated to the movement of the ossicles, when an organ pipe of 256 cycles per second was sounded with a 3-gram weight on the tensor tympani thread. This created tension on the eardrum and constrained the movement of the ossicles, approximating the assumed influence that attention would have on the tensor tympani. When they ran the experiment again with an organ pipe of 1,024 cycles per second, they found that the displacement of the stirrup thread was less. When they pulled on the stapedius muscle, further constraining the movement of the ossicles, the displacement of the stirrup thread was reduced further still. Mach and Kessel believed that this demonstrated that changed tension on the eardrum resulted in changed transmission of the sound waves through the ossicles.[26]

Next, Mach and Kessel attached both organ pipes via rubber tubes to the single rubber tube in the outer ear canal. A Lissajous vibration microscope was also attached to the system and set up to project two-dimensional images of the stirrup-thread displacement.[27] First, with no weight on the tensor tympani thread, the low pipe was sounded. The image projected by the vibration microscope can be seen in the first of the set of figures shown in figure 4. When the same pipe was sounded while there was tension on the tensor tympani (weight was added), the second image was seen. When just the higher pipe was sounded, with no weight on the tensor tympani, the third image was seen. When both pipes were sounded simultaneously, again with no weight on the tensor tympani, the fourth image resulted. They then tried sounding both pipes and weighting the tensor tympani. The image that resulted was the third, the same image as when the higher pipe sounded with no weight. Changing the tension on the tensor tympani while two tones were sounding simultaneously altered the movement of the ossicles to appear as if only the higher tone was sounding. It appeared that Mach and Kessel had located the accommodation mechanism; it was, as suspected, the tensor tympani.[28]

Mach and Kessel then attempted to replicate this series of experiments on a living ear. They of course could not cut away bone in order to attach little pulleys and weights on a living person, and so they instead constructed an "ear mirror" (*Ohrenspiegel*) with which to observe the displacement of gold flecks on the outside of the eardrum. The vibration microscope images of the displacement of the eardrum for

[26] Mach and Kessel, "Versuche über die Accommodation des Ohres," *Sitzungsberichte der kaiserlichen Akademie der Wissenschaften* 66 (1872): 337–43.

[27] The vibration microscope, first developed in 1855 by Jules Lissajous, who would later become the scientific consultant on Napoleon III's commission to establish a standard pitch, consisted of a small lens attached by an arm to a vibrating object, traditionally a tuning fork. Objects (a light beam or stylus mark on another tuning fork) viewed through this lens would thus appear to vibrate, tracing out a one-dimensional path of oscillation. If the object viewed through the lens also vibrated perpendicular to the plane of the primary vibrating tuning fork, then a two-dimensional curve would appear to be traced out. These Lissajous curves or figures could then be analyzed relative to the known frequency of the primary tuning fork. For his accommodation experiments, Mach viewed the oscillating stirrup thread through the vibrating lens of the vibration microscope and then projected the two-dimensional path onto a screen. See Steven Turner's article on the development of Lissajous apparatuses, "Demonstrating Harmony: Some of the Many Devices Used to Produce Lissajous Curves before the Oscilloscope," *Rittenhaus* 11, no. 2 (1997): 33–51.

[28] Mach and Kessel, "Versuche über die Accommodation des Ohres" (cit. n. 26).

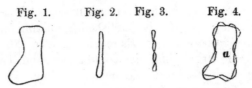

Figure 4. The Lissajous microscope images. Mach and Kessel, "Versuche über die Accommodation des Ohres" (cit. n. 26), 339.

lower tones were the same as those for the prepared ear. They were not, however, the same for higher tones. Mach and Kessel could not recreate the mechanism of accommodation in a living ear. This discrepancy between results on a nonliving ear and a living one, Mach claimed, indicated that listening and hearing were not the same thing. Attention and accommodation could not be replicated by merely adding a weight-and-pulley system to the tensor tympani.[29] The psychophysical nature of accommodation in hearing limited its study to individuals' subjective observation of the phenomenon. Its mechanism could not be located by outside observers.

MACH AS SELF-APPOINTED BRIDGE BETWEEN PHYSICS AND MUSIC

Mach relied heavily on music in his science. He employed musical examples to demonstrate the phenomenon of accommodation to his audience, and he turned to music-theoretical arguments to support his assertion that the phenomenon did, in fact, exist. While his efforts to locate the accommodation mechanism of sound sensation remained fruitless, music was a means of demonstrating and discussing the otherwise inaccessible psychophysical phenomenon. Music was a proxy scientific language. For Mach, it was not simply that sound and music were interchangeable, but rather, that it was critical that they be interchangeable, for both demonstrating and testing certain phenomena of the sensory perception of sound required employing music as sound. As a consequence, the search for the accommodation mechanism of hearing was also a study of musical aesthetics informed by Mach's relationship with music.

While he toiled away with Kessel in the laboratory, Mach simultaneously offered up his findings to the music world. Further, Mach was in regular communication with his close friend, the Viennese music critic Eduard Kulke, about his work, as Kulke, too, was struggling with issues surrounding the individual's subjective experience of sound. Mach's use of music in his search for the accommodation mechanism in hearing was not simply a consequence of his musical milieu. It was critical that he use music *as* music to study accommodation in hearing. For one, music was the best way to convincingly demonstrate the existence of the phenomenon of accommodation. And, in turn, locating the mechanism of accommodation could potentially buttress efforts to explain and validate individuals' subjective experience of sound.

Mach moderated and moved with ease through the worlds of natural science and music, actively engaging both socially and intellectually. He used musical examples in his scientific writings, and he reached out to the music world with the results of his research on sound sensation. Mach saw his work and himself as a bridge between

[29] Ibid., 342.

physics and music, overlapping and intersecting with several issues important to music theoreticians at the time.[30] In 1866, he published *Einleitung in die Helmholtz'sche Musiktheorie*, a reworking of Helmholtz's much-heralded 1863 text *Die Lehre von den Tonempfindungen* in terms a musician could understand. Throughout the 1860s and 1870s, he wrote articles for music and musicology journals, and also contributed to the *Musikalisches Conversations-Lexikon*, a twelve-volume music encyclopedia published between 1870 and 1879 (as well as an 1888 supplemental volume). In all of these writings but most explicitly in *Einleitung in die Helmholtz'sche Musiktheorie*, Mach sought to clarify scientific concepts in the service of musicians and music theorists. He explained that he hoped the text would result in cooperation between musicians and physicists and that it would help some musicians overcome their initial reticence about studying Helmholtz's technical work. Mach was making a self-conscious effort to place himself at the intersection of the natural science and music worlds.[31]

An 1867 review of Mach's reworking of Helmholtz in the *Leipziger allgemeine musikalische Zeitung* claimed to welcome any work that would help the musician befriend the work of Helmholtz through further illumination.[32] The author hesitantly allowed that it remained to be seen, however, whether Mach's text would provide such illumination, especially the section on harmony. Three months later the same journal ran an extended article titled "Zur Theorie der Musik: Die Physiker und die Musiker" that explored the overlap between the intellectual circles of Helmholtz's and Moritz Hauptmann's physiological acoustics and that of music philosophy. A third of the text was devoted to a discussion of Mach's efforts. The editors recommended Mach's book to anyone hoping to gain insight into Helmholtz's theories.[33] But the editors also explained that, while Mach's work was both instructive and convincing, it missed Helmholtz's main points.[34] So in 1867, although the musicologists cautiously welcomed Mach's presence in the music world, they were also willing to question and challenge his theories.

Mach's participation in the *Musikalisches Conversations-Lexikon* indicates the extent to which his prestige had grown since his initial efforts to engage the scholarly music community with his text on Helmholtz. Mach was now regarded as an expert. The "Mach" entry in *Musikalisches Conversations-Lexikon* described him as one of the "most worthy scholars of the science of music." His text on Helmholtz was declared extraordinarily worthwhile.[35] In 1887, one of the leading musicology journals, *Vierteljahrsschrift für Musikwissenschaft*, published a lengthy excerpt of Mach's *Analyse der Empfindungen*. Mach's efforts to construct a bridge of ideas between physicists and musicians appeared to have been successful.

Perhaps the most concrete example of Mach's exchange of ideas with the world

[30] Mach, *Einleitung in die Helmholtz'sche Musiktheorie* (Graz, 1866), v–4.
[31] Ibid., vii.
[32] Unsigned review of *Einleitung in die Helmholtz'sche Musiktheorie*, by Mach, *Leipziger allgemeine musikalische Zeitung*, no. 7 (1867): 58.
[33] "Zur Theorie der Musik: Die Physiker und die Musiker," *Leipziger allgemeine musikalische Zeitung*, no. 21 (1867): 165–9.
[34] These main points were, apparently, that Helmholtz's "physical-psychological" theory was internally consistent and that this theory was in full harmony with the art of music. Ibid., 166–7.
[35] *Musikalisches Conversations-Lexikon*, vol. 7, ed. Hermann Mendel (Berlin, 1875), s.v. "Mach, Ernst."

of music is found in his friendship with Eduard Kulke. They met in 1863 when Mach was drawn into a lively discussion among musicians over the nature of musical tones at a Viennese café.[36] When later describing his initial encounter with the group, Mach recalled that he had chosen to side with Kulke due to Kulke's more sober, more *wissenschaftlich* position on sound sensation.[37] Kulke was part of a circle of Viennese Wagnerians that included Anton Bruckner and Peter Cornelius, among others. A moving performance of Wagner's *Tannhäuser* in 1854 had prompted what Kulke described as his "aesthetic heresy" (*ästhetische Ketzerei*). He found Wagner's music to be pleasing, but his personal reaction was countered by acoustic and music-theoretical arguments and accusations that no one of taste or aesthetic education could possibly find Wagnerian music beautiful. Kulke was troubled by the accepted belief that there could be correct and incorrect taste despite overwhelming evidence that musical tastes varied greatly among individuals. The Wagnerian opera motivated a lifelong effort by Kulke to reconcile the music-theoretical analyses that condemned Wagner's harmonies as ugly with his own, individual enjoyment of Wagnerian music.[38]

Mach and Kulke's friendship and the intellectual exchange with which it began on that day in the café illustrate the extent to which the individual, subjective experience of sound sensation presented issues common to both psychophysics and music.[39] Kulke sought to develop an aesthetic of music that would explain the individual's subjective appreciation of music. Correspondingly, Mach worked to locate and explain accommodation in hearing, the locus of an individual's subjective experience of sound. An individual could, for example, by focusing on certain tones, hear the opening chords of *Tannhäuser* as quite beautiful and moving. Another, directing her attention to different harmonies, could find Wagner's opera to be jarring and dissonant. It should be little wonder, then, that Mach and Kulke began such a close friendship over a discussion of musical tones that day at the café; they were working out similar problems.

From the very beginning of his psychophysical studies of accommodation in hearing, Mach had relied on music to practice his science. He employed the chord

[36] These rowdy and informal café gatherings were apparently a regular occurrence, and Mach continued to frequent them. Ludwig Karpath wrote: "A close relative has told me a lot about a Vienna coffeehouse which I can no longer remember, the 'Café Elephant,' which was located in a narrow passage between Stephen's Place and the Graben. Every day, scholars, artists, and doctors of medicine and law would gather together. The regulars (*Stammgaesten*) included such later famous people as Professor Mach, Lynkeus (Popper), a group of Wagner-oriented musicians: Peter Cornelius, Heinrich Proges, the music critic Grad, the composer Goldmark, and many others. People wandered in around 2 p.m. and stayed until 2 in the morning, that is, some were always leaving while others were arriving. Unbroken wit and argument on philosophical, scientific, and artistic matters kept the discussion sharp and stimulating. To a certain extent the young Dozent Ernst Mach presided over the gathering. His profound understanding and reflective manner impressed everyone. According to my relative he was one of the first to occupy himself deeply with the recently published work of Helmholtz on tone perceptions about which he formed many interesting and instructive conclusions." This letter is part of the private collection of Ernst Anton Lederer (Mach's grandson) at his home in Essex Fells, N.J. Quoted and translated in John Blackmore, *Ernst Mach: His Work, Life, and Influence* (Berkeley, Calif., 1972), 23.

[37] Mach, foreword, in Kulke, *Kritik der Philosophie des Schönen* (Leipzig, 1906), x–xi.

[38] Kulke, foreword (written in 1896), in *Kritik der Philosophie des Schönen* (cit. n. 37), vii–viii.

[39] I discuss Mach and Kulke's lifelong correspondence more extensively in Hui, *Psychophysical Ear* (cit. n. *).

demonstration of the accommodation phenomenon discussed at the beginning of this article both in his writings and in public lectures.[40] In addition to mobilizing music (and the music training of his audience) to demonstrate the phenomenon of accommodation in hearing, Mach also regularly employed music to explain the phenomenon, as best seen in his popular lectures on musical acoustics, "Über die Cortischen Fasern des Ohres" and "Die Erklärung der Harmonie."

In the first, "Über die Cortischen Fasern des Ohres," Mach employed the piano as a model for the concept of sympathetic vibration.[41] Standing two pianos next to each other, lifting the dampers on one piano (by pressing the sostenuto pedal), and striking a key on the other resulted in the same note ringing on the undamped piano. Similar results occurred in response to a major triad, and so on. The undamped piano, Mach explained, separated the sounded tone in the air into its individual component parts, performing a spectral analysis of sound.[42]

According to Mach, the ear had the same ability to perform a spectral analysis of sound. Mobilizing the same analogy as employed by Helmholtz, Mach directed his audience to imagine the undamped piano as, instead, the cortical fibers of the ear. Just as a string on the undamped piano would sound in response to a tone sounded by the other piano, so too would a single cortical fiber be thrown into vibration. The large number of cortical fibers—one per piano string—allowed for accommodation and thus an appreciation of music. A listener could, for example, pick out the melodic lines of a Bach fugue. Or she could distinguish separate tones of simultaneous impressions (*Eindrücke*), not just a harmonious chord but any combination of tones, by directing her attention.[43]

Mach elaborated on the role of attention in an individual's experience of music in "Die Erklärung der Harmonie," using the chord demonstration discussed above. An individual would hear the harmonic sequence of the two chords differently depending on which tones—the roots of the chord, which changed, or the upper tones, which remained unchanged—she directed her attention toward. The art of music composition, Mach asserted, therefore lay in guiding the listener's attention. There was also an art of hearing, he continued, which was not the gift of every person.[44] Only through extensive practice could one develop the ability to further differentiate a single tone into its fundamental tone and harmonic overtones.[45] Attention combined with the accommodation mechanism allowed the individual to distinguish harmonic overtones, which, according to Helmholtz, were the root of Western harmony. Mach's work

[40] Mach presented this example in his 1865 lecture "Die Erklärung der Harmonie," reprinted in *Zwei populäre Vorlesungen über musikalische Akustik* (Graz, 1865), 18–31, and in his article of the same year, "Bemerkungen über die Accommodation des Ohres" (cit. n. 2).

[41] According to Mach, groups of sonorous bodies behaved similarly—individual tones within the group only sounded when their particular note (*sein Eigenton*) was struck. Mach gave other illustrations of his point: a dog's response to "Phylax," his name, and utter indifference to such other heroic names as "Hercules" or "Plato," or the unified throb of two hearts in love. Mach, "Über die Cortischen Fasern des Ohres," reprinted in *Zwei populäre Vorlesungen* (cit. n. 40), 10–1.

[42] Ibid., 27.

[43] Ibid., 28.

[44] Mach, "Erklärung der Harmonie" (cit. n. 40), 37.

[45] In a clear reference to Helmholtz's work on tone sensation, Mach explained that these overtones played an important part in the formation of musical timbre as well as the consonance of sound. Mach developed a fuller, generally celebratory, discussion of Helmholtz's theories in his *Einleitung in die Helmholtz'sche Musiktheorie* (cit. n. 30).

suggested that Western musical aesthetics both shaped and were the product of the accommodation mechanism in hearing.

The pervasive use of musical examples and arguments in Mach's investigations of accommodation in hearing suggests that he made the three following assumptions: First, he assumed that music was a valid avenue through which to understand the sensation of sound because music and sound were equivalent or at least closely related. Second, he assumed that his audience was well versed in music theory and had extensive musical experience—enough to follow and be convinced by the musical passages written on the page. Third, Mach's frequent use of musical examples also assumed the validity of his audience's individual experience of music. An experience of the accommodation phenomenon could only be subjective. By attempting to root sound sensation in accommodation, Mach tied it to the attention of the individual and his or her experience. Individual attention was of course part of individual aesthetics, and specific to local time and space. It was historicist, at least for a single individual.

MACH'S HISTORICITY OF HEARING AND THE EPISTEMOLOGICAL CONSEQUENCES

Let us turn now to some of the consequences of Mach's use of music, and by association his musical aesthetics, in his studies of accommodation in hearing, and how these consequences might relate to his increasingly evolutionary worldview. Although Mach had read Darwin's *Origin of the Species* soon after its publication in 1859 and later integrated many elements of it into his theory of "economy of thought," his conception of evolution was very much rooted in Lamarckian inheritance of acquired traits, as was typical at the time. It was Lamarck's secondary law of inherited acquired traits that gained a new popularity—in isolation from his full program—during the latter half of the nineteenth century.[46] Darwinian and Lamarckian conceptions of evolution were often tangled at the end of the nineteenth century. Individuals who described themselves as Darwinians—or have since been described as Darwinians, like Herbert Spencer—frequently gave Lamarckian inheritance equal if not greater weight in their evolutionary perspectives. Mach was no different.

Further, Mach embraced Ewald Hering's work on the inheritance of memory. It was Hering's elaboration on Lamarckian inheritance to include Gustav Fechner's psychophysics that motivated Mach to expand his understanding of evolutionary inheritance to include psychological traits as well.[47] Hering argued that the inheritance of acquired traits included not only physical traits but psychical ones as well—after all, they were the same, according to Fechner.[48] Memory, Hering claimed, was an

[46] Peter Bowler explains that neo-Lamarckism was a diverse movement, varied both geographically and disciplinarily but unified by a perception of the shortcomings of the Darwinian evolutionary mechanism of natural selection. Neo-Lamarckians appreciated that the inheritance of acquired traits was directed, so that there was no issue of unfit individuals being lost in the fitness struggle. In this sense neo-Lamarckism was appealing because it was moral; organisms were not at the mercy of their environment and could retain control over their own destiny. Bowler, *Evolution: The History of an Idea*, 3rd ed. (Berkeley, Calif., 2003), 236–8.

[47] Hering presented his lecture "On Memory as a General Function of Organized Matter" before the Imperial Academy of Sciences in Vienna on 30 May 1870. Reprinted in Samuel Butler, ed., *Unconscious Memory*, 2nd ed. (New York, 1911), 80–3.

[48] In his 1860 text *Elemente der Psychophysik*, Fechner presented a monistic understanding of the world in which psychical and physical experiences were two different perspectives on the same event, two sides of the same reality. He offered the example of a circle differing depending on whether one stood inside or outside of it. According to Fechner, it was impossible, when standing on the plane of

inherited germ that was the unifying force of consciousness, powering the development of individual beings, and therefore the universal function of organized matter.
This view of memory implied that "the development of one of the more highly organized animals represents a continuous series of organized recollections concerning the
past development of the great chain of living forms."[49]

Mach similarly saw both physical and psychical traits as being transmitted through
generations. In an 1875 article, "Bemerkung über die Function der Ohrmuschel,"
Mach explained how Darwin's discussion in *Descent of Man* of a statue of Puck in
which he was given the pinna (the tips of the outer ear) of an animal prompted Mach
to an understanding that the outer ear acted as a resonator for higher notes in humans
as in animals.[50] Certainly animals gained advantage from fine determination of the
direction of sound and changes of timbre—the crackle of grass or leaves—and some
of the residual traces of this function likely still remained in humans for similar reasons.[51] The ability to distinguish subtle sonic differences as well as invoke organized
recollections for the sake of survival persisted in the ears of listeners at the end of the
nineteenth century.

Mach had combined this evolutionary understanding of sensory perception with
aesthetics from very early on. In an 1867 lecture, "Why Has Man Two Eyes?" Mach
first addressed the question as one of survival in which the two eyes were required for
depth perception. But his discussion soon turned to a survey of the different visual
aesthetic traditions of ancient cultures. Mach explained that if you "change man's
eye . . . you change his conception of the world. We have observed the truth of this
fact among our nearest kin, the Egyptians, the Chinese, the lake-dweller."[52] Evolving
physiology—the changing of the sight organ itself—explained the variety of visual
aesthetic traditions across time and place.

In 1871, while he was particularly engrossed in his search for the accommodation
mechanism in hearing, Mach presented his belief that repetition of sensory stimulation was the key to "agreeableness."[53] Symmetry was the most agreeable stimulus
because it conditioned repeated sensations, most noticeable in the visually pleasing
effect of regular figures, especially straight vertical and horizontal lines.[54]

Continuing, Mach explained that human appreciation for symmetry was deeply
rooted in the physiology of the sensory perception organ.[55] He claimed that human

---

the circle, to simultaneously experience both the convex and concave sides. Fechner, *Elemente der
Psychophysik*, vol. 1 (1860; repr., Amsterdam, 1864), 2–3.

[49] Hering, "On Memory as a General Function of Organized Matter" (cit. n. 47), 83.

[50] Mach, "Bemerkung über die Function der Ohrmuschel," *Archiv für Ohrenheilkunde* 9 (1875):
72–6.

[51] Ibid., 76.

[52] Mach, "Why Has Man Two Eyes?" in *Popular Scientific Lectures*, trans. Thomas McCormack
(Chicago, 1898), 66–88, on 82.

[53] Mach, "Ueber die physikalische Bedeutung der Gesetze der Symmetrie," *Lotos* 21 (1871): 139–47.

[54] Attempting to locate symmetry in sound sensation, Mach performed experiments in which he
played a series of moving notes and chords on a piano while looking in a mirror, then with the sheet
music reflected vertically and then horizontally in the mirror. Melodies became unrecognizable and
the chords, when mirrored, were reversed (the series of major key intervals, when reflected across a
horizontal axis, sounded like a series of minor key intervals and vice versa). Because sound sensation
was temporal, rather than spatial, like visual sensation, Mach concluded that there was no symmetry
in the sonic realm other than an intellectual one in the transposition from a major to a minor key in the
Western harmonic system. Ibid., 145–7.

[55] This was, Mach explained, why visual symmetry was still appreciated by those who had, for example, lost an eye. Ibid., 144.

beings' notions of beauty could very well be different if their physiology were different. With that, Mach then further asserted that conceptions of beauty could be modified by culture, "which stamps its unmistakable traces on the human body."[56] The idea of eternal beauty was thus mistaken; had not, in the recent past, all musical beauty been restricted to a five-toned scale?

So aesthetics were psychophysical. And an individual's aesthetics were subject to Hering/Lamarckian inheritance. It is likely that Mach understood this inherited knowledge to be quite broadly defined, extending to both the psychophysics of sensation and aesthetics. In an 1872 letter to Kulke, Mach asked if the people of their country could presently hear what the Greeks had heard and the Slavs still hear. And could the answer lie in attention (*Aufmerksamkeit*) only? He mused that a developmental history of melody, harmony, and rhythm would be very interesting.[57] In a later letter, Mach asked why the Germans and the Slavic people phrased their melodies differently. Kulke replied that the question was a historical one.[58] It was also, Kulke continued, an issue of aesthetics and a further application of Darwinian evolution to the arts. For both Mach and Kulke, musical aesthetics were subject to variation and inheritance, much like physiological attributes.[59]

By 1885 the discussion of the accommodation mechanism of sound sensation had all but disappeared from Mach's writings. He never did find it. But, between his psychophysical studies and his ongoing dialogue with Kulke, Mach's historicist understanding of both sound sensation and musical aesthetics was fully formed. In *Die Analyse der Empfindungen*, Mach asked how the development of modern music and the sudden appearance of great musical talent—genius, he claimed, seemingly appeared in a single generation rather than through the slow accumulation of ancestors' efforts—contributed to the survival of the species. Why did humans possess such fine discrimination of pitch, sense of intervals, and sensitivity to acoustic coloring that so far exceeded necessity or even usefulness? After all, according to Mach, music "satisfie[d] no practical need and for the most part depict[ed] nothing." He concluded that individuals developed their discriminating sensation of tone—of pitch, of harmony—much as they developed their visual ability to distinguish lines, "as a sort of collateral product of [their] training, a sense for the agreeable combination of lines."[60] By 1885, the ability to create and appreciate music was for Mach a byproduct of evolution:

> To deny the influence of pedigree on psychical dispositions would be as unreasonable as to reduce everything to it, as is done, whether from narrow-mindedness or dishonesty, by

[56] Ibid.

[57] Mach to Eduard Kulke, 30 May 1872, no. 4, Ernst Mach Papers, Dibner Library of the History of Science and Technology, Smithsonian Institute Special Collections, Washington, D.C.

[58] Eduard Kulke to Mach, 26 Oct. 1872, Ernst Mach Nachlass, Deutsches Museum Archives, Munich.

[59] Fifty years later, folklorist and composer Béla Bartók maintained a similarly imprecise but also similarly evolutionary and psychophysiological position, explaining that "peasant music is the outcome of changes wrought by a natural unconscious; it is impulsively created by a community of men who have had no schooling; it is as much a natural product as are the various forms of animal and vegetable life." Bartók, *The Hungarian Folk Song*, ed. Benjamin Suchoff, trans. M. D. Calvocoressi (Albany, N.Y., 1981), 3.

[60] Mach, *Analysis of the Sensations and the Relation of the Physical to the Psychical*, trans. C. M. Williams (New York, 1959), 307–8. Originally published as *Beiträge zur Analyse der Empfingdungen* (Jena, 1886).

modern fanatics on the question of race. Surely everyone knows from his own experience what rich psychical acquisitions he owes to his cultural environment, to the influence of long vanished generations, and to his contemporaries. The factors of development do not suddenly become inoperative in post-embryonic life.[61]

For a while Mach viewed musical aesthetics as an alternative entry point to investigating sound sensation, because culture "stamps its unmistakable traces on the human body." This is evidenced in his decades-long use of music to explore the psychophysics of accommodation in hearing. But once he determined that sound sensation was, not just for the individual, but for the human species as a whole, historically and culturally contingent, the mechanism of accommodation in hearing became irrelevant and he ended his pursuits.

These early psychophysical investigations of sound sensation of course had philosophical consequences for Mach; namely, an increasingly phenomenological and positivist approach to the world.[62] In *Die Analyse der Empfindungen*, Mach described how the superfluity of the Kantian "thing in itself" abruptly dawned on him: "On a bright summer day in the open air, the world with my ego suddenly appeared to me as *one* coherent mass of sensations."[63] This realization fueled his belief that one should approach the world as if it were made up entirely of elements of sensation—reality, appearances, and one's self in the world all one buzzing mass. Man did not surrender as an individual into nature, but rather, was already part of it. When one conceived of the world, as Mach did, as a single coherent mass of sensations, the individual was one perspective within a complete psychophysical whole. Fechner had eliminated the causal connection between the physical and the psychical, replacing it with mathematical dependence only. With his development of psychophysical monism, Mach dispensed with Fechner's two-sided nature in favor of complete unification.

[61] Ibid., 309. In this last sentence Mach was making an oblique reference to the work of his former student, the Viennese music critic and comparative musicologist Richard Wallaschek, *Anfänge der Tonkunst* (Leipzig, 1903).

[62] A historiographical note: I have aimed to push the origins of Mach's historicism back to 1863, the year in which he first met Kulke and Kulke's dilemma, and the year in which he first began his search for the accommodation mechanism of sound sensation. Already in 1867 Mach was thinking of the sense organs as well as visual aesthetics within an evolutionary, historicist framework. Certainly by 1871 he was. If any doubt remains about the early date of 1863, it should be noted that Mach himself described his work since 1863 as "historico-physical investigations." Mach, *Analysis of the Sensations* (cit. n. 60), 30.

[63] Ibid.; translator's emphasis. The passage in the original German reads: "An einem heitern Sommertage im Freien erschien mir einmal die Welt sammt meinem Ich als eine zusammenhängende Masse von Empfindungen." Mach, *Die Analyse der Empfindungen und das Verhältnis des Physischen zum Psychischen*, 4th ed. (Jena, 1903), 24. The operative verb here, *erscheinen*, is usually translated as "to seem" or "to appear" but it can also be translated as "to strike." All of these suggest an inseparability of nature and the experiences of the mind that likely sounds familiar to scholars of pragmatism and William James's radical empiricism. Indeed, James and Mach had known each other since the 1870s, corresponded extensively, and pored over each other's work. Historian Gerald Horton has carefully documented the extent of the two men's intellectual exchange, emphasizing the significant impression Mach's writings made on James. Judith Ryan has further argued that it was the 1886 edition of Mach's *Analyse der Empfindungen* that was the missing intellectual link between the evolving versions (the 1884 manuscript versus the 1890 chapter in *Principles of Psychology*) of James's essay "The Stream of Thought." Horton, "Ernst Mach and the Fortunes of Positivism in America," *Isis* 83, no. 1 (1992): 27–60; Ryan, "American Pragmatism, Viennese Psychology," *Raritan* 8 (1989): 45–55.

MAX PLANCK'S ACCOMMODATION EXPERIMENTS

Mach might have been the first to admit that these phenomenological consequences of his psychophysical work were not inevitable. A study of accommodation in hearing did not necessarily a positivist make. Let us turn now to the accommodation study of the Berlin physicist Max Planck.[64] To be sure, Planck's investigation was not nearly as extensive as Mach's, nor did Planck actively place himself at the intersection of the music and natural science worlds. But he was no less steeped in a rich world of music. He played several instruments and was a frequent concert attendee. Planck was also well versed in the sound-sensation writings of his Berlin colleagues Hermann von Helmholtz and Carl Stumpf. And—connecting additional threads between music and experimental studies of sound sensation—Planck attended several evenings of music at the Helmholtz household, where he heard the violinist Joseph Joachim perform his *Hungarian Dances* as well as Marianne Brandt and Rudolf Oberhauser sing selections from Wagner's *Walküre*.[65]

Soon after Planck's assignment to the Berlin Institute for Theoretical Physics in the early 1890s, the institute received a large harmonium designed by Carl Eitz, built for the Prussian state government by the Schiedmayer piano factory in Stuttgart. Eitz, originally from Halberstadt, taught primary school in Eisbaden until his retirement in 1911, but was better known as a music pedagogue and acoustician.[66] Halberstadt's most famous musical resident at this time was the town organ, completed by organ builder Nicholas Faber in 1361. The third manual of the Halberstadt organ contained the earliest form of a keyboard of seven keys in succession within which an additional five were arranged in groups of three and two—the arrangement predominantly seen on keyboard instruments today. Harry Partch would later describe the Halberstadt organ as the apple in Eden.[67] Eitz developed various scientific instruments, including an apparatus that rendered sound waves visible, as well as a number of musical instruments. Of these, most notable was a harmonium with 104 pure tones per octave. He developed a *Tonwort* solmization method for training students to sing according to just intonation, which was later implemented in the Bavarian school system.[68] An early exploration of quarter tones, the system employed 31 separate syllabic note names for each diatonic, chromatic, and enharmonic degree of the untempered scale. He later further extended this system to support the harmonium that had been developed two decades earlier by R. H. M. Bosanquet, with 53 mathematically pure tones per octave.[69]

[64] Planck is best known for his research on black-body radiation in which he extended the principles of mechanics (employing Ludwig Boltzmann's definition of entropy in a gas) to radiation, in turn establishing some of the initial groundwork for quantum physics.

[65] Planck, "Persönliche Erinnerungen aus alten Zeiten," in *Vorträge und Erinnerungen von Max Planck*, 5th ed. (Stuttgart, 1949), 7.

[66] Halberstadt is currently home to the ASLSP/John-Cage-Orgel-Kunst-Projekt in which, beginning in 2000, John Cage's piece "As Slow as Possible" is being performed very slowly over 639 years in the church of St. Burchardi.

[67] Partch, *Genesis of a Music* (cit. n. 3), 373–4. Partch also, incidentally, mentioned Eitz's harmonium (ibid., 438).

[68] Eitz, *Das mathematisch-reine Tonsystem* (Leipzig, 1891).

[69] Bosanquet's "generalized keyboard harmonium" was first presented at the Musical Association in London in May of 1875. *Grove Music Online*, s.v. "Eitz method," by Bernarr Rainbow, 15 Oct. 2011, http://www.oxfordmusiconline.com/.

Planck was tasked with examining the Eitz harmonium's untempered scale. Eitz's instrument consisted of four and a half octaves, beginning with F and ending with the C three octaves above. Each octave contained 104 tones (the keyboard familiar to most readers has 13 tones per octave). Planck published a summary of this investigation in 1893, "Ein neues Harmonium in natürlicher Stimmung nach dem System C. Eitz," noting that the instrument was relatively easy to play with a little practice.[70]

In the course of studying Eitz's instrument Planck became quite interested in the role of the untempered, "natural" scale in modern, unaccompanied vocal music. He tested both himself as well as a number of vocalists and players of string instruments (instruments without fixed tuning systems), comparing aesthetic impressions of both tempered and untempered scales, with the goal of determining the extent to which the untempered scale played any practical role in contemporary music making. Planck claimed to find, somewhat to his surprise, that the tempered scale was far more pleasing. Continuing, he explained that even in the harmonic major triad, the natural third sounded feeble and inexpressive when compared with the tempered third.[71] He ascribed this unanticipated preference for tempered tuning "to a habituation through years and generations," because "before J. S. Bach, the tempered scale had not been universally known at all."[72]

In his 1893 article "Die natürliche Stimmung in der modernen Vokalmusik," published in the musicology journal *Vierteljahrsschrift für Musikwissenschaft*, Planck argued that accommodation in hearing may very well have been at the root of this preference for temperament.[73] He explained that the relaxed, inattentive ear, hearing a major triad in which the third was pure but the fifth was slightly tempered, would have the impression of consonance. Only with proper and careful attention would the listener determine the fifth to be tempered. Planck explained that even the early nineteenth-century acoustician Ernst Friedrich Florens Chladni had noted something similar, that an interval only ever so slightly off from pure could be heard as pure.[74]

So Planck believed that accommodation in hearing did not simply bring out the tones focused on more distinctly, but actually altered the impression of their tuning. Accommodation for Planck meant that the listener heard different sounds. Taking Mach's chord example from the beginning of this article, if the demonstration had been performed on an equal-tempered piano, the listener, when instructed to simply listen to the chords (the first time the progression was played), would have, according to Planck, heard them as just tuned. The second and third times through the chord progression, when asked to focus his or her attention on the high and low pitches, the listener would have heard the chords as equal-tempered. Perhaps we can borrow the nineteenth-century distinction that listening was active and hearing was passive for

[70] The 104 tones are determined by the formula $2^p\, 3^q\, 5^r$, if p, q, and r are close for any positive or negative integers or zero sets. The syntonic comma (81:80), the Pythagorean comma (approximately 74:73), and the schism (about 887:886) can be given on the instrument by an interval that is still too large by about 301:300, almost the natural seventh (7:4). Planck, "Ein neues Harmonium in natürlicher Stimmung nach dem System C. Eitz," *Fortschritte der Physik* 49 (1895): 557–8.

[71] Planck, *Max Planck: Vorträge, Reden, Erinnerungen* (Berlin, 2001), 61.

[72] Ibid.; translation mine.

[73] Planck, "Die natürliche Stimmung in der modernen Vokalmusik," *Vierteljahrsschrift für Musikwissenschaft* 9 (1893): 418–40. The article was republished the following year as a pamphlet by Breitkopf and Härtel (Leipzig). I must extend a special thank-you to Erwin Hiebert for bringing this article to my attention nearly a decade ago upon hearing of my interest in Mach's studies of accommodation.

[74] Ibid., 422.

further clarification: for Planck, listeners listened in natural tuning and heard in equal temperament.

Planck's understanding of accommodation—that it altered the sound heard—bore similarities to what Stumpf termed "music consciousness" (*Musikbewusstsein*). Highly skilled, musically trained listeners, Stumpf explained, would hear impure intervals as pure out of musical habit, biased by their music consciousness.[75] Helmholtz had previously made a similar claim, though about the sound generated by highly skilled musicians rather than the sound heard.[76] Helmholtz argued that pure intervals, with their coincident overtones, were heard as more consonant and that this was the reason that the very best musicians naturally migrated to pure intervals. He also saw this as further evidence of the superiority of untempered tuning systems (the natural tuning Planck referred to).

Planck, thirty years later, to his admitted surprise, found the opposite: both listeners and musicians migrated to tempered intervals.[77] Now, Planck did not link this accommodative ability to musical skill. In fact, he thought that it was for the most part involuntary.[78] Explaining that a sustained sounded interval would, over time, become smaller, smoother, and softer, with a correlated sense of decreasing tension, Planck implied that the phenomenon was psychophysical.[79] It was not, however, a timeless effect. Planck explained that the fact that temperament was the aural resting state (that focus and accommodation were required to hear pure, natural intervals, which would then slip back to tempered intervals when the listener relaxed) showed the "habituation of our ear to tempered tuning" as a consequence of its predominance.[80]

Planck suspected that this phenomenon of accommodation in hearing—and the corresponding phenomenon in which individuals would, when relaxed and unfocused, hear in tempered tuning—was linked to various effects in Western musical composition. It contributed, he believed, to the power of the leading tone, the fermata, and the compositional return to the tonic key, all of which were dependent on the subtle and sustained aftereffects created by the habituation of a listener constantly exposed to equal temperament.[81] Accommodation amplified musical aesthetics.

In pursuit of experimental support of this theory beyond studies of his own ear with the Eitz harmonium, Planck took advantage of the abundance of musical groups in

[75] Stumpf introduced this concept (he also used the term "music-infected consciousness"—*musik-infizierten Bewusstsein*) in the course of a fierce debate in the early 1890s with Wilhelm Wundt and Wundt's student Carl Lorenz over tone-differentiation data collected by Lorenz from both musically trained and musically untrained experimental subjects. See Alexandra Hui, "The Bias of Music Consciousness: The Aesthetics of Listening, in the Laboratory and on the City Streets of Fin-de-Siècle Berlin and Vienna," *Journal for the History of the Behavioral Sciences* 28, no. 3 (2012): 236–50.

[76] According to Helmholtz, it was impossible to detect "any false consonances" in the music performed by the very best instrumentalists (on bowed and wind instruments, where the intonation can be controlled by the players themselves) because they naturally played in just intonation and they "know how to stop the tones they want to hear, and hence do not submit to the rules of an imperfect school." Hermann von Helmholtz, *On the Sensations of Tone as a Physiological Basis for the Theory of Music*, 4th ed., trans. from the 4th German ed. (London, 1877), 324–5.

[77] Samples of the passages Planck tested, in both equal temperament and just intonation, can be heard in audio 2 (300KB; MP3) and audio 3 (300KB; MP3) in the electronic version of this article. Both samples were recorded by Peter Pesic using Finale music-notation software, Santa Fe, N.M., 2012. Special thanks to Alexei Pesic and William Sethares.

[78] Planck, "Natürliche Stimmung in der modernen Vokalmusik" (cit. n. 73), 423.

[79] Ibid., 424, 431–2.

[80] Ibid., 425.

[81] Ibid., 425–6.

Berlin. His studies of various a cappella choirs arrived at similar conclusions: with only a few exceptions, singing was unambiguously tempered.[82] He asked the choirs to perform a series of sequences of sustained chords that would sound distinctly different when sung in natural tuning, with pure intervals, than when sung tempered. Even the very best choirs slid into tempered tuning and could only sustain pure intervals with repeated practice and focused attention. Modern vocal music, Planck concluded, was performed almost without exception in tempered form. This sliding was, of course, not due to poor singing or laziness but to instinct buttressed by accommodation.[83]

PLANCK'S HISTORICITY OF HEARING AND
THE EPISTEMOLOGICAL CONSEQUENCES

So, was Helmholtz wrong? Did the musically trained passively hear and generate tempered intervals rather than pure ones? Planck made no attempt to explain his contradictory findings other than to note his astonishment. And also to suggest that perhaps accommodation in hearing was similar to other stimuli that affect sensory perception, and correspondingly, aesthetics, in subtle ways.[84] Helmholtz had performed his *Tonempfindungen* research three decades prior. Perhaps the Berliners' ears had changed.[85]

Indeed, the implication of Planck's conclusions was that the hearing organ was remarkably sensitive and malleable. Though tempered tuning had only recently been introduced on a large scale, Planck's studies of accommodation in hearing suggested that, unless pushed to do otherwise, not only did listeners hear tempered intervals but musicians generated tempered intervals. Through habituation, Berliners' ears (or at least Planck's and those of members of the Royal College of Music) had become accustomed to both creating and receiving tempered tones in only a few generations—a very Lamarckian time scale.

It is worth noting that Planck did not mobilize his findings toward a discussion of Western musical aesthetics. He ventured that tempered hearing and accommodation might explain the power of some musical features, but he stopped there. Certainly Planck did not, as Helmholtz did, see and lament the expansion of equal temperament as the harbinger of poorly performed and poorly composed music.

Let us now attempt to situate Planck's concept of sound sensation within his broader understanding of the world. Planck had understood his task as a scientist, from the very beginning of his career, to be the pursuit of a unified conception of all the forces of nature. The interconnection of laws of mechanics, conservation, least action, quantum physics, and relativity drew on more general Wilhelmian values em-

---

[82] He worked mostly with the Royal College of Music in Berlin, under the direction of Adolf Schulze. Ibid., 430–1.

[83] Ibid., 434–7, 439–40.

[84] Ibid., 430.

[85] One could also point here to Stumpf's *Musikbewusstsein* debate with Wundt and his student Lorenz. Stumpf argued that the musically trained overwhelmingly heard pure intervals, in contrast to the untrained—essentially the exact opposite findings from Planck. If Stumpf's claims were based only on Lorenz's data, then perhaps the Leipzig musicians heard differently from the vocalists of Berlin that Planck worked with. If Stumpf's claims were, however, rooted in some of his earlier *Tonpsychologie* research in Berlin, then he and Planck may very well have employed the same musicians.

phasizing an ideal of unity among political, cultural, and intellectual life.[86] At the beginning of the twentieth century, Planck began to articulate an epistemological framework made up of no less than three worlds: the world of the human senses, the real world, and the world of physics. This latter world, because it was advanced by the finite human mind, was subject to change in relation to the other two worlds.

The world of physics sought to determine the laws connecting the world of sense with the real world. Planck described a historicity in this program, explaining that depending on the stability of physics at a particular time, one worldview would dominate.[87] There had been overall, however, a steady march of the physics world toward the real world. As the world of physics had moved away from the world of sense, according to Planck, it had lost much of its former anthropomorphic character and become progressively more abstract.[88] Planck advocated an ultimate—though admittedly unattainable—dual goal of complete domination of the world of sense and total understanding of the real world by physics. It was therefore the duty of scientists to continue to purge all anthropomorphic elements and no longer admit "any concepts based in any way upon human mensuration."[89]

Planck's conception of sound sensation fit into this widening space between the world of physics and the sense world based on human mensuration. For one, this increasing distance reinforced Planck's framework by revealing the deceptiveness of the senses—a single listener could hear the same interval differently depending on her use of accommodation, and a vocalist could only maintain a pure interval with careful and sustained concentration. In his 1908 Leiden lecture "Die Einheit des physikalischen Weltbildes" (The unity of the physical universe), when surveying the progress of physics away from the sense world, Planck explained that the fields of physical acoustics, optics, and thermodynamics had all eliminated their dependence on the immediate perceptions of the senses.[90] Though accommodation in hearing could, with effort, overcome the tempered hearing acquired through habitual exposure to the recent explosion of equal temperament, the use of frequency and wavelength to measure sound waves eliminated the human element and was certainly more reliable.[91]

Further, the implication of Planck's studies of accommodation among vocalists underscored his position that the sense world and the real world were indeed separate. If passive hearing could shift significantly in just a few generations due to material and aesthetic shifts in the music world with the spread of equal temperament

[86] John Heilbron, *The Dilemmas of an Upright Man: Max Planck and the Fortunes of German Science* (Berkeley, Calif., 1996), 4.

[87] Planck offered the examples of the second half of the nineteenth century, during which the physical world was very stable and many believed that a complete understanding of the real world was within reach, and conversely, the interwar period, a time of change and instability in physics, when many turned to positivism. Planck, *Das Weltbild der neuen Physik* (Leipzig, 1931), trans. W. H. Johnston as *The Universe in the Light of Modern Physics*, 2nd ed. (London, 1937), 12–3.

[88] Planck, *Universe in the Light of Modern Physics* (cit. n. 87), 14–5.

[89] Ibid., 49.

[90] Planck, "Die Einheit des physikalischen Weltbildes" (lecture, University of Leiden, 9 Dec. 1908), trans. R. Jones and D. H. Williams as "The Unity of the Physical Universe," in *A Survey of Physical Theory: Max Planck* (London, 1993), 1–25.

[91] Such sentiments and efforts to eliminate human judgment from acoustics continued through the twentieth century, as seen in the rebuffing of Sheridan Speeth's psychoacoustic approach to international nuclear arms control, discussed by Axel Volmar in this volume.

(that is, very much as a result of human activity), and yet a listener could, with accommodation, toggle between tuning systems in her sensory perception of sound, certainly the sense world and the real world were not one and the same. For Planck, physicists could know more than just the world of direct experience. This conception of physics as oscillating between the worlds of sense and reality but also asymptotically approaching unity with the real world is a rather delightful solution that allowed Planck room to historicize scientific thought while maintaining an antipositivist stance.

Much has been written by historians of science on the epistemological clash between Planck and Mach, and going into too much detail here would distract from the larger points of this article.[92] It should be noted, though, that Planck was deeply disturbed by the popularity of Mach's phenomenology and believed it to be a threat to exact science. Mach's system, according to Planck, was not science and "evade[d] the most convenient criterion of all scientific research—the finding of a *fixed* world picture independent of the variation of time and people." The goal of science, Planck continued, was not the adaptation of our ideas to our perceptions, but rather, "the complete liberation of the physical picture from the individuality of the separate intellects."[93] Planck attacked Mach's phenomenology as misguided self-alienation when applied to the sciences and ended his 1908 Leiden lecture by decrying Mach as a false prophet.

CONCLUSION

I would like to conclude with a brief discussion of the work of Harry Partch, the early twentieth-century American composer and musical-instrument maker, perhaps best known for his monophonic system based on a forty-three-note octave. To perform his monophonic pieces, he built the Ptolemy harmonium.[94] In contrast to Bosanquet's or Eitz's harmonium (he mentioned Bosanquet's in his 1949 text *Genesis of a Music*), which were built as acoustic and music-theoretical instruments, Partch explained that his harmonium was built for a composer and his musical creations. Extending the aspirations of Arnold Schönberg, Alexander Scriabin, and Henry Cowell to expand the triad basis of tonality, Partch's Ptolemy ended the double tyranny of temperament and the diatonic scale.[95]

I include Partch here for two reasons. First, as the earliest composer to seriously

[92] The most thorough treatment is Heilbron's *Dilemmas of an Upright Man* (cit. n. 86), but see also Lawrence Badash, "The Completeness of Nineteenth-Century Science," *Isis* 63, no. 1 (1972): 48–58; John Blackmore, "Ernst Mach Leaves 'The Church of Physics,'" *British Journal for the Philosophy of Science* 40, no. 4 (1989): 519–40; Steve Fuller, "Retrieving the Point of the Realism-Instrumentalism Debate," *Proceedings of the Biennial Meeting of the Philosophy of Science Association* 1994, vol. 1: *Contributed Papers*, 200–8; Nadia Robotti and Massimiliano Badino, "Max Planck and the 'Constants of Nature,'" *Annals of Science* 58 (2001): 137–62; and Richard Staley, "On the Co-creation of Classical and Modern Physics," *Isis* 96, no. 4 (2005): 530–58.

[93] Planck, "Unity of the Physical Universe" (cit. n. 90), 38–9.

[94] Partch explained that whereas the basis of tonality in the contemporary music system was the triad chord (composed of a fundament, the third, and the fifth interval), his monophonic system continued past the fifth through the eleventh overtone in fourteen of its tonalities and through the ninth overtone in the other ten. Partch, "A New Instrument," in *Enclosure 3: Harry Partch*, ed. Philip Blackburn (St. Paul, Minn., 1997), 49. The article was originally published in *Musical Opinion* (1935): 764–5.

[95] Ibid.

pursue microtonal music, right down to the development of his own microtonal musical instruments, Partch drew on the psychophysical studies of sound sensation of the nineteenth century. Indeed, he described his discovery of an English translation of Helmholtz's *Tonempfindungen* at the Sacramento public library as "the key for which I had been searching."[96] Helmholtz's discussion of the various tuning systems of Western music encouraged Partch to abandon equal temperament.

I also see Partch as a resolution of the contrasts between Planck and Mach. His realization of the historicity of tuning systems liberated him to develop both a new tuning system and new instruments with which to perform music composed in this new system. As a consequence his music sounded new; for most of us, a piece composed in the forty-three-tone scale will include tones and intervals never before heard. One could say that Partch wanted to break the habituation to tempered tuning discovered by Planck. His criticism of musicological training and pedagogy was that it rehashed old compositional styles, techniques, and sounds. Recall Planck's assertion that listeners' ears must have changed through habituation over time to hear—when unaccommodated—tempered rather than pure intervals because "before J. S. Bach, the tempered scale had not been at all universally known." Partch quipped in response: "There is, thank God, a large segment of our population that never heard of J. S. Bach."[97]

In the early 1920s, Partch had become increasingly dissatisfied with the limits of traditional music training. In particular, he was troubled by the emphasis on technique and skill in playing an instrument and, correspondingly, the belief that polished technique was the equivalent to good performing. Further, Partch was frustrated by the popular assumption that the present represented progress in comparison to the past. By 1928 he had drafted his monophonic principles, which were based on his belief that the individual's spoken words were the most potent and intimate tonal ingredients available and therefore "the juice of a given identity in the tonal world."[98]

Following his intuitive break with the aesthetic expectations of the Western tonal system, Partch then set about justifying his innovative sounds through critical and historical analysis. In *Genesis of a Music: An Account of a Creative Work, Its Roots, and Its Fulfillments*, he documented four thousand years of music history in terms of a shift, with noteworthy exceptions, away from "instinctive Corporeal attitudes" to "an Abstract character."[99] The antiphonal singing that replaced the Chinese and Greek chant traditions liberated music from language and forced the listener to think of music itself as conveying meaning. By the eighteenth century, music had transcended language, space, moment, and the human body (musical morality in fact denied the human body) as the mass expression of pure form. Certainly there were

[96] Quoted in Jules Joseph, "Harry Partch Uses 43 Tone Scale to Preserve Natural Word Rhythm," *University of Wisconsin–Madison Daily Cardinal*, 30 Jan. 1945. Cited in Bob Gilmore, *Harry Partch: A Biography* (New Haven, Conn., 1998), 48.

[97] Partch, *Enclosure 3* (cit. n. 94), 93.

[98] Partch, *Genesis of a Music* (cit. n. 3), 5–6, quotation on 7.

[99] Partch offered sung or chanted stories, poems recited or intoned, early seventeenth-century Florentine dramas, and music intended for dances in which a story or situation is described as examples of corporeality. Abstract character could be seen in such musical forms as "songs with words that are intended not to convey meaning but simply to set the mood of the music; songs or dramas with words that do not convey meaning because of the style of composition; . . . all purely instrumental music." Ibid., 9.

recent exceptions—Modest Mussorgsky, Arnold Schönberg—but for the most part, contemporary conductors, performers, and composers were all conditioned by the "steam-pressure exploitation of mediocrity."[100]

Not that Partch wanted to return to an ancient corporeal music. Certainly the phrases of the past were not timeless in their meaning, or they would be abstractions. Partch's monophonic music was instead a new individual expression. But it was also "frankly and extremely Corporeal," seeking the intimacy of one voice and one instrument.[101] Monophonic music was liberated from the shackles of habituation and conditioning through its explicit prioritization of the individual's expression and experience of sound. It had meaning (if only to the individual) precisely because it was so firmly subjective.

* * *

One of the larger goals of this article has been to examine why and how music was appropriated for psychophysical studies of sound sensation and the consequences of this appropriation. I have shown that for Mach, deeply steeped in the music world, the use of music and musical instruments in his search for the mechanism of accommodation in hearing was a natural and uncomplicated impulse. Planck studied an a cappella choir to resolve questions raised by what had begun as an investigation of the acoustics of microtones. Both men turned to music to better understand psychophysical phenomena without hesitation or self-conscious defense.

The use of music to explore psychophysical phenomena required both men to accept the validity of individual, subjective experiences of sound. Or, when sound was music, to accept the validity of individual, subjective musical tastes. This they also appeared to do without hesitation, indicating a confidence in their own musical tastes—a further indication of their comfort in the world of music.

Both Mach and Planck were unsuccessful in achieving their initial respective goals of locating the accommodation mechanism and establishing the root of natural tuning in modern vocal music. Instead they were only able to reconcile the subjective experience of accommodation in hearing with more universal claims about musical aesthetics by conceiving of hearing itself as changeable, historical. If an individual's experience of sound was psychophysically bound to the individual's musical milieu, then not only were her musical tastes changeable but so were her ears. It was precisely at this time that Western musical aesthetics were also changing radically. The sonic upheaval fueled by changing tuning systems and the introduction of non-Western music reinforced Mach's and Planck's shared belief in the historicity of hearing.

The central role of subjectivity in accommodation in hearing revealed by both Mach's and Planck's studies buttressed and, in Mach's case, directly contributed to their opposing epistemological positions. Mach was eventually able to reconcile psychophysical laws of sound sensation with historically and culturally contingent musical aesthetics by rejecting the universality of sound-sensation processes and coming to the belief that hearing itself was historical.

Planck's work on accommodation, though far less extensive than Mach's, also essentially came to the conclusion that hearing was historical. For Planck, the ear

[100] Ibid., 60.
[101] Ibid., 61.

could, through habitual exposure, begin to hear in different tuning systems. And yet they mobilized their findings to different ends: Mach to show that knowledge itself was historical and that the only reality was that indicated by sensations; Planck to argue that knowledge was only exact and valuable when it moved away from the world of sensation toward a separate reality.

Partch presents a third path. His monophonic music unified the subjectivity of creating sound with the subjectivity of sensing sound. For Partch, music was not the means of finding answers to psychophysical questions. Music was the answer.

The Audiovisual Field in
Bruce Nauman's Videos

by Armin Schäfer*

ABSTRACT

This article investigates the audiovisual field in some of Bruce Nauman's early videos. Historians of science and anthropologists have pointed to the necessity of considering seeing and hearing as faculties that are acquired. I argue that media technology plays a crucial role in that process. Nauman's videos explore the relation of perception, the human body, and media technology, bringing to the fore the extent to which this seemingly self-evident relationship is artificial and constructed. In particular, Nauman's *Violin Tuned D.E.A.D.* employs techniques of musical training that emerged in the late nineteenth century from the physiology of movement and fatigue.

In an interview, American artist Bruce Nauman recalled the 1960s, when he attended graduate school in fine arts at the University of California, Davis: "I was working very little, teaching a class one night a week, and I didn't know what to do with all that time. . . . There was nothing in the studio because I didn't have much money for materials. So I was forced to examine myself, and what I was doing there. I was drinking a lot of coffee, that's what I was doing."[1]

Since that time, Nauman has explored bodily and cultural techniques such as walking, pacing around, jumping, or playing a musical instrument while recording these performances on film and video. The recorded activities are presented in a way that strips them of their self-evidence and brings the conditions that are imposed on visual and auditory perception by media technology to the fore. When one looks at and listens to the performances, media technology's effect upon those activities is itself what is seen and heard. Screen and loudspeaker intervene in perception. Seeing and hearing, although grounded in psychophysiological principles, are not anthropological universals but must be learned, similar to laboratory-based skills discussed by historians and sociologists of science over the past twenty-five years.[2] By examin-

* Department of German Literature and the History of Media Culture, University of Hagen, Universitätsstraße 33, D-58084 Hagen, Germany; armin.schaefer@fernuni-hagen.de.

I wish to thank the editors of this volume, Mara Mills, and John Tresch for comments on an earlier version of this article, as well as the participants in the preparatory Berlin conference.

[1] Willoughby Sharp, "Interview with Bruce Nauman, 1970," in *Please Pay Attention Please: Bruce Nauman's Words; Writings and Interviews*, ed. Janet Kraynak (Cambridge, Mass., 2003), 111–30, on 117–8.

[2] For the historical construction of vision, see Jonathan Crary, *Techniques of the Observer: On Vision and Modernity in the Nineteenth Century* (Cambridge, Mass., 1992), 1–25; for the historical construction of listening and hearing, respectively, see Jonathan Sterne, *The Audible Past: Cultural Origins of Sound Reproduction* (Durham, N.C., 2003), 87–136, and Veit Erlmann, *Reason and Reso-*

© 2013 by The History of Science Society. All rights reserved. 0369-7827/11/2013-0008$10.00

ing the role of enculturation in seeing and hearing, media studies can enrich histories of science interested in the importance of skill acquisition and representation. This article intends to contribute to such studies by investigating activities that can be seen and heard in Nauman's videos: the body techniques displayed in his videos are learned. Perceiving them in turn exerts a specific feedback on the spectator. In order to grasp the problem that is posed by Nauman's videos, it is not enough simply to differentiate between a perceiving subject and the perceived object. Nauman's videos explore the relations between learned, habitual techniques of the body and perception.

In the following, Nauman's early video performances will be traced back to nineteenth-century experimental physiology. His art stages simple situations that, upon second glance, are rooted in those experimental explorations of the body, which used registration devices to analyze the movements of the body. Nauman performed simple activities, such as walking or playing a few notes on a violin, and recorded them on videotape with the technology of the 1960s. While the artist performed these activities in his studio under precisely defined conditions, as if he were investigating them, he left the analysis to the spectator. Yet his use of media technology introduces a second mode of artistic experimentation that will be shown to relate his explorations of body techniques to an exploration of his spectators' perception. The spectators who watch the video are involved in an experimental setup that forces them to question their habits of perception. They experience the extent to which perception is shaped and formatted by such habits when they enter what I will call in the following the "audiovisual field" of Nauman's videos. In this experience, what we usually accept as self-evident about our perception turns out to be produced to a great extent by media technology.[3] The following analysis of the audiovisual field centers on Nauman's video *Violin Tuned D.E.A.D.* This piece questions a history of registration devices by confronting it with musicians' instrumental training. The registration on video differs from the scientific inscription that emerged in physiological laboratories of the nineteenth century. The video does not aim directly at formalizing and quantifying, whereas an inscription yields standardized semiotic data.[4] As will be shown, the interrelation between art and science has migrated to the artist's studio, turning it into a laboratory where the human body and perception are explored.

The article will proceed in three steps. Throughout, concepts derived from analyses of the body and mind in the history of philosophy, physiology, anthropology, and psychology will be deployed as illuminating precursors and foils for Nauman's interventions. The first part introduces the notion of body techniques, in order to grasp the artificiality of Nauman's activities in front of the camera. His videotape *Slow Angle Walk (Beckett Walk)* is an example of Nauman carrying out the body technique of

nance: A History of Modern Aurality (New York, 2010); for the historical construction of the scientific observer, see Lorraine Daston and Peter Galison, *Objectivity* (New York, 2007), 191–251. For social studies of skill, see Michel Polanyi, *Personal Knowledge: Towards a Post-critical Philosophy* (New York, 1958); Harry M. Collins, "The TEA Set: Tacit Knowledge and Scientific Networks," *Science Studies* 4 (1974): 165–86; Trevor Pinch, Collins, and Larry Carbone, "Inside Knowledge: Second Order Measures of Skill," *Sociological Review* 44 (1996): 163–86. For historical studies of skill, see Kathryn M. Olesko, "Tacit Knowledge and School Formation," *Osiris*, 2nd ser., 8 (1993): 16–29; Myles W. Jackson, *Spectrum of Belief: Joseph von Fraunhofer and the Craft of Precision Optics* (Cambridge, Mass., 2000).

[3] Simon Schaffer, "Self Evidence," *Critical Inquiry* 18, no. 2 (1992): 327–62.

[4] See Robert M. Brain, "Standards and Semiotics," in *Inscribing Science: Scientific Texts and the Materiality of Communication*, ed. Timothy Lenoir (Stanford, Calif., 1998), 249–84.

walking in a specific way. For simplicity, this analysis will concentrate on the visual aspects in Nauman's construction of the audiovisual field.

In the second step, the video *Violin Tuned D.E.A.D.* will be discussed to consider how Nauman carries out the culturally specific technique of playing the violin and how this specificity is made visible and audible in the video: screen and loudspeaker constitute an audiovisual field, whose consistency cannot be sufficiently explained as merely an opposition to everyday perception or as a defamiliarization of the latter. When a viewer looks at the video, seeing merges into the imagination of something seen; hearing in turn infers a source from the sound and provokes, referring to previous experience, an imagination of violin playing. Seeing and hearing thus conceal bits of imagination that are constantly feeding back into perception.

I shall show in the third step how the audiovisual field in Nauman's video pushes the viewer to continue this process of imagination, leading eventually to a semantics of exhaustion. Nauman's video art explores auditory perception as well as musical practice by involving the spectator in a situation in which the violin is decoupled from the habitual inventory of movements. His performance equally questions the role of training and instrumental practice. The video shows the violin player in a situation that does not grant him access to the habitual body techniques, threatening the failure of conventional training. The spectator follows this movement up to a point where either the playing would have to be interrupted or new techniques of tone production would have to be invented, but at this point the video ends.

Playing a musical instrument and presenting his performance on videotape, Nauman explores culturally constructed habits. In the video, the technology of film and video, the object of perception, and the execution of techniques of the body form a circular cohesion. Even though this circle involves movements of the human body, it is a thoroughly artificial product that does not even create the appearance of running smoothly or automatically. The body is certainly not just one object of perception among others. Rather, it is the point of reference for perception in general, which guarantees that the field of perception synthesizes spontaneously into one homogeneous unity. It would be misleading, however, to infer that this homogeneous unity is a given, nor is there a natural bond between its elements or between the visible and the audible. The spectator connects sound and image, synthesizing them into an audiovisual field, because this is what he or she has learned to do. The audiovisual field in Nauman's art is, in addition, produced artificially and with the help of media technology. Rather than embedding its objects into the habitual audiovisual field of the spectator, it creates disjunctive syntheses that relate seeing and hearing only through disconnecting them. In this process, the automated, unreflected, and self-evident manner that is inherent to everyday perception can be actively experienced. The double exploration of body techniques and perception thus does not result in creating scientific knowledge, but it creates experiences this knowledge refers to.

BODY TECHNIQUES

In 1968, Nauman made a video entitled *Slow Angle Walk (Beckett Walk)*.[5] No formatting was added to that produced by the basic technical conditions of the time. The

---

[5] See Coosje van Bruggen, *Bruce Nauman* (New York, 1988), 235–40; Kathryn Chiong, "Nauman's *Beckett Walk*," *October* 86 (Fall 1998): 63–81.

video is black-and-white because this was the technological standard in the 1960s, and it lasts one hour because this was the length of a standard tape. The camera has a fixed position and thereby determines the frame. This technical feature, however, introduces a first disorientation for the spectator. Part of the activity Nauman performs takes place within this frame, and part of it outside. The viewer cannot see everything and misses some parts of the walk.

Nauman described his walking protocol as follows:

> These steps are made by raising the leg, without bending the knee, until it is at a right angle to the body, then swinging 90 degrees in the direction indicated in the diagram. . . . The body then falls forward onto the raised foot and the other leg is lifted to again make a straight line with the body (which now forms a T over the support leg). The body swings upright with the non-support leg swinging through the vertical and into the 90-degree position, as at the start, and proceeds into the next 90-degree position, as at the beginning. Three step-turns to the right and then three step-turns to the left will advance you two paces—each three steps advances you one step.[6]

This way of walking was inspired by Samuel Beckett's novel *Watt* (1953); hence the video's title.[7] Beckett's protagonist Watt walks like this:

> Watt's way of advancing due east, for example, was to turn his bust as far as possible towards the north and at the same time to fling out his right leg as far as possible towards the south, and then to turn his bust as far as possible towards the south and at the same time to fling out his left leg as far as possible towards the north, and then again to turn his bust as far as possible towards the north and to fling out his right leg as far as possible towards the south, and then again to turn his bust as far as possible towards the south and to fling out his left leg as far as possible the north, and so on, over and over again, many many times, until he reached his destination, and could sit down. So, standing first on one leg, and then on the other, he moved forward, a headlong tardigrade, in a straight line. The knees, on these occasions, did not bend. They could have, but they did not. No knees could better bend than Watt's, when they chose, there was nothing the matter with Watt's knees, as may appear. But when out walking they did not bend, for some obscure reason.[8]

Nauman imitates this manner of walking, transposing it into a physical activity, which is recorded and presented on videotape; however, Nauman's *Beckett Walk* does more than realize the way of walking Beckett invented. The video also translates the way in which this gait is textually represented. Beckett stripped the body technique of walking of its self-evidence and described it like a mechanically executed program. Although a person is walking here, this person—Watt—does not seem to walk by himself, but the movement seems to be driven by some unknown mechanism. The movement of legs in Watt's gait is described almost geometrically; the natural gait, in contrast, defies calculation.

From its very beginning, scientific research into human locomotion focused on the calculus of movement. But as it soon turned out, it is extremely difficult to describe what exactly happens while walking. In the eighteenth century, ways of walking were explored in the Prussian military. Tactical issues guided the search for an efficient

[6] Van Bruggen, *Bruce Nauman* (cit. n. 5), 115.
[7] See Samuel Beckett/Bruce Nauman, *Kunsthalle Wien*, ed. Michael Glasmeier (Vienna, 2000), 161.
[8] Samuel Beckett, *Watt*, new ed. (London, 1998), 28–9.

way of walking, resulting in a walking style with straight knees. This matched the
interest of the Prussian infantry in a regular and efficient way of movement that could
be implanted in the body through drilling and exercise. Later, the military success of
Napoleon's army incited a new interest in walking. While the Prussian military had
forced the soldiers' bodies into the desired movement, now the spontaneous physiology of walking was used for developing efficient marching styles. Investigating these
issues, physiologists encountered the natural self-regulation in walking movements.[9]

In 1836, Wilhelm and Eduard Weber postulated in their groundbreaking book, *Mechanics of the Human Walking Apparatus: An Anatomico-physiological Investigation*, that it is impossible to give a theory of human movement. "It can perhaps be
questioned," the Webers wrote, "whether a theory of walking and running can be
provided at all, since we are not walking machines, and these movements can be altered in many ways by our free will."[10] The daily use of our legs happens unconsciously and automatically. One does not know what one is doing while walking:
"Man binds his movements to certain rules, even if he cannot express these rules in
words. These rules are based totally on the structure of his body and on the given external conditions."[11]

In their experiments, the two brothers found a very complicated but regular movement of human legs that one can perform even if tired or dead. They wrote:

> Even in a corpse, this swinging movement can be generated by pushing the legs, given
> that the stiffness caused by rigor mortis has passed or the stiffened muscles have been
> cut through. . . . From these circumstances it becomes evident that the equal duration of
> these oscillations is caused by gravity automatically, without involving our will.[12]

The brothers demonstrated that the movement of the leg while walking is not that of
a single pendulum, but that of a double pendulum swinging both from the hip and
from the knee. This, however, was a mechanical figure that defied calculation even
if it was governed by laws of mechanics. Although the human will could alter the
regular movement of the human leg, humans were apparently unable to perceive the
mechanics of the leg as a double pendulum. Therefore it was impossible to understand—whether through external or internal observation—precisely how the movement happens.

During the nineteenth century, physiological research continued to investigate the
human gait, now recording the movement with the help of technical media such as
chronophotography and, later, film. Chronophotography and slow-motion cinematography altered our understanding of movements.[13] In the nineteenth century, physi-

[9] Mary Mosher Flesher, "Repetitive Order and the Human Walking Apparatus: Prussian Military
Science versus the Webers's Locomotion Research," *Annals of Science* 54 (1997): 463–87.

[10] Weber and Weber, *Mechanik der menschlichen Gehwerkzeuge: Eine anatomisch-physiologische
Untersuchung* (Göttingen, 1836), v; translation mine; see also Weber and Weber, *Mechanics of the
Human Walking Apparatus*, trans. Paul Maquet and Ronald Furlong (New York, 1992); Friedrich Kit-
tler, "Man as a Drunken Town Musician," *Modern Language Notes* 118, no. 3 (2003): 637–52.

[11] Weber and Weber, *Mechanik der menschlichen Gehwerkzeuge* (cit. n. 10), vi.

[12] Ibid., 19.

[13] See François Dagognet and Etienne Jules Marey, *A Passion for the Trace* (New York, 1992); Marta
Braun, *Picturing Time: The Work of Etienne Jules Marey (1830–1904)* (Chicago, 1992); Andreas
Mayer, "Autographien des Ganges: Repräsentation und Redressement bewegter Körper im 19. Jahrhundert," in *Kunstmaschinen: Spielräume des Wissens zwischen Ästhetik und Wissenschaft*, ed. Mayer
and Alexandre Métraux (Frankfurt am Main, 2005), 101–38.

ologist Emil du Bois-Reymond told his colleague Wilhelm Weber that he had observed a peculiar effect in a chronophotographic registration of human walking:

> The place where one rests for a short time on both feet certainly looks completely as painters have always portrayed walking people, except that in the middle of the step, where the so-called moving leg swings past the standing leg, the most strange and even ludicrous sight appears: like a drunken town-musician, he seems to trip over his own feet, and no one has ever seen a walking man in such a position.[14]

As Du Bois-Reymond watched the recorded motion, walking changed into something completely different from its everyday perception.

In the course of this research, the human gait lost its self-evidence. Physiology discovered in the movement of the legs a phenomenon that was unfamiliar to the point of seeming unreal. Soon after, sociologists and ethnographers began questioning the distinction between natural and artificial ways of performing human activities. Eventually, in a lecture given at the Société de psychologie in Paris in 1934, French ethnographer and sociologist Marcel Mauss defined his concept of "techniques of the body" as follows: "I use the term 'body techniques' in the plural. . . . By this expression I mean the ways in which, from society to society, men know how to use their bodies."[15] This idea had come to Mauss, he said, while he was reading the article on swimming from the 1902 edition of the *Encyclopaedia Britannica*. As Mauss observed, swimming styles had changed rapidly over the previous thirty years:

> Our generation has witnessed a complete change in technique: we have seen the breast-stroke with the head out of the water replaced by the different sorts of crawl. Moreover, the habit of swallowing water and spitting it out again has gone. In my days, the swimmers thought of themselves as a kind of steamboat. It was stupid, but in fact I still do this: I cannot get rid of my technique.[16]

The historical change in swimming styles and the introduction of the front crawl in Europe disturbed the common understanding of the body as behaving in a natural way and performing movements naturally. This invited Mauss to speculate also on the formative powers that are capable of altering the body's techniques. As one example, he suggested the influence of movies on the unconscious conception and execution of movement:

> I was ill in New York. I wondered where I had seen girls walking the way my nurses walked. I had time to think about it. At last I realized that it was in movies. Returning to France, I noticed how common this gait was, especially in Paris; the girls were French and they too were walking in this way. In fact, American walking fashions had begun to arrive over here, thanks to the movies. This was an idea I could generalize. The positions of the arms and hands while walking form a social idiosyncrasy—they are not

[14] Du Bois-Reymond, "Naturwissenschaft und bildende Kunst: Zur Feier der Leibniz-Sitzung der Akademie der Wissenschaften zu Berlin am 3. Juli 1890 gehaltene Rede," in *Reden von Emil du Bois-Reymond in zwei Bänden*, vol. 2, 2nd rev. ed., ed. Estelle du Bois-Reymond (Leipzig, 1912), 390–425, on 407, quoted after Kittler ("Man as a Drunken Town Musician" [cit. n. 10], on 650), trans. Jocelyn Holland.

[15] Mauss, "Techniques of the Body," in *Incorporations*, ed. Jonathan Crary and Sanford Kwinter, Zone 6 (New York, 1992), 455–77, on 455.

[16] Ibid., 456.

simply a product of some purely individual, almost completely psychic, arrangements and mechanisms.[17]

For Mauss, the "natural" state of the human body was no longer evident; although difficult to change, the techniques of the body were not anthropological universals—hence the necessity for human beings to have a mimetic model when learning to walk.

PERCEPTION AS A TECHNIQUE OF THE BODY

It is impossible to decide in Nauman's *Beckett Walk* whether the movement should be seen as a natural or artificial activity or technique of the body. Furthermore, it is impossible to reduce the video to a mere presentation or documentation of the performance. Rather, Nauman introduced certain features into his use of video technology that made the view unfamiliar to the spectator of the late 1960s. He mounted the camera in such a way that the vertical axis was rotated by ninety degrees. With reference to French philosopher Maurice Merleau-Ponty, the effect of the rotated camera can be explained as follows:[18] Because we are bodies—not because we have a body—our perception is always oriented. There is a great difference between a face that is seen in normal position and a face that is seen upside down; "for the subject of perception the face seen 'upside down' is unrecognizable." The subject perceives an object as dependent on its own and the object's position in space. Therefore the meaning of an object changes when it is not oriented in the usual way with relation to the body. Thinking, in contrast, abstracts from the actual position of the object in space, as Merleau-Ponty explains: "For the thinking subject a face seen 'the right way up' and the same face seen 'upside down' are indistinguishable."[19] He thereby distinguishes the object, which remains the same, from the object's perception, whose meaning can get lost when its orientation changes. He even claims: "To invert an object is to deprive it of its significance. Its being as an object is, therefore, not a being-for-the-thinking subject, but a being-for-the-gaze which meets it at a certain angle, and otherwise fails to recognize it."[20] To see a face "is to take a certain hold upon it, to be able to follow on its surface a perceptual route with its ups and downs." And this general orientation in space is "not a contingent characteristic of the object, it is the means whereby I recognize it and am conscious of it as an object."[21] Obviously, it is possible to recognize a face when it is turned upside down or to bend the head to the side without the world becoming unrecognizable; however, Merleau-Ponty argues that the perception of objects is irreducibly anchored in spatial orientation, and everything we perceive refers to this orientation. Orientation is not just an additional aspect of our perception, it is a capacity human beings never lose, even when perceiving an inverted or rotated object: after a while things readjust.

In Nauman's works for video, the body techniques, their staging, and the presentation on a video screen intermingle. Both perception and body techniques relate to orientation in space, but in Nauman's videos, this relation is interrupted. Nauman's

---

[17] Ibid., 457–8.
[18] Nauman denied that he ever read Merleau-Ponty. See Lorraine Sciarra, "Bruce Nauman, January, 1973," in Kraynak, *Please Pay Attention Please* (cit. n. 1), 166.
[19] Merleau-Ponty, *Phenomenology of Perception*, trans. Colin Smith (New York, 2003), 294.
[20] Ibid., 294–5.
[21] Ibid., 295.

videos present their objects in such a way as to elude "natural" orientation or, more precisely, the habitual orientation of the body. To begin with, Nauman executes the body techniques in such a way as to disorient the viewer. The exaggerated, geometrical movements of the Beckett Walk do not correspond to the habitual placement of the walking body in space. Turning the camera at an angle of ninety degrees makes the body's movements appear doubly transformed. What the spectators see does not easily translate into a normal view of the walking body. Still, seeing a body walking evokes the desire to understand these movements as connected to the ground. Nauman thus walks for an altered gaze: the meaning of walking is disturbed. The camera position transforms the principle of walking into a mode of perception. The body's technique of walking has, as the Weber brothers showed, its fundamental condition in gravity; wherever a body is located and in whatever way the body is positioned, it always feels gravity. Nauman is working against this condition: the legs do not swing like a pendulum; they are moved intentionally. In addition, through the rotation of ninety degrees, the vertical and horizontal axes change places. Nauman's way of walking without bending the knees emphasizes vertical and horizontal lines. He works against gravity with his walking, and the camera takes this up and continues it by disorienting the perception of the viewer, again by ninety degrees.

At this point, it will be helpful to introduce the notion of a *sensorimotor schema*, which was prepared by Henri Bergson, introduced as a term by Jean Piaget, and further developed by Gilles Deleuze.[22] Whoever learns to see also learns that perception is continued into movement and that movement follows perception. When one person performs an activity, the perception of another person watching this activity is connected to that activity by a sensorimotor schema. The notion of sensorimotor schema, again, does not denote a natural connectivity between perception and movement, but their habitual coordination. This coordination must be learned; it also forms the backdrop for other perceptions that are not related to movements of the perceiver. The sensorimotor schema is an elastic band that does not break, even when the viewer is standing still and the movement viewed makes reference only to a virtual, hypothetical movement in the viewer's own body.

When Nauman's video confronts the spectator with a movement that defies immediate correlation to the viewer's perception, her perception gets correlated to such a virtual movement. Because the video disorients the viewer's perception, it also disturbs her sensorimotor schema. The viewer actively strives to establish the connection between her own perception and the viewed movement. Perception is thus not merely embodied, but is itself a technique of the body that, like any other technique of the body, is not naturally given but must be learned.

PERCEPTION OF THE AUDIOVISUAL FIELD

In a ten-minute black-and-white video that Nauman made in 1967–8, *Playing a Note on the Violin while I Walk around the Studio*, he walks to and fro, incessantly sawing at a violin. The camera rests in a fixed position, image and soundtrack run

22 See Bergson, *Matter and Memory*, trans. Nancy Margaret Paul and W. Scott Palmer (New York, 1929), 153–69; Piaget, *The Origins of Intelligence in Children*, trans. Margaret Cook (1953; repr., New York, 2001); Deleuze, *Cinema 1: The Movement-Image*, trans. Hugh Tomlinson and Barbara Habberjam (Minneapolis, 1986), 155–9, 197–204; Deleuze, *Cinema 2: The Time-Image*, trans. Tomlinson and Habberjam (London, 2005), 1–12.

asynchronously, and now and then Nauman disappears from the frame. He reported in an interview, "In one of those first films, the violin film, I played the violin as long as I could: I don't know how to play the violin, so it was hard, playing on all four strings as fast as I could as long as I could. I had ten minutes of film and ran about seven minutes of it before I got tired and had to stop and to rest a little bit and then finish it." When asked by the interviewer, "But you could go on longer than the ten minutes?" Nauman replied, "I would have had to stop and rest more often. My fingers got very tired and I couldn't hold the violin any more."[23]

In *Violin Tuned D.E.A.D.* of 1969, Nauman tackled the relation of time and bodily movement in his violin playing from a different angle: "I wanted to set up a problem where it wouldn't matter whether I knew how to play the violin or not. What I did was to play as fast as I could on all four strings with the violin tuned D, E, A, and D, rather than the customary G, D, A, and E."[24] This video lasts for one hour. The visible effort of the player, Nauman explained, conveys to the viewer the sincerity of his activity: "If you are honestly getting tired, or if you are honestly trying to balance on one foot for a long time, there has to be a certain sympathetic response in someone who is watching you."[25] The extended duration of the performance not only exposes the viewer to a sensorimotor schema, but it also leads to the player's exhaustion, which in turn becomes visible and thereby even more strongly appeals to the viewer to comprehend the sensorimotor schema. In the video, Nauman turns his back to the camera, the axis of which is again rotated by ninety degrees. As he explained to the interviewer, the violin strings are tuned to the pitches D, E, A, and D, and he plays the open strings from the lowest to the highest in one down-bow, that is to say, in the particular order of the letters D, E, A, D, producing the sound in one movement.

In the video, Nauman is seen only from the back. According to the phenomenological analysis of perception, a viewer is able to imagine the side of an object that is hidden from view. The object's reverse side is made "co-present," by "analogous apperception," as this process was described by Edmund Husserl: "An appresentation occurs even in external experience, since the strictly seen front of a physical thing always and necessarily appresents a rear aspect and prescribes it for a more or less or determined content."[26] Alfred Schütz explained the ability to appresent an object as follows:

> When we apperceive an object of the external world, we see, strictly speaking, only the front aspect of it. This perception of the visible front also contains the analogous apperception of the unseen rear aspect. The latter is, however, only an empty expectation of what we would perceive if we would turn the object around or walk behind it. This anticipation is learned and based on our past experiences of normal objects of this kind.[27]

Nauman dealt with the problem inherent to appresentation in a short, untitled text that sketches a hypothetical performance that is supposed to detach the gaze from its grounding in the body schema.

---

[23] Sharp, "Interview with Bruce Nauman" (cit. n. 1), 142.

[24] Ibid., 147.

[25] Ibid., 148.

[26] Husserl, *Cartesian Meditations: An Introduction to Phenomenology*, trans. Dorion Cairns (Dordrecht, 1999), 109.

[27] Schütz, "Symbol, Reality and Society," in Schütz, *Collected Papers I: The Problem of Social Reality*, ed. Maurice Natanson (Boston, 1982), 287–311, on 294.

A person enters and lives in a room for a long time—a period of years or a lifetime.
One wall of the room mirrors the room but from the opposite side: that is, the image room has the same left-right orientation as the real room.
Standing facing the image, one sees oneself from the back in the room, standing facing a wall.
There should be no progression of images: that can be controlled by adjusting the kind of information the sensor would use and the kind the mirror wall put out.
After a period of time, the time in the mirror room begins to fall behind the real time—until after a number of years, the person would no longer recognize this relationship to his mirrored image. (He would no longer relate to his mirrored image or delay of his own time.)[28]

Similar to the situation in René Magritte's painting *Réproduction interdite*, the person in the room would see his own back view in a mirror, which in addition would follow his own movements. He would thus see himself as seen by others. He would by no means become disembodied, but his visual perspective would be placed outside of his own body, and he would have to learn to see himself from outside and to regain consciousness of his own body's coordinates.

The function undertaken by the mirror room is accomplished by the technically produced audiovisual field in *Violin Tuned D.E.A.D.*: the audiovisual field detaches perception from its anchoring in the body, only to place the—now unfamiliar—perception back in the body again. Turning his back to the camera, Nauman creates an expectation of how his front would appear. The camera positioning disorients perception and unsettles the meaning of what can be seen. The back view of the player stimulates the appresentation of his front view. As long as the viewer sees the source of the sound, that is to say, the violin player, the habitual image of a violin player in a frontal position is automatically coproduced. The ear, in turn, processes the information that comes out of the loudspeaker: although it does not hear the sound of the original source, but the recorded sound, it has learned to infer from the latter to the former. The extent to which the body shapes musical notes became evident when electronic music confronted listeners with an alternative. In electronic music the production of sound does not have to make its way through the body of the musician to become audible. The electronic music of the 1950s and '60s allowed the perception for the first time of a new class of audible objects that were not a part of the circle that connected the listeners' bodies, via sound, with the musicians' bodies. Electronic music thereby revealed the hidden anthropomorphism in any sound produced by the dispositive of player and instrument.[29] The loudspeaker further defines the spatiality of hearing because it is not just a neutral transmission device: any sound that is heard will thus be constructed, resulting from a complicated psychophysical calculus. While sound transmission disconnects the sound from the situation where it originated, hearing engages in reconstructing this situation, binding it back to a source and the source's position.

Nauman's violin video dissolves the correlation of bodily movement, hearing, and seeing as it has been established in everyday experience. Although viewers think that

---

[28] Nauman, "Untitled, 1969," in Kraynek, *Please Pay Attention Please* (cit. n. 1), 55.

[29] See Pierre Boulez, "À la limite du pays fertile," in Boulez, *Points de repère*, vol. 1, *Imaginer*, new ed., ed. Jean-Jacques Nattiez and Sophie Galais (Paris, 1995), 315–30; Julia Kursell and Armin Schäfer, "Kräftespiel: Zur Dissymmetrie von Schall und Wahrnehmung," *Zeitschrift für Medienwissenschaft* 2, no. 1 (2010): 24–40.

the sound source is visible, they can only speculate about the act of sound production; although they hear the sound, they have to infer from the loudspeaker to the sound source. Nauman's visible body, however, complicates these inferences. The viewers will relate what they see to what they hear, correlating sound and movements by referring them to their own body. Here again Nauman's work can be seen to engage with the history of the experimental study of bodily and mental processes. Starting in the second half of the nineteenth century, physiology and psychology investigated the perception of time. As was found out, the perception of time is actively coupled to the perception of one's bodily actions: a listener measures acoustic events by comparing them to the movements of his or her own body. The perception of temporal relations is carried out through so-called registering movements of one's body.[30] The perception of the sequence of notes D, E, A, D in Nauman's film thus takes two steps. The listener continues the perceived sequence in movements produced by his or her own body to perceive the difference between the perceived sequence and his or her own movements. This difference becomes the actual referent of the temporal perception.

Tuning the violin to D, E, A, D instead of the usual G, D, A, E, Nauman creates an indefinite, but intense, semantics. He does not apply the common bow technique, which he has not learned, but produces the sounds with a high expenditure of energy. Although the resulting sounds are shaped by a bodily gesture, they remain far removed from the ideal of the beautiful tone. This exposes the body in a new way. In the video, he repeats the same bowing gesture over and over for a seemingly endless time. This is a standard situation in pedagogical practice; that of the pupil who repeats the "correct" movement in order to appropriate it. In traditional violin pedagogy, special attention is paid to the generation of the tone: students learn to construct and control a "beautiful tone," considered to be the result of a specific movement that must be acquired and trained.[31] Players are urged to control their movements with the ear, which means that they have to build a feedback loop incorporating the body's appearance, movement, and hearing. The player's body and the instrument produce a tone whose quality in turn allows a listener to infer back to the way in which the tone's production occurred. Nineteenth-century instrumental pedagogy developed an exercise routine that decomposed playing into elementary movements and drilled the movements until they were automatic.[32] If instrumental lessons had to surmount difficulties in order to discipline the body, this was not so much because obstinacy or the lack of assiduity in the pupils compromised the effort, but because fatigue jeopardized the playing as a whole. The disciplining of the body finds its greatest enemy in fatigue, which cannot simply be passed over or broken. The pupil has to accommodate his or her own body and its unavoidable fatigue. Nauman, however, takes his body to the limit of discipline and fatigue, which is exhaus-

[30] Friedrich Schumann, "Über die Schätzung kleiner Zeitgrößen," *Zeitschrift für Psychologie und Physiologie der Sinnesorgane* 4 (1893): 1–69. See also (for temporal and spatial relations) Brian Massumi, *Parables for the Virtual: Movement, Affect, Sensation*, Post-contemporary Interventions (Durham, N.C., 2002), 61–2.

[31] See Myles W. Jackson, *Physicists, Musicians, and Instrument Makers in Nineteenth-Century Germany* (Cambridge, Mass., 2006), 255–9; Wolfgang Scherer, *Klavier-Spiele: Die Psychotechnik der Klaviere im 18. und 19. Jahrhundert* (Munich, 1989); Grete Wehmeyer, *Carl Czerny und die Einzelhaft am Klavier* (Kassel, 1983). For pedagogy and training of the body in physics, see Andrew C. Warwick, *Master of Theory: Cambridge and the Rise of Mathematical Physics* (Chicago, 2003).

[32] See Otakar Ševčík, *The Little Ševčík: An Elementary Violin Tutor* (Miami, 1988), 5–6 (1st ed. Miami, 1901).

tion. The movements are repeated endlessly, become recursive, and form a rhythm of autoreflexivity. Inasmuch as his movements lack a destination, they do not obey a rule that would indicate the point at which they must end.

PLAYING ON UNTIL EXHAUSTION

In the 1860s, Italian physiologist Angelo Mosso was the first to investigate fatigue systematically.[33] He invented the so-called ergograph, a device that recorded the working performance of experimental subjects, and he was able to demonstrate that one of the most specific characteristics of any individual's life is the way he or she gets tired. The phenomenon of fatigue is governed by natural laws; it is inevitable that everyone gets tired, but everyone gets tired in an individual way. He also showed that fatigue alters the personality once a certain threshold is transgressed.

> Extreme fatigue, whether intellectual or muscular, produces a change in our temper, causing us to become more irritable; it seems to consume our noblest qualities—those which distinguish the brain of civilized from that of savage man. When we are fatigued we can no longer govern ourselves, and our passions attain to such violence that we can no longer master them by reason. Education, which is wont to curb our reflex movement, slackens the reins, and we seem to sink several degrees in the social hierarchy.[34]

Mosso's investigations awakened the interest of psychiatry. German psychiatrist Emil Kraepelin and his assistant William Halse Rivers began to experiment with fatigue and its consequences. They discovered that fatigue was unavoidable and that it protected an organism against exhaustion: "No doubt, fatigue begins at the same time as the action itself. To avoid the occurrence of fatigue would mean to renounce work itself. Even without any work, we could not avoid getting tired."[35] Whatever you are doing or not doing, you will get tired, for it is not work alone that is tiring, but idleness, too. The first sign of fatigue Kraepelin and Rivers determined was a rise in the number of mistakes made by an experimental subject performing some task. It is possible to counteract fatigue to a certain degree by resting, eating, drinking, and sleeping, on the one hand, and by regular training and effort of will, on the other. The signs given by the body should by no means be disregarded. Fatigue that eventually leads to sleep is a way for the body to protect itself from the more extreme and potentially injurious state of exhaustion. Normally, the body is capable of governing itself.

> Prolonged work produces fatigue and with it difficulty of further application. Up to a certain degree, this fatigue, which may be considered as a safeguard against overwork, may be overcome by an increased exertion of will power, which in long and fatiguing work gives rise to a feeling of "increased effort."[36]

[33] See Anson Rabinbach, *The Human Motor: Energy, Fatigue, and the Origins of Modernity* (Berkeley, Calif., 1992), 133–42; Philipp Felsch, *Laborlandschaften: Physiologische Alpenreisen im 19. Jahrhundert* (Göttingen, 2007).

[34] Mosso, *Fatigue*, trans. Margaret Drummond and W. B. Drummond (New York, 1904), 238.

[35] W. H. R. Rivers and Emil Kraepelin, "Ueber Ermüdung und Erholung," *Psychologische Arbeiten* 1 (1896): 669; my translation.

[36] Emil Kraepelin, *Clinical Psychiatry: A Text-book for Students and Physicians*, abstracted and adapted from the 7th German ed. of Kraepelin's *Lehrbuch der Psychiatrie* by A. Ross Diefendorf, new ed. (New York, 1915), 148.

The self-protection of the working body can be suspended. Effort of will can compensate for fatigue and does not lead to exhaustion. This explanation, however, was partly motivated by Kraepelin's desire to avoid concluding that effort of will might in fact lead to exhaustion, because will was considered at the time to be within the framework of reasonable behavior.

> While the increased exertion of the will can for a time balance the effects of fatigue through an increased expenditure of power, the effects of fatigue ultimately gain the upper hand and force one to cease work. The first indications of exhaustion are when, under certain conditions, the increased exertion of will continues for some time in spite of the uncomfortable feeling of fatigue. This is what happens when work is performed under intense emotional excitement. The signs of fatigue, which call for relaxation, either do not appear or are overwhelmed, and work is prolonged beyond a permissible degree.[37]

Passion, in contrast to will, drives one's work level over the threshold and leads to exhaustion and self-damage. Therefore, the performing arts such as theater and music can be dangerous when practiced with great passion.[38]

Scientists gave a number of different explanations for the psychophysiological processes of fatigue and exhaustion. Fatigue was seen as resulting from the consumption of a substance or as a self-intoxication by the products of metabolism. Exhaustion was defined as the destruction of the foundations of psychic processes due to excessive consumption or insufficient recreation. Notwithstanding the different explanations given by the scientists, they all agreed on the danger that results from exhaustion. Exhaustion causes a permanent reduction in working energy. Psychiatrists warned that increasing fatigue was the first step toward the self-destruction of the nervous system through its own activity. Therefore, they considered the exhausted person to be in a dangerous state that they compared to mental illness, especially to psychosis. The consequences of exhaustion are, in the wording of the psychiatrists, a dissociation of the personality, a loss of personality, and an abolition of the self. Exhausted subjects are only a mechanically driven bundle of functions. They are in danger of serious damage and present a risk to their environment. The symptoms of exhaustion are not apathy, withdrawal from activity, and extinction of movements, but mere action. The danger for the exhausted lies in the fact that they continue their activity like an overheated machine performing idle motion or an idiot performing repetitive stereotypical movements.

Whereas the tired person is able to resume activity in a predictable way, it is uncertain whether and how the exhausted one can ever do so. Pausing or sleeping enables the tired to take up their activity anew or to continue their thoughts; they do not quit the activity, but suspend or defer it. Therefore, lying down prepares the tired to fall asleep, and sleeping, in turn, prepares them to resume the interrupted activity. Yet exhaustion is something different. Deleuze discussed the role of exhaustion in Beckett's novels and plays and observed that "lying down is never the end, the last word, but rather the penultimate, and there is too much risk of being rested enough, if not to get up, at least to roll or to crawl."[39] The exhausted subject is beyond any calculus

[37] Ibid.
[38] See, e.g., James Kennaway, *Bad Vibrations: The History of the Idea of Music as a Cause of Disease* (Farnham, 2012).
[39] Gilles Deleuze, "The Exhausted," trans. Anthony Uhlmann, *SubStance* 24, no. 3, issue 78 (1995): 3–28, on 6.

of activity; she would do anything to continue an activity rather than stop for recreation. The tired person will rely on her habit and perform the body technique in the way it has been learned. She will abandon her movements at some moment, but the exhausted person goes on. The tired person, according to Deleuze, "is no longer prepared for any possibility (subjective): he therefore cannot realize the smallest possibility (objective). . . . The tired has only exhausted realization, while the exhausted realizes all of the possible."[40]

Exhaustion is not just an amplification of tiredness, but a different state. The tired person wants to continue his activity and therefore rests in order to take it up again in the habitual way. The automatism in body techniques guarantees continuation: one who performs in a certain way can rely on automatism and not be forced to change the manner of movement. The exhausted person will perform an activity even if she makes mistakes and loses control or will perform the activity in an unpredictable way. When, in the state of exhaustion, the performer begins to lose control of the activity, the activity will be carried out in an indefinite way, and the economy of means and ends begins to be undermined. When arms and fingers begin to hurt, and the violin threatens to fall, even acquired techniques can no longer be executed. If the player still wants to continue, she has to think of some new technique. Playing the violin in a new way does not mean here that one exceptional occurrence is to be singled out, but rather that playing as a whole is being transformed. At any moment, the next movement is the only aim. The exhausted player cannot rely on acquired body techniques because the self-regulation of the body has broken down. Therefore, new forms or ways of movement must be invented in order to continue. As Beckett's Unnamable puts it, "That the impossible should be asked of me, good, what else could be asked of me?"[41]

Exhaustion is an ambivalent state. It opens up the possibility of doing something new, but it also contains a lethal danger. In *Violin Tuned D.E.A.D.*, Nauman attains a state where violin playing becomes exhausting and where something unforeseen can occur. The new does not appear at a moment that can be indicated, at a distinct point in the notes he plays, or in the form of a new combination of tones, but it is something that interrupts the self-evidence of the playing. The automatism of executing a self-evident movement guarantees continuation in the playing of an instrument. A trained violinist who is able to continue can entrust continuation to her acquired automatism, even when improvising; it then becomes unnecessary to make decisions about whether to stop or to continue. This automatism does not include any rule governing when to stop; the aim of practicing is to continue. In this mode of automatic playing anything new that occurs consists of combination and variation of the acquired and the known.

Nauman's violin playing, in contrast, seems to be constantly endangered: at any moment, he may have to stop. Continuation is not just endangered; it becomes completely unclear in what way it can be possible for him to continue. Nauman cannot refer to acquired techniques. Rather, the repeated action shown in the video takes part in his process of movement acquisition. The principle that constitutes this way of playing lies at every moment in the actual present. This is emphasized by the use of sound: the long reverberation in the empty studio points, in this case, to the

[40] Ibid., 3.
[41] Samuel Beckett, *Molloy, Malone Dies, the Unnamable* (New York, 2003), 340.

synchronicity of what can be seen and heard. From this a kind of music emerges that does not have to make its way along the circular path through the body—the path that involves acquired abilities, techniques, and automatisms. Rather, the video registers an individual activity that provokes the viewer to observe whether and how long Nauman will be able to go on with his unruly and uneconomic practice of violin playing. This is an experimental use of musical body techniques that results in an inconceivable new piece of music.

However, the exhaustion leading to this music cannot be understood as a desirable state that eventually allows creativity to come into being. Exhaustion does not always lead to something new. More often, it leads to death. The title that Nauman chose for his violin performance unmistakably points to this: *Violin Tuned D.E.A.D.* Even though the arbitrary letters that name the tones turn into a meaningful word here, this pun does not acquire meaning from the symbolic operation alone, but also from Nauman's specific way of playing the violin. The title points to the disjunctive synthesis in the audiovisual field: just as perception synthesizes the heterogeneous orders of vision and audition by forcing them together into one field, so the title of the video connects unrelated orders. And so also the media technology of video generates a disjunctive synthesis, binding the heterogeneous orders of sound and image with a technical bond. The title thus creates a meaning by connecting the heterogeneous orders of notes, letters, and linguistic code. Nauman's title acts like a *symbolon*. This Greek word denotes a specific operation of concatenation, of aligning and fusing parts together. Originally, it was used for a ring of clay that could be broken into two parts but later joined into one whole again. The specific materiality of the clay and the way in which the ring was broken guaranteed that only the parts of the original ring could be fitted together with each other. Nauman's title, however, does not join fitting parts, as did the Greek *symbolon*, but heterogeneous parts. He offers no explanation of how cohesion might be guaranteed among the orders of sound, image, perception, and the body, which are forced together in his videos; instead, he points, however ironically, to the activity of the artist: *The True Artist Helps the World by Revealing Mystic Truths*, as the title of another of his works has it. These mystic truths are hidden in a ring of media technology, perception, and techniques of the body, which Nauman does not break in two, but rather shatters into innumerable splinters.

In his video *Violin Tuned D.E.A.D.* Nauman appears to be an anonymous figure whose identity—encoded, again, in the front view—remains in suspense. Turning his back to the camera, he avoids the classic position of the virtuoso. This posture, in combination with the tilted camera that immediately prostrates Nauman in the position of a tired and even exhausted man, interrupts the continuity of vision and audition. The sounds he produces in one way or another do not refer to a personal style but become detached from the body in a strange manner. The emergence of the new requires a paradoxical individuation: no matter how self-obsessed the player may be, he does not refer to himself while he plays. He plays something that points beyond himself, something that, however, does not make him bigger: playing is not intended to amplify the self.

Nauman's videos present body techniques in a way that strips them of their self-evidence while constituting an audiovisual field. The videos dissolve the correlation of bodily movement, hearing, and seeing as it has been learned by the viewers under specific, given conditions and reconfigure this interrelationship in a new way. In this process, the pedagogical calculus that typically adjusts the movements of playing the violin to hearing is undermined. Both Nauman's playing and the video technology efface the necessary points of reference for coordinating bodily movement and sound, seen and heard perception. The videos construct a disjunctive synthesis of sound and image. Vision and audition are disconnected from their anchoring in the body with the eventual result that the cultural semantics of violin playing is unsettled. This audiovisual field in Nauman's videos can no longer be conceived of as the retroactive defamiliarization of natural perception and its subsequent aesthetization. Rather, media technology, body techniques, and the habits of perception determine how vision and audition interact in the videos.

Camera Silenta:
Time Experiments, Media Networks, and the Experience of Organlessness

by Henning Schmidgen*

ABSTRACT

In order to communicate with isolated test subjects in physiological and psychological laboratories in the late nineteenth century, scholars such as Wilhelm Wundt, Edward W. Scripture, and Hendrik Zwaardemaker used modern technologies, in particular telegraphy. In a similar vein, Marcel Proust equipped his apartment with a soundproof room and a network of cables and switches in order to conduct his famous "research" on lost time. The combined use of the *camera silenta* and advanced communication technologies turned time experts around 1900 into spiders: without ears, eyes, or nose, they were waiting at the edge of an extended web of simultaneities for the slightest vibrations their bodies could receive. With Gilles Deleuze and Félix Guattari, one could say that they experienced states of "organlessness."

> Weil das Gehör auf die Zeit einschlägt, so begleitet es alle Verstandesvorstellungen vom Object, bringt aber keine Vorstellung des Objects hervor, also ist bei ihm nichts als Empfindung und Form der Veränderung, nicht aber die Erscheinung eines Gegenstandes.
>
> —Immanuel Kant[1]

Anybody who knows John Cage knows the story. When Cage came to Cambridge, Massachusetts, at the beginning of the 1950s, he visited an anechoic chamber at Harvard University. He entered the room expecting to experience absolute silence. To his surprise two sounds became noticeable: one high and one deep. Cage asked the engineer responsible for the room what the origin of these sounds might be. The answer was, "The high one was your nervous system in operation, the low one was your blood in circulation."[2]

---

* Institut für Information und Medien, Sprache und Kultur (I:IMSK), Universität Regensburg, Universitätsstr. 31, 93053 Regensburg, Germany; henning.schmidgen@ur.de.
Unless otherwise credited, all translations are my own.

[1] Kant, "Reflexionen Kants zur kritischen Philosophie," in *Reflexionen Kants zur Anthropologie*, ed. Benno Erdmann (Leipzig, 1882), 82. The passage might be translated as follows: "Since the ear responds to time, it accompanies all our rational ideas of the object, although it itself creates no idea of the object. There is nothing in hearing but the feeling and the form of change, not the image of an object."

[2] Cage, "How to Pass, Kick, Fall and Run," in *A Year from Monday: New Lectures and Writings* (Middletown, Conn., 1967), 133–41, on 134. See also Cage, "Composition as Process," in *Silence: Lectures and Writings* (Middletown, Conn., 1961), 18–34, on 32, and Cage and Daniel Charles,

© 2013 by The History of Science Society. All rights reserved. 0369-7827/11/2013-0009$10.00

OSIRIS 2013, 28 : 162–188

162

Cage drew the conclusion from this that there was no such thing as absolute silence: "Something is always happening that makes a sound."[3] It was shortly after this experience that he composed the piece that everyone who is familiar with John Cage also knows: 4'33". On 19 August 1952, it was premiered by David Tudor in Woodstock, New York.

While this story is well known, it is still uncertain in precisely which room it took place. At that time Harvard University was equipped with two anechoic rooms. One had been set up in 1943 and belonged to the Cruft Laboratories, a research lab for physics founded in 1913 where, in the early 1940s, the legendary Mark I computer began its work. Frederick Vinton Hunt used this anechoic room to test acoustic devices for their usability in military contexts (see fig. 1). The other anechoic room was in the Psychoacoustics Laboratory located in Harvard's Memorial Hall. The laboratory had been set up in 1940 as an extension of the research sites for psychology, which had existed at the university since the days of Hugo Münsterberg, who worked at Harvard from 1892 to 1916. The anechoic room set up there was used by Stanley Smith Stevens to investigate the psychoacoustics of hearing and human communication (see fig. 2).

Which of the rooms was Cage in? The fact that the answer to his question was that the sounds he heard were caused by the circulation of his blood and the working of his nervous system suggests that he was in the Psychoacoustics Laboratory, but the fact that it was an engineer who answered his question would seem rather to point to the Cruft Laboratories.[4]

The decisive point here, however, is not where Cage had his experience of "roaring silence," but that it became the foundation for his later work as an experimental composer. In the early 1940s Cage had at first based his work on a dialectical concept of silence, which he used above all to penetrate the structures of musical material. At this point sound and silence were mutually exclusive phenomena for Cage: their succession determined essentially what music was. Not much later he developed a spatial idea of silence that diverged from his original idea. It was his involvement with the music of Erik Satie that played a critical role here. Silence related now above all to surrounding noises, which were heard by certain listeners in concrete situations.

It was only after his Harvard experience that Cage came to associate silence with unpredictability and, to that extent, with time. After his visit to the anechoic room the idea emerged for him that the sounds that filled the silence were associated with each other by the absence of intention. The sounds of silence had in common that they followed no defined direction, determination, or meaning. As a result, fluctuating noise became the zero state both of hearing music and of making music, a state in which there was a constant openness with respect to what happened next—which also meant that silence and living time had moved closer to each other: "The common

---

"Third Interview," in *For the Birds: John Cage in Conversation with Daniel Charles* (Boston, 1981), 101–20, on 115–6.
    [3] John Cage, "45 Minutes for a Speaker," in *Silence* (cit. n. 2), 146–93, on 191. See also ibid., 152: "Silence, like music, is non-existent. There are always sounds."
    [4] David Revill, *The Roaring Silence: John Cage, a Life* (New York, 1992), 162–86 (there is also an illustration here of the anechoic room of the Cruft Laboratories [fig. 9]). A description of the soundproof room of the Psychoacoustics Laboratory can be found in Stanley S. Stevens and Edwin G. Boring, "The New Harvard Psychological Laboratories," *American Psychologist* 2 (1947): 239–43.

Figure 1. *Interior of the anechoic room at the Cruft Laboratory of Physics, Harvard University, 1949.* Education, Bricks and Mortar: Harvard Buildings and Their Contribution to the Advancement of Learning *(Cambridge, Mass., 1949), 61.*

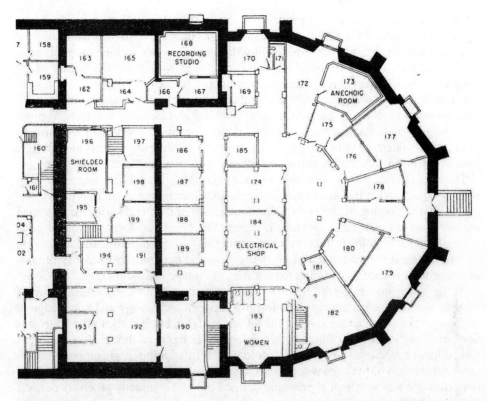

Figure 2. *Plan of the Harvard Psychology Laboratory with anechoic room (no. 173, top right), 1947. Stevens and Boring, "New Harvard Psychological Laboratories" (cit. n. 4), 241.*

denominator [of music and life] is zero, where the heart beats (no one *means* to circulate his blood)."[5]

In recent years the spaces in which science and art are produced and received have come increasingly to occupy the interest of scholars in sociology, history, and cultural studies.[6] Whereas the *camera obscura* and the various optical media associated with it have been the subject of numerous studies,[7] to date no comparable investigations have been devoted to its acoustical counterpart, the *camera silenta* and similar

[5] Cage, "Erik Satie," in *Silence* (cit. n. 2), 76–82, on 80. See also the article by Eric de Visscher, "'So etwas wie Stille gibt es nicht . . .': John Cages Poetik der Stille," *MusikTexte* 40/41 (1991): 48–54; Daniel Charles, "Über die Nullzeit-Poetik von John Cage," in *Musketaquid: John Cage, Charles Ives und der Transzendentalismus* (Berlin, 1994), 95–118.

[6] See, e.g., Brian O'Doherty, *Inside the White Cube: The Ideology of the Gallery Space* (Santa Monica, Calif., 1986); Douglas Crimp, *On the Museum's Ruins* (Cambridge, Mass., 1993); Caroline A. Jones, *Machine in the Studio: Constructing the Postwar American Artist* (Chicago, 1996); Hans-Jörg Rheinberger, Michael Hagner, and Bettina Wahrig-Schmidt, eds., *Räume des Wissens: Repräsentation, Codierung, Spur* (Berlin, 1997); Peter Galison and Emily Thompson, eds., *The Architecture of Science* (Cambridge, Mass., 1999); Thompson, *The Soundscape of Modernity: Architectural Acoustics and the Culture of Listening in America, 1900–1933* (Cambridge, Mass., 2002).

[7] See, among others, Svetlana Alpers, *The Art of Describing: Dutch Art in the Seventeenth Century* (Chicago, 1983); Jonathan Crary, *Techniques of the Observer: On Vision and Modernity in the Nineteenth Century* (Cambridge. Mass., 1990); Friedrich Kittler, *Optische Medien: Berliner Vorlesung, 1999* (Berlin, 2002).

devices—although their history probably points to similarly productive assemblages. Anechoic, soundproof, and sound-insulated rooms play a role in the history of prisons and clinics,[8] and they are associated with the emergence of the bourgeois *intérieur*.[9] But they are also to be found—and perhaps above all—in the history of the laboratory. To a certain extent these rooms can even be seen as laboratory rooms par excellence since—similar to bubble chambers and respiration rooms—they reproduce within the laboratory the architectural shielding and technical connections modern scientific undertakings use to separate themselves from everyday environments in order to connect up with them by other means: with the help of texts and images, but also by means of circuits, pipes, and other infrastructures.[10]

The experiences that people have in such laboratory fractals are experiences of *organlessness*. What becomes perceivable in anechoic rooms is an amorphous, undifferentiated flowing of the body, a humming, which is at odds with the common organization of the organs. As a result, reports like Cage's contain an implicit criticism of oculocentrism. In fact, many contributions from cultural studies focus on vision, the most power-related of all the senses, and explain other sensory organs, instruments, and media according to its model or "image" (*Vor-Bild*).[11]

At the same time, however, reports such as Cage's subvert a persisting figure of thought that refers the changing relationship between technology and knowledge to the "projection" and "extension" of the human body or, conversely, to the "introjection" and "embodiment" of human technology. This figure of thought—as Georges Canguilhem has shown—is deeply tautological.[12] In addition, it presupposes a certain organization of the organs (or tools), taking the body (or technology) as a kind of irreducible statue, which at precisely defined points has equally precisely defined functions built into it. As a consequence, one does not take into account that individual and collective bodies are embedded in processes of becoming and that experiments, understood as technoepistemic production assemblages, can create bodies that are only defined by axes and vectors, gradients and thresholds, so that organs and tools appear and function here only as intensities.

Such "bodies without organs," as Gilles Deleuze and Félix Guattari have explained, are not scenes; they are not places—not even supports upon which something comes to pass. Rather, they indicate borderlines, states, intensities equal to zero, which are bound to nothing more than sets of practices.[13] As fields of immanence or planes of

[8] See here the historical documents in Olaf Arndt and Rob Moonen, *Camera silens: Ein Projekt von Moonen und Arndt; Parochial-Kirche, 3. bis 24. November 1995*, 2nd expanded and revised ed. (Hamburg, 1995).

[9] Wolf Lepenies, *Melancholie und Gesellschaft*, 2nd ed. (Frankfurt am Main, 1981), 115–54.

[10] On the bubble chamber, see Peter L. Galison, *Image and Logic: A Material Culture of Microphysics* (Chicago, 1997), 313–431, and Andy Pickering, *The Mangle of Practice: Time, Agency, and Science* (Chicago, 1995), 37–67. On respiration rooms, see Frederic L. Holmes, "The Formation of the Munich School of Metabolism," in *The Investigative Enterprise: Experimental Physiology in Nineteenth-Century Medicine*, eds. William Coleman and Frederic L. Holmes (Berkeley, Calif., 1988), 179–210. More generally, see Hans-Jörg Rheinberger, "Wissenschaft zwischen Öffentlichkeit und Labor," in *Jahrbuch 2000 des Collegium Helveticum der ETH Zürich* (Zurich, 2001), 159–75.

[11] See here the detailed analysis by Jonathan Sterne, *The Audible Past: Cultural Origins of Sound Reproduction* (Durham, N.C., 2003), 1–29.

[12] Canguilhem, "Machine and Organism," in *Knowledge of Life*, trans. Stefanos Geroulanos and Daniela Ginsburg (New York, 2008), 75–97. Canguilhem's text was originally published in 1952.

[13] See Deleuze and Guattari, *A Thousand Plateaus: Capitalism and Schizophrenia*, trans. Brian Massumi (Minneapolis, 1987), 150. On the question of the dynamic development of technical media, see, e.g., Joseph Vogl, "Medien-Werden: Galileis Fernrohr," *Archiv für Mediengeschichte* 1 (2001): 115–24.

consistency tied to experimentation these bodies arise during the constructive functioning of those subjective-objective production assemblages that combine parts of machines, theories, human beings, animals, and other components to create epistemic or aesthetic effects of innovation. Bodies without organs are, therefore, produced, created in experiments, and this can—to quote Deleuze and Guattari—"take place in very different social formations through very different assemblages (perverse, artistic, scientific, mystical, political)."[14]

This article historicizes the experience of organlessness in the camera silenta. It argues that once time experiments, as they were carried out in physiological and psychological laboratories in the nineteenth century, began to involve the use of sound-insulated, soundproof, and/or anechoic rooms, they created a roaring silence in test subjects and other human actors. Against this background, the article discusses the experience of organlessness reported by artists such as Cage as tied to the encounter with sound-insulated rooms. The first section introduces the topic of organlessness with reference to the time experiments of physiologist Hermann von Helmholtz and literary writer Marcel Proust. Although working independently of one another, Helmholtz and Proust shared an interest in the problem of experiencing "lost time," or *temps perdu*, and of the lack of human organs that directly and accurately sense time. In addition, Helmholtz and Proust worked in technological contexts that were highly networked through telegraphy, the telephone, and pneumatic tubes.

The second section of this article is devoted to the emergence of the camera silenta as an architectural structure in physiological and psychological laboratories. In the wake of Helmholtz's pioneering measurement of the propagation speed of nervous impulses, laboratory scientists of the 1880s began to investigate reaction times in human beings. In order to protect their test subjects from acoustical distraction, they distributed their experimental setups over various rooms of their laboratories, eventually locking test subjects in soundproof rooms. The third section shows that this spatial distribution of time experiments became possible through the use of telegraph technology. Beginning in the 1860s, physiologists and psychologists increasingly relied on electromagnetic precision timers that had been developed (and sold) in the context of time telegraphy. This technological and economical context was also crucial for installing time networks in European cities, as the fourth section explains.

From this background the final section of this article shifts back to the architectural structure of the camera silenta and the corresponding experience of organlessness. Around 1900, psychophysiologists such as Edward Scripture and Hendrik Zwaardemaker and literary writers such as Proust were sitting in their respective research rooms like spiders in their webs. Eyes, nose, and mouth shut, they were waiting at the edge of an extended web of urban synchronicities for the slightest vibrations that their bodies could detect. Locked in the camera silenta, they experienced intense states of organlessness. According to Cage, however, this experience was nothing other than the experience of duration.

TEMPORALITY AND ORGANLESSNESS

There is no organ that directly senses time, and therefore, it is not surprising that time experimenters more than anyone else have referred to the organlessness of the human

[14] Deleuze and Guattari, *Thousand Plateaus* (cit. n. 13), 157.

body. In December 1850 Hermann von Helmholtz began his lecture "On the Methods of Measuring Very Small Portions of Time" before the Physical-Economic Society in Königsberg as follows: "The perception of small differences of time by means of our senses and without artificial help is not very fine." This judgment prepared his famous statement about the deplorable errors of the human "tool of vision."[15]

Given the lack of adequate time organs in our body, the Königsberg physiologist dreamed of a "microscopy of time,"[16] and he found that this dream had at least partly been fulfilled by the electromagnetic timing devices developed in the 1840s and 1850s by Claude Pouillet, Werner Siemens, Charles Wheatstone, and others. Helmholtz's metaphor was rather convincing. Even some forty years later, Hugo Münsterberg would compare the Hipp chronoscope and other devices for measuring short time intervals that he collected at the Psychological Institute at Harvard University with the same extension and amplification of the human eye: "Our clocks have somewhat the same function as the microscope of the anatomist. With his *microscopes* he can distinguish the thousandth part of a millimeter; with our *chronoscope* we can measure the thousandth part of a second."[17]

It was the French physiologist Etienne Jules Marey who, in 1880, spelled out the prerequisites that gave rise to such comparisons. According to Marey the sensory organs were unreliable because they delivered perceptions "too slowly." Marey's "graphic method"—his extensive use of inscription devices for registering the functions of the living organism (respiration, blood flow, muscle contractions, etc.)—was indeed intended, above all, to aid the human senses in their condition of temporally caused limitation: "The recording devices [of the graphic method] measure infinitely small intervals of time. The fastest and the weakest movements, as well as the least variation in forces, do not escape them."[18]

Marey's *physiologie cinématique* is not the only place in which organlessness and time enter into such an instructive association. Marcel Proust, at the beginning of the twentieth century, had similar experiences with the "inscription device" he used for pursuing his *Recherche* concerning lost time. It is true that Proust mainly used pen and paper. At the same time, he integrated products and effects of the media into the writing process, from newspaper clippings to photographs and telephone calls. However, in contrast to Helmholtz's and Münsterberg's devices, Proust's novel-writing machine was supposed to function not as a microscope but as "a telescope focused on

[15] Helmholtz, "On the Methods of Measuring Very Small Portions of Time, and Their Application to Physiological Purposes," *London, Edinburgh and Dublin Philosophical Magazine and Journal of Science* 4 (1853): 313–25, on 313. The later statement referred to is the following: "Now it is not too much to say that if an optician wanted to sell me an instrument which had all these defects [i.e., standard features of the human eye such as an irregular curvature of the cornea and the radial formation of the lens], I should think myself quite justified in blaming his carelessness in the strongest terms, and giving him back his instrument." Helmholtz, "The Recent Progress of the Theory of Vision," in Helmholtz, *Science and Culture: Popular and Philosophical Essays*, ed. David Cahan (Chicago, 1995), 127–203, on 141.

[16] Hermann von Helmholtz, "Ueber die Methoden, kleinste Zeittheile zu messen, und ihre Anwendung für physiologische Zwecke," *Koenigsberger naturwissenschaftliche Unterhaltungen* 2 (1850): 169–89, on 177. The English translation ("On the Methods," cit. n. 15, on 318) omits this sentence.

[17] Münsterberg, "The New Psychology, and Harvard's Equipment for Teaching It," *Harvard Graduate's Magazine* 1 (1893): 201–9, on 205–6; emphasis mine.

[18] Marey, *La méthode graphique dans les sciences expérimentales et principalement en physiologie et en médecine* (Paris, 1878), iii.

time." Proust called it "an organ that presents to consciousness unconscious phenomena that, completely forgotten, lie far back in the past."[19]

But the organlessness that the narrator of *À la recherche du temps perdu* notes—for example, when he kisses Albertine for the first time[20]—in no way corresponds, as has often been noted,[21] to a lack of memory nor to some kind of difficulty in remembering on the part of the author. Like Helmholtz, Proust was interested in time relations that are *present* in the living organism.

This can, perhaps, already be inferred from the French title of Proust's novel, since the expression *temps perdu* can also be found in French translations of texts by Helmholtz.[22] But it can be seen concretely in the fact that in his practical investigation of time, Proust, similar to Helmholtz and other time experimenters of the nineteenth and early twentieth centuries, relied on and referred to the labor of synchronizing society to which factory owners, newspaper publishers, telegraph builders, and watchmakers, each in their own way and often independently of each other, made their contribution.[23]

In the case of Helmholtz, a clear reference to this situation can be found in the examples he presented when explaining his experiments on the transmission speed of stimuli in the motor nerves of the frog. In a letter from April 1850 Helmholtz referred not only to the individual errors in astronomical determinations of time but also to the physiological difficulties involved in the synchronization of clocks: "[It is impossible] to determine whether the ticking of two quiet pocket watches occurs at the same time or in succession when one holds one watch to each ear, while nothing is easier than to determine this when one holds both watches to the same ear."[24]

Similarly, Proust's work on his novel was associated with a simultaneity that had become precarious, although in this case the materiality of simultaneity was a different one. In his drafts for the opening scene of the *Recherche*, Proust first has his narrator Marcel determine the time his brain needs to identify as his own the text that he had written for *Le Figaro*: "For a moment my thoughts, swept on by the impetus of this reaction, . . . continue to believe it isn't."[25] Then the more general question arises of how it is possible that by means of a newspaper, that product of a "mysterious process of multiplication," "thousands of wakened attentions" can emerge at the

[19] Proust to Camille Vettard, March 1922, in *Marcel Proust: Correspondance générale*, vol. 3, *Lettres à M. et Mme Sydney Schiff, Paul Souday, J.-E. Blanche . . .*, ed. Robert Proust, Paul Brach, and Suzy Mante-Proust (Paris, 1932), 195.

[20] "That man, a creature clearly less rudimentary than the sea-urchin or even the whale, nevertheless lacks a certain number of essential organs . . ." Proust, *The Guermantes Way*, trans. Mark Treharne (London, 2002), 362.

[21] Georges Poulet, "Marcel Proust," in *Études sur le temps humain*, vol. 4, *Mesure de l'instant* (Paris, 1968), 299–335; Gilles Deleuze, *Proust and Signs*, trans. Richard Howard (Minneapolis, 2000); Félix Guattari, "Les ritornelles du temps perdu," in *L'inconscient machinique: Essais de schizo-analyse* (Paris, 1979), 237–336.

[22] On this point, see more generally Henning Schmidgen, *Die Helmholtz-Kurven: Auf der Spur der verlorenen Zeit* (Berlin, 2009).

[23] Steven Kern, *The Culture of Time and Space, 1880–1918* (Cambridge, Mass., 1983), 67–81; Peter Galison, *Einstein's Clocks, Poincaré's Maps: Empires of Time* (New York, 2003); Jimena Canales, *A Tenth of a Second: A History* (Chicago, 2009). See also the classic paper by Edward P. Thompson, "Time, Work-Discipline, and Industrial Capitalism," *Past and Present* 38 (1967): 56–98.

[24] Quoted according to Leo Koenigsberger, *Hermann von Helmholtz*, vol. 1 (Brunswick, 1902), 124.

[25] Proust, *By Way of Sainte-Beuve*, trans. Sylvia Townsend Warner (London, 1958), 47.

same time, that is, on a given morning: "At this moment each sentence that I extorted from myself flows not into my own mind, but into the minds of thousands on thousands of readers who have just woken up and opened *Figaro*."[26]

In other words, inscribed into one of the most famous scenes of the *Recherche* is amazement concerning the fact that the "marvelous ideas" of one author can penetrate "*at the same moment* all the brains" that are occupied with reading a newspaper.[27] It is precisely this connection over time of spatially separate events that is investigated in the novel in new situations from the perspective of the narrator. And as the final version of the novel's first chapter makes clear, the problematic character of such connections is measured more in "seconds," "instants," and "moments" than in years or decades.

As two of the most obvious sources of societal synchronization, clocks and newspapers set the technological framework within which the time experiments of the late nineteenth and early twentieth centuries were carried out. However, these experiments gained their epistemic potential only by pointing out nonsimultaneity, deviations, differences. Only by distancing themselves from the dominant simultaneities were experimenters such as Helmholtz, Münsterberg, or Proust able to conduct productive time experiments. The reason for this is that when time becomes the only dimension of investigation, an exact assignment of anatomical, physiological, or psychological facts to the object being investigated is only possible if one acts within clearly limited and disruption-free rooms—which also means that watches are not worn and even pendulum clocks have been "silenced."[28]

Helmholtz already recognized that the coherence of his time measurements on human beings was endangered by any "slight indisposition" or "tiredness" of the experimental subject, and even by distractions of all kinds: "If at the time of the perception of the signal the experimental subject's thoughts are occupied with something else and his mind then has to recall the movement to be carried out, it [i.e., the reaction] takes much longer."[29]

Moreover, workers in physiological and psychological laboratories paid attention not only to the experimental subjects' difficulty concentrating but also to the distractions themselves. In the 1880s they started to focus on preventing all conceivable distractions arising in the environment of their experimental subjects, by removing time-measuring devices that made noise, by building the research sites in quiet areas, and by setting up soundproof rooms.

Proust also set up a sound-insulated room in order to carry out his investigations of lost time without being disturbed, as far as possible. Using curtains, roller blinds, and cork sheets he created a room in his apartment on the Boulevard Haussmann that, in the words of Marcel, worked as a "whole of which I was only a small part and whose

---

[26] Ibid., 48.

[27] Ibid., 49 (translation modified; emphasis mine).

[28] Proust, *Swann's Way*, trans. Lydia Davis (New York, 2003), 8.

[29] Helmholtz, "Mittheilung an die physikalische Gesellschaft in Berlin betreffend Versuche über die Fortpflanzungsgeschwindigkeit der Reizung in den sensiblen Nerven des Menschen," Archive of the Berlin-Brandenburg Academy of Science, NL Helmholtz 540, 4. (A reproduction of this document including a transcription can be found on the "Virtual Laboratory" of the Max Planck Institute for the History of Science at http://vlp.mpiwg-berlin.mpg.de/lise/lit15976/index.html [accessed 24 Jan. 2013].)

insensibility I would soon return to share."[30] This "insensibilty" was one of the decisive conditions for the precise investigation of lost time.[31]

The architectural separation of the time experimenters from noise-ridden environments was, however, subverted by technical connections, taken in the widest sense. The resulting contacts were strangely filtered: by means of newspapers and letters, as well as telegraphs and telephones. Spatially separated from each other, time researchers sent telegraphic signals through their wired apartments and laboratories, used pipe systems to convey messages, and established connection with events that occurred at other locations by means of telephones. Characteristic of time experiments around 1900 are heterogeneous assemblages that remind one of a spiderweb, with the experimenter as the spider: separated, on the one hand, from the noise of time; connected, on the other, with the materiality of the simultaneous.

As a consequence, time researchers reacted much more sensitively in their protected rooms to vibrations in their webs, which still reached them through padded doors and double walls, leaping on acoustic and optical stimuli, temperature differences, and voices of actresses or professors over the telephone as if they were prey. And what they discovered here, to their horror or their joy, were not only uncommon organizations of their organs, but also novel lines of flight concerning time: discontinuities, distances, noise.[32]

ROOMS WITHIN ROOMS

Two Swedish physiologists, Robert Tigerstedt and Jakob Bergqvist, were the first to draw attention to the laboratory effects resulting from the close contact of precise time-measuring devices and organic individuals. Their work led to the construction, ten years later, of the first soundproof rooms in physiological and psychological laboratories.

In 1883 Tigerstedt and Bergqvist published a paper "On the Duration of Apperception in Compound Visual Representations." In this paper they expressed considerable doubt as to the times for conscious perceptions that had been measured in earlier experiments at the Institute for Experimental Psychology in Leipzig. The two Swedish researchers found the apperception times that had been obtained by Wilhelm Wundt's student Max Friedrich to be clearly too long. They asked the rhetorical question, "How would it be . . . possible to read or write at all if the perception of every letter and every digit were to take several tenths of a second?"[33]

For the two scientists from Stockholm, this cultural argument was the point of departure for a detailed criticism of the experimental practice that had provided the foundation for Friedrich's publication in the first volume of Wundt's journal *Philosophische*

[30] Proust, *Swann's Way* (cit. n. 28), 4.

[31] See in this sense also Lepenies, *Melancholie und Gesellschaft* (cit. n. 9), 145: "In Proust the *intérieur* becomes a place from which—in life as in the novel—'the world' can be regained."

[32] On the image of an organless body as a spider in its web, see Deleuze, *Proust and Signs* (cit. n. 21), 181–2. On the experience of noise in general, see Sabine Sanio, ed., *Das Rauschen: Aufsätze zu einem Themenschwerpunkt im Rahmen des Festivals "Musikprotokoll'95 im Steirischen Herbst"* (Hofheim, 1995); Andreas Hiepko and Katja Stopka, eds., *Rauschen: Seine Phänomenologie und Semantik zwischen Sinn und Störung* (Würzburg, 2001).

[33] Tigerstedt and Bergqvist, "Zur Kenntniss der Apperceptionsdauer zusammengesetzter Gesichtsvorstellungen," *Zeitschrift für Biologie* 19/new ser. 1 (1883): 5–44, on 18.

Studien in 1883. Friedrich not only neglected the physiological time that the human eye needed to adapt to sudden bright-light stimuli; the very setup of his experiment interfered with the times he measured. The two scientists referred above all to the fact that during the experiments carried out by Friedrich the experimental subject and the experimenter were in the same room, in which, additionally, the stimulation device and the measuring instruments were also located:

> This means that the reacting subject necessarily was disturbed by the noise of the apparatus used during the investigation—that is, by the rattling of the Hipp chronoscope and the noise of the contact breakers in the induction devices, of which at least the latter is quite loud. Further, the presence of several people in the experimental room, the continual changing of the object, etc., must have had an influence on the mind of the reacting subject.[34]

In other words, the closer to the experimental subject the measuring instruments were placed, the longer the time that was to be measured became.

The strategy that the two physiologists suggested for getting around this kind of uncertainty relation was correspondingly simple: the subject being experimented upon was to be separated spatially from both the experimenter and the measuring equipment, but was to be reconnected with both by technical means (see fig. 3). Tigerstedt and Bergqvist profited here from the fact that their device for measuring time—similar to the chronoscope used by Friedrich—used electromagnetism. The "reacting subject" (i.e., the experimental subject) and the "recording subject" (i.e., the experimenter) could be placed in different rooms, which were connected by wires. Thus the experimental subject was kept "completely free of disrupting influences."[35]

The results obtained in this manner differed significantly from those obtained in Leipzig. With precision to three places after the decimal the two physiologists measured apperception times between 0.014 and 0.035 seconds, whereas Friedrich had obtained values between 0.290 and 1.595 seconds. Their general conclusion was "that the true apperception time for a compound representation is so short that it can at most be a few hundredths of a second."[36]

The reaction from Leipzig was not long in coming. In 1888 Wundt's brilliant assistant Ludwig Lange published his paper titled "New Experiments on the Process of Simple Reactions" in *Philosophische Studien*. Lange attempted to get around the problem of physiological adaptation to time by shifting—in contrast to Friedrich as well as Tigerstedt and Bergqvist—to acoustical stimuli. Since the early 1860s, when Adolphe Hirsch, an astronomer from Halberstadt who in 1859 became the director of the newly established observatory in Neuchâtel, Switzerland, conducted his chronoscopic investigations on physiological time, the ear had apparently been considered largely unproblematic in this regard (as opposed to the eye).

After Tigerstedt and Bergqvist's criticism of Friedrich's experimental setup, Lange very clearly went to great lengths to remove every conceivable source of disturbance from his investigations. At least for Leipzig his study set the standard for the correct execution of reaction-time experiments.

[34] Ibid., 17.
[35] Ibid., 20.
[36] Ibid., 42.

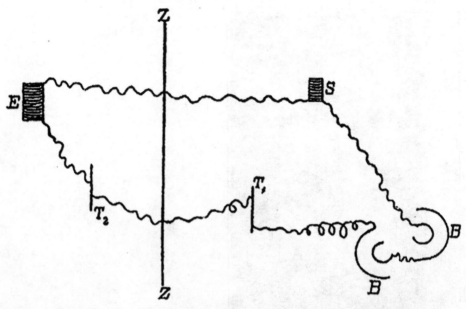

Figure 3. *Schematic drawing of the experimental setup used by Tigerstedt and Bergqvist. The experimental subject is to be imagined on the left, the experimenter on the right. E refers to the electromagnet inside the box by means of which the stimuli were presented to the test subject. S is the electric signal by means of which time is recorded. T1 and T2 are the switches handled by the experimental subject and experimenter. The B's are the batteries required for the running of the experiment. Tigerstedt and Bergqvist, "Zur Kenntniss der Apperceptionsdauer" (cit. n. 33), 21.*

Lange's main interest was in making certain that "the noises issuing from the equipment for measuring time would not be disruptive" since these would otherwise "contaminate" the required experimental conditions.[37] Following his Swedish colleagues, he placed the experimental subject and the device producing the stimulus in one room and the experimenter and the measuring equipment in another. The two rooms were connected by electric lines. Further, the experimental subject was given strict instructions to "avoid all distracting thoughts" in order not to cause any unnecessary "deviations from the normal" in the reaction times.[38]

Following Lange another Wundt student drew attention to the "ambient errors" that endangered precise work in psychological laboratories. In his lecture at the second annual meeting of the American Psychological Association in 1893, Edward W. Scripture claimed that "distracting noises [are] probably the worst source of errors" in psychological time experiments. Like Tigerstedt, Bergqvist, and Lange, Scripture was convinced that spatial separation of the experimental subject from the measuring equipment and the experimenter was the proper means of overcoming such disturbances. For Scripture, who had himself worked in the Leipzig laboratory at the end

[37] Lange, "Neue Experimente über den Vorgang der einfachen Reaction auf Sinneseindrücke," *Philosophische Studien* 4 (1888): 479–510, on 481.
[38] Ibid., 485.

Figure 4. Subject under experimentation placed in an isolated room within
the psychological laboratory at Yale
University, 1895. Scripture, Thinking,
Feeling, Doing *(cit. n. 40), 41.*

of the 1880s, it was self-evident that "the experimenter, the recording apparatus, and
the stimulating apparatus are in a part of the building distant from the person experimented upon."[39]

Scripture even went a step further than his Europe-based colleagues. In his laboratory at Yale University he set up an isolation room especially for the experimental
subject: "To be rid of all distraction the person experimented upon is put in a queer
room, called the 'isolated room,' whose thick walls and double doors keep out all
sound and light. When a person locks himself in, he has no communication with the
outside world, except by telephone"—and by a telegraph key that was connected
with measurement and recording devices located where the experimenter was (see
fig. 4).[40] Variations on this arrangement, which was described as a "room within a

---

[39] Scripture, "Accurate Work in Psychology," *American Journal of Psychology* 6 (1893): 427–30,
on 429.
[40] Scripture, *Thinking, Feeling, Doing* (New York, 1895), 41.

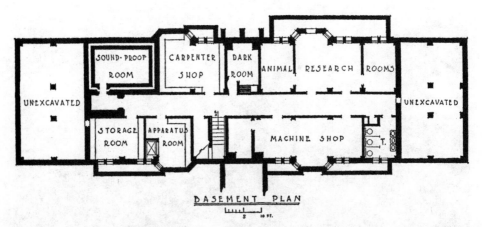

Figure 5. *Soundproof room in the psychological laboratory at Princeton University, 1926. Herbert S. Langfeld, "Princeton Psychological Laboratory,"* Journal of Experimental Psychology 9 (1926): 259–70, on 264.

room," were used in the following years in laboratories at the universities in Leipzig, Utrecht, Princeton, and Austin, among others (see fig. 5).

SYNCHRONIZING CLOCKS—AND HUMANS

With the introduction of telegraph technology for connecting the spatially separated parties in reaction-time experiments, researchers turned to a technological system that, since the 1860s, had defined both the horizon and the limits of these experiments.[41] This technological system was used above all for the telegraphic transmission of time.[42] The setup of the reaction experiment with the Hipp chronoscope was adopted by the Leipzig psychologists from Hirsch. In his *Principles of Physiological Psychology* Wundt made reference to a corresponding publication by Hirsch, which was facilitated by the fact that a German translation of Hirsch's article had appeared not in an astronomical journal, but in a physiological periodical. The research documented in "Chronoscopic Experiments on the Speed of the Various Sense Impressions and Nerve Conduction," which Hirsch had carried out at his observatory in October and November 1861, had been motivated by the attempt to determine the individual errors made in astronomical time measurements in order to be able to reduce them. Referring to this investigation Wundt suggested in his *Principles* that the chronoscope be used in reaction-time experiments. In comparison with other procedures for measuring short periods of time, it had the advantage "that its use is very comfortable and that the reading of the two dials immediately gives the absolute time."[43]

[41] On the relationship between technological and epistemic things, see Hans-Jörg Rheinberger, *Toward a History of Epistemic Things: Synthesizing Proteins in the Test Tube* (Stanford, Calif., 1997), 28–31.

[42] Henning Schmidgen, "Time and Noise: The Stable Surroundings of Reaction Experiments, 1860–1890," *Studies in History and Philosophy of Biological and Biomedical Sciences* 34 (2003): 237–75.

[43] Wilhelm Wundt, *Grundzüge der physiologischen Psychologie* (Leipzig, 1874), 772; Adolphe Hirsch, "Chronoskopische Versuche über die Geschwindigkeit der verschiedenen Sinneseindrücke und der Nerven-Leitung," *Untersuchungen zur Naturlehre des Menschen und der Thiere* 9 (1865):

Figure 6. *The Neuchâtel observatory, ca. 1860. Département de l'instruction publique, Ob-*
servatoire cantonal neuchâtelois *(cit. n. 44), 6.*

In contrast with the large observatories in Greenwich or Vienna, the canton observatory at Neuchâtel (see fig. 6) was almost exclusively devoted to time service.
This service comprised the astronomical determination of time, as well as the investigation and evaluation of marine chronometers, pocket chronometers, and pendulum clocks. Clock manufacturers in Neuchâtel such as Grandjean, Bertschinger, and
Breitling brought their products to the observatory, where they were put into different
positions and exposed to different temperatures, in accordance with a clearly defined
program. When a clock had passed the tests, Hirsch issued certificates attesting to the
quality of the clock as an observatory chronometer. In addition, the results of these
tests were published annually, which allowed a direct comparison of the clocks and
motivated competition among the clockmakers.[44]
   Further, the observatory headed by Hirsch also supplied numerous clock and watch
manufacturers in Neuchâtel with time signals. Once a day Hirsch sent a time signal
to the workshops in the Jura mountains. This was done using the observatory's telegraph, which made use of the existing telegraph network. Every day at 1:00 p.m.
a time signal was sent to the participating sites, without interruption of the normal
telegraph service. It was this part of the time service provided by the observatory
that expanded rapidly. By 1859 there were already connections from the observatory
to the telegraph office in Neuchâtel, and to the villages of La Chaux-de-Fonds, Le

183–99. For a recent translation of this article into English, see Serge Nicolas, "'On the Speed of Different Senses and Nerve Transmission' by Hirsch (1862)," *Psychological Research* 59 (1997): 261–8.
   [44] On the observatory in Neuchâtel, see Edmond Guyot, "L'observatoire cantonal de Neuchâtel,
1858–1938: Son histoire, son organisation et ses buts actuels," *Bulletin de la Société neuchateloise
des sciences naturelles* 63 (1938): 5–36; Département de l'instruction publique, ed., *L'observatoire
cantonal neuchâtelois, 1858–1912: Souvenir de son cinquantenaire et de l'inauguration du Pavillon
Hirsch* (Neuchâtel, 1912).

Locle, Les Ponts, and Fleurier. The following year the clocks in all the telegraph offices throughout Switzerland were connected to the time signal of the Neuchâtel observatory, and a short time later this time signal was transmitted to the public clocks in certain regions of Switzerland, such as the canton of Waadt and the city of Biel.

It is hardly surprising, then, that in a contemporary travel guide the authors, referring to the observatory, claimed with a certain degree of pride, "By means of electric wire time is sent from there to all of Switzerland."[45] In fact, it was from the Neuchâtel observatory that there arose "in Switzerland for the first time a standard for uniform time over a large geographic area."[46]

Hirsch played a significant part in these developments. As cofounder and secretary of the Swiss Geodetic Commission, as secretary of the International Earth Measurement office, and as the Swiss delegate to conferences in Rome and Washington that immediately preceded the international standardization of the latitudes and of time,[47] the astronomer was an active proponent of the dissemination of time and length standards over large areas.

Hirsch received practical support in his efforts to standardize and communicate time from Matthäus Hipp, the head of the federal telegraph factory in Bern, who was already well known as the inventor of the chronoscope and other devices that were on the borderline between clock mechanisms and electromagnetism. From 1860 onward the skilled watchmaker and mechanic had had his own telegraph factory in Neuchâtel. Hipp had also installed in the observatory the telegraphic equipment that ensured the transmission of the time signal. In Geneva, Neuchâtel, Zurich, Rome, and other European and American cities Hipp had installed systems of electric clocks, which guaranteed the transmission of the time over geographic areas both small and large—from individual buildings to whole cities.

As the manufacturer of the chronoscope, Hipp was also directly involved in Hirsch's experiments on the speed of sense impressions. Not only did he supply his business partner and friend with two chronoscopes made in his factory, but as a member of the Society for Natural Research in Neuchâtel, Hipp also took part as an experimental subject in Hirsch's experiments. In other words, the fact that twenty years later, around 1880, the contact between experimental subjects and experimenters in psychological laboratories could be conducted by means of telecommunications was due to the connections that had already long existed between precision electromagnetic time-measurement devices and the technological system of conveying time by telegraph.

A decisive factor in carrying out precise work in observatories and psychological laboratories was the noise of the time device. Astronomers like Hirsch had to listen carefully to the ticking of their astronomical pendulum clocks in order to coordinate them exactly with the passage of stars they were observing; psychologists such as Lange and Scripture paid attention to the sound of the small fly springs that regu-

[45] Louis Favre and Louis Guillaume, *Guide du voyageur à Neuchâtel, Chaumont et le long du lac, accompagné de l'indication de quelques courses dans le canton de Neuchâtel* (Neuchâtel, 1867), 23.

[46] Jakob Messerli, *Gleichmässig, pünktlich, schnell: Zeiteinteilung und Zeitgebrauch in der Schweiz im 19. Jahrhundert* (Zurich, 1995), 74.

[47] See Ian R. Bartky, *One Time Fits All: The Campaigns for Global Uniformity* (Stanford, Calif., 2007), and Clark Blaise, *Time Lord: Sir Sanford Fleming and the Creation of Standard Time* (London, 2000).

lated the chronoscopes to be certain of the proper functioning of their instruments. These acoustic aspects of the materiality of time, however, resulted in astronomers and psychologists being easily affected by all kinds of noises that existed or occurred in the area around their research sites. However satisfactory the technical outfitting of his laboratory at Harvard University might have been, in Münsterberg's eyes—or rather, ears—its geographic location was completely unsatisfactory:

> Whoever has undertaken psychological investigations on the corner of *Harvard Square*, at a place where electric cars cross from four directions, and where the hand organs of the whole neighborhood make their *rendezvous,*—out of his soul will not vanish the wish that a new laboratory may sometime arise in a more quiet spot.[48]

Ten years later, with the construction of the spacious Emerson Hall in the middle of the Harvard campus, it seems that this quieter site was found. At any rate, the complaints about disruptive noise stopped.[49]

In the psychological institutes of German universities as well, one was by no means removed from ambient noise. In fact, in the protocols of the experimental subjects that Wundt and his pupils studied, mentions of distraction by unexpected acoustic events can repeatedly be found—for example, in Götz Martius's investigation on the relationship between reaction time and attention. One of Martius's experimental subjects says, for example, that he had focused his attention "quite well" before the experiment began, but he was so distracted by a noise from outside that the reaction time measured was probably wrong. In the experimental protocol there is a laconic statement as to the reason: "Disruption due to a church clock."[50] Only after the Leipzig institute had moved into a new university building in the fall of 1896 did it have a "quiet room" at its disposal (see fig. 7), which was protected against noise from both outside and within the laboratory by large double windows, a "mattress door," and "double walls filled with rubble."[51]

TIME AND THE CITY

The close relationship between time and noise had already occupied Hirsch. Before the construction of the observatory, in an expert report that he had written for the council of state of Neuchâtel, Hirsch presented his requirements for the instrumental outfitting of a time observatory and for the geographic location of the observatory. Hirsch explained that when choosing the best site for construction, one would have to look for a horizon as wide as possible. In addition, the observatory building had to be protected against all kinds of impacts and vibrations, as these could disturb the precise work of the astronomical instruments. Hirsch then emphasized that there had to be a "profound tranquility" so that the astronomer could "always hear his pendulum clock":

---

[48] Münsterberg, "New Psychology" (cit. n. 17), 206.

[49] The new laboratory is described in Hugo Münsterberg, "Emerson Hall," *Harvard Psychological Studies* 2 (1906): 3–39.

[50] Martius, "Ueber die muskuläre Reaktion und die Aufmerksamkeit," *Philosophische Studien* 6 (1890): 167–216, on 202.

[51] Wilhelm Wundt, "Das Institut für experimentelle Psychologie," in *Festschrift zur Feier des 500jährigen Bestehens der Universität Leipzig*, ed. Rektor and Senat (Leipzig, 1909), 118–33, on 125.

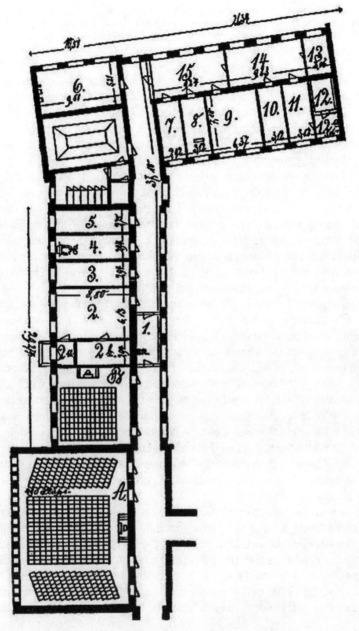

Figure 7. Quiet room (no. 12, top right) in the Institute for Experimental Psychology at Leipzig, 1909. Wundt, "Institut für experimentelle Psychologie" (cit. n. 51), 123.

> The neighborhood of church towers, of busy sections of the city and, above all, of large streets has to be avoided, so that the noise of the bells, of the vehicular traffic and of the whistle of locomotives does not disturb the astronomical observations and the vibrations of the ground are not transferred to the observatory's foundation and instruments.[52]

In other words, Hirsch found that his attempts to determine time exactly and to transmit it by telegraphic means were endangered by a different, older system for communicating time. The "language of the bells" could affect the concentrated coordination of eye and ear, which was necessary for precisely checking the astronomical pendulum clock against the passage of the stars.[53] As mentioned above, the psychologists who were measuring time faced similar problems.

Hirsch's concerns were apparently the result of experiences he had had while working for Urbain Jean Joseph Le Verrier at the Paris Observatory. In 1853 Le Verrier was chosen to head the observatory in the French capital. Among the expressed goals of the newly named director were the improvement of the precision of astronomical time measurement, the installation of equipment that would make the telegraphic distribution of the time possible, and the determination of the longitudes in cooperation with other observatories. Further, Le Verrier was interested in making the Paris observatory a center for meteorological research.[54]

The steps required for the implementation of these projects were not facilitated by the location of the observatory in the center of the city. Le Verrier and his colleagues saw the value in the proximity of other scientific institutions and of the factories of the instrument makers, but the vibrations of the ground coming from the traffic in the immediate vicinity of the observatory, the dust and haze in the center of the city, and the nighttime lighting of the streets often rendered precise astronomical work impossible.[55]

The measurement of time was interrupted by the bells of religious establishments that rang more or less continuously in the vicinity of the observatory. It was only in the 1860s that these clocks were better synchronized so that, as a contemporary observer said, "it no longer happened that an astronomer would hear the same hour rung for over half an hour by different bells."[56] When Hirsch wrote his observatory report in 1858 for the state council in Neuchâtel, he no doubt had this chaos of bells still ringing in his ears.

In contrast with the Neuchâtel projects, Le Verrier's plans for the telegraphic distribution of observatory time in the French capital were slow to be fully implemented. A connection between the pendulum clock at the Paris observatory and the telegraph office had been established in the mid-1850s, but it would take more than another twenty years before one could speak of an effective distribution of time in the capital.

At the end of the 1870s two corresponding projects began to be implemented that were more or less independent of each other. In 1880 the city of Paris installed a first

[52] Hirsch, "Rapport de M. le Dr Hirsch sur le projet de fonder un observatoire cantonal à Neuchâtel," 1858, State Archive of Neuchâtel, Neuchâtel, Switzerland, Archive of the Industry Department, Observatory, box 83, 5.

[53] On this, see also Alain Corbin, *Village Bells: Sound and Meaning in the Nineteenth-Century French Countryside*, trans. Martin Thom (New York, 1998).

[54] Le Verrier, "Rapport sur l'observatoire impérial de Paris et projet d'organisation," *Annales de l'observatoire de Paris* 1 (1855): 1–68.

[55] David Aubin, "The Fading Star of the Paris Observatory in the Nineteenth Century: Astronomers' Urban Culture of Circulation and Observation," *Osiris*, 2nd ser., 18 (2003): 79–100.

[56] Émile Bourdelin, "L'observatoire de Paris," *Le monde illustré* 6, no. 251 (1862): 70–1, on 71.

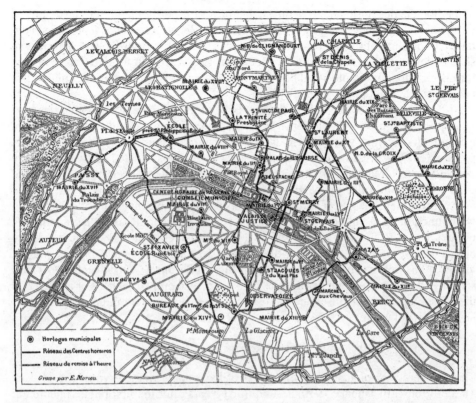

Figure 8. *Networks of unified time in Paris, 1881. Niaudet, "Unification de l'heure à Paris" (cit. n. 57), 101.*

system of electric clocks. Official institutions such as city halls and schools in the different arrondissements were supplied by telegraph with a time signal, which came from the observatory. Two circuits for supplying time were set up: an eastern one, which connected the city halls of the sixth and twelfth arrondissements, among other locations, and a western one, which connected the city halls of the fifth, tenth, and eleventh arrondissements, as well as the Hôtel de Ville (see fig. 8). The plan was to expand this network of electric clocks; however, the extension was left up to the initiative of interested private persons and institutions within the different city districts, as well as to the city halls of the neighboring arrondissements.[57]

This self-organization of the electric clock system does not seem to have been very successful, in particular as it was in competition with another network for the distribution of time. In 1878, at the universal exposition in Vienna, Austria had introduced a system of pneumatic clocks, which had already been working with success for some time in the Austrian capital. In the same year interest developed in Paris in introducing a comparable system of clocks.[58] At the end of 1878 the city gave approval

[57] Alfred Niaudet, "L'unification de l'heure à Paris," *La Nature* 9 (1881): 99–102.
[58] On the pneumatic clocks at the exposition in Vienna, see Alphonse F. Noguès, "Les horloges pneumatiques à l'Exposition universelle," *La Nature* 6 (1878): 161–2. On the system of pneumatic clocks that distributed the time in the French capital, see Edouard Hospitalier, "Les horloges pneumatiques:

to the Société des horlogers to begin such a project. Shortly thereafter, the details of
its technical implementation were discussed in the *Revue chronométrique*. After a detailed description of the hydropneumatic clocks of Eugène Bourdon and a somewhat
shorter description of the pneumatic clock that Eadweard Muybridge (better known
for his chronophotography) had built for San Francisco,[59] the Paris Clockmaker Society decided to adopt the Vienna system.

At the beginning of the 1880s, the newly founded Compagnie générale des horloges pneumatiques began the construction of a corresponding facility. Working from
a central clock, which was adjusted daily via telegraph according to observatory time,
pressurized air was distributed throughout the city in pipes that had partly been newly
laid and partly already existed (the public pneumatic mail had been functioning since
1879), which controlled the clocks connected to the system.

In March 1880, the first reports were published about the pneumatic clocks, which
were functioning at public sites in Paris, as well as at the homes of a "large number"
of subscribers to the service (see fig. 9). With satisfaction it was claimed that the
clocks that were installed on high pillars in the streets of Paris made the time plainly
visible: "The time can be seen clearly at a sufficient distance." The advantages of sub-
scribing to time within one's own four walls were also emphasized: "For a sou a day
the subscriber receives observatory time with precision to the minute, without ever
having to wind up, set, or take care of the clocks."[60]

PROUSTIAN INSCRIPTION DEVICES

There is a precise point at which the clocks of Paris make their entry into the work of
Marcel Proust. In January 1910, around thirty years after the first public pneumatic
clocks were set up in the city, but still three years before publication of the first part of
the *Recherche*, Paris was hit by long and heavy rainfalls. The rain was so heavy that
the Seine overflowed its banks. In the night of 22 January all the pneumatic clocks
stopped at ten past eleven: the cellar in which the pressurized air was produced had
been flooded.

Proust apparently did not subscribe to the pneumatically communicated time, because he only learned about the historic stoppage of the clocks of Paris two days after
it had occurred. In a letter dated 25 January, his friend Lionel Hauser reported the
significant event to him—without relating it, however, to Proust's work on his novel.
Hauser was one of the writer's financial advisers, and the main purpose of his letter was to keep Proust from investing more money in industrial stocks. It was in this
connection that he referred to the dangers that industrial facilities in Paris such as the
Paris Compressed Air Company were exposed to.[61]

In his answer to Hauser, Proust did not mention the stoppage of the pneumatic
clocks, but he did express his concern that the pneumatic mail system, which he often

La distribution de l'heure à domicile," *La Nature* 8 (1880): 407–9; "Das Betriebssystem der pneuma-
tischen Uhren in Paris," *Deutsche Uhrmacherzeitung* 6 (1882): 24–5, 31–2, 39–40.

[59] Benjamin Bourdon, "Horloge à tube flexible et son moteur hydro-pneumatique," *Revue chronométrique* 24 (1879): 202–7; "Horloge pneumatique de M. E.-J. Muybridge de San Francisco," *Revue
chronométrique* 24 (1879): 285–6.

[60] Hospitalier, "Horloges pneumatiques" (cit. n. 58), 407.

[61] Hauser to Proust, 25 Jan. 1910, in *Marcel Proust, Correspondance*, vol. 10, *1910–1911*, ed. Philip
Kolb (Paris, 1983), 36–9, on 36.

Figure 9. *Pneumatically driven clock in Paris, Place de la Madeleine, within walking distance of Proust's apartment at 102 Boulevard Haussmann. Hospitalier, "Horloges pneumatiques" (cit. n. 58), 409.*

used to send letters across the city, might not be working again. The intensive correspondence carried out by the writer, who lived in seclusion, enabled him to maintain connection with a wide network of friends, relatives, and partners. This included the sending not only of handwritten papers but also of photographs, especially portraits, which Proust collected for use as visual raw material for his writing.[62]

Additional contact with the outside world was mediated by telephone; a set had been installed in Proust's apartment on Boulevard Haussmann. This was used, for example, to call taxis or the husband of his housekeeper in a café nearby to ask him

[62] Brassaï, *Proust in the Power of Photography*, trans. Richard Howard (Chicago, 2001), 17–27.

Figure 10. *Central office of the Paris* théâtrophone, *1892. Mareschal, "Théâtrophone" (cit. n. 64), 57.*

to run errands.[63] Further access to the simultaneities of the Parisian metropolis was achieved through newspapers. Proust wrote articles for *Le Figaro* and other Paris papers, whose business sections he read very attentively, as he was a stockholder.

In addition, beginning in 1913 Proust had a subscription to the *théâtrophone*, a service introduced in Paris in the 1890s, which enabled subscribers to listen to theatrical performances and concerts over the telephone (see fig. 10). This expanded the written and visual raw material that Proust received in his apartment through the mail by a whole spectrum of acoustic materials. A contemporary description shows the effectiveness of the *théâtrophone*:

> As for the subscribers to the telephone network, they have it best since they can believe they have been transported to their favorite theater without leaving their apartment and even while they are lying in bed. If it's a play they're familiar with they actually see the stage sets and the actors before them, since they recognize their voices, and they don't miss a single note or a single syllable.[64]

Not only visits to the theater and concert hall, but also daily movements in the large apartment Proust lived in were conveyed by means of a network of wires. The

[63] On these and the following biographical details, see Ronald Hayman, *Proust: A Biography* (London, 1990); Céleste Albaret, *Monsieur Proust: Souvenirs recueillis par Georges Belmont* (Paris, 1973).
[64] Georges Mareschal, "Le théâtrophone," *La Nature* 20 (1892): 55–8, on 58.

author's combined bedroom and study was connected with the other five rooms of his apartment by electric lines. Next to the table on which Proust kept his photographs there was a switchbox, which allowed him not only to turn the lights on and off but also to call his housekeeper and to operate an electric kettle—"even while he was lying in bed."

If Proust's bedroom-study was connected by technological systems to the synchronicity of the surrounding city, it was simultaneously separated architecturally from these surroundings. Because he was sensitive to dust, noise, and light, the writer had everything possible done to isolate his room—which was up two flights of stairs and faced the street—from the outer world. The shutters on the two large windows were always kept closed, as were both the drapes and the curtains, which were filled with cotton. Large sheets of cork that covered the walls and the ceiling as of July 1910 were supposed to provide additional soundproofing.

The idea for this came from a woman friend who was a writer, Anna de Noailles, who herself had gotten the idea from the playwright Henry Bernstein, who was extremely sensitive to noise. That Proust had already been familiar with this means of sound insulation for some time and even approached it with a certain irony can be seen from the fact that in 1908, in a *pastiche* about Flaubert, he had written about people who covered their walls with cork in order to protect themselves from loud neighbors.[65]

The silence that Proust obtained for his investigations was not, however, complete. The author did indeed find himself in a room in which "each sound serves only to make the silent apparent by displacing it."[66] We can also assume that Proust—just like his narrator Marcel—became able, in this room, to twitch and to jump "like a machine with the brake on running in neutral,"[67] which probably contributed to his productivity as a writer. At any rate, in the novel, isolation and deprivation often turn into the precondition for an enhanced capacity for perception.

In his secluded room Marcel succeeds in taking part even more intensively than usual in the daily events of the outer world:

> It was above all inside myself that I heard with delight a new sound struck from the inner violin. Its strings are tightened or slackened by simple variations in temperature, in exterior light. Within our being, that instrument which the uniformity of habit has reduced to silence, melody springs from these changes, these variations, which are the source of all music: the weather on particular days makes us move immediately from one note to another. . . . Only these inner changes (though they came from outside) brought the outer world alive again for me. Connecting doors, long walled up, were opening again in my brain.[68]

Under a musical sign, time (*temps*) detaches itself from outer conditions such as clocks and newspapers and expands itself to weather (*temps*), a periodically variable phenomenon that is both spatially and temporally determined, which takes hold of the whole body, redefining it as "time-space."

[65] Henri Raczymow, *Le Paris littéraire et intime de Marcel Proust* (Paris, 1997), 41.

[66] Proust, "Days of Reading (I)," in *Against Sainte Beuve and Other Essays*, trans. John Sturrock (London, 1988), 194–226, on 202.

[67] Proust, "The Prisoner," in *In Search of Lost Time*, vol. 5, *The Prisoner/The Fugitive*, trans. Carol Clark and Peter Collier (London, 2002), 1–384, on 19.

[68] Ibid., 18.

Even if Marcel's experience here approaches the modes of a music of silence reminiscent of Satie (and thus also of Cage), Proust was forced to recognize that despite the transformation of his bedroom-study into a soundproof room, freedom from disruption still remained an unreachable goal. After the cork sheets had been installed, for example, the author suffered a series of asthma attacks, which he suspected were caused by the cork. A later attempt to overcome noises that continued to disturb him by using earplugs led to a middle-ear infection. Perhaps Proust's inscription device was above all driven by the impossibility of achieving an enduring condition of living quietude.

The time researchers in their laboratories had basically the same experience. For example, the soundproof room that Hendrik Zwaardemaker set up in 1904 at the Physiological Institute of the University of Utrecht quickly proved to be inadequate. In his investigations of the physiology of the sense of smell Zwaardemaker had developed a "zero-method" that he also wanted to use in the area of psychological acoustics.[69] His goal here was to isolate the experimental subject from all external distractions. Within this isolation, he wanted to reconstruct the complexities of acoustical experience by starting with two stimuli that were so simple that in combination they would be perceived as neutral. Thus, the point of departure for his investigations in the soundproof room was a kind of minimal acoustic pair. While he was successful in creating relatively easily a state without smells with this method, achieving a zero degree of noise by this method was not so easy. The deep noises that were caused by heavy vehicles driving past in the street in front of the laboratory meant that the silence in the isolated room could not be considered complete.

Zwaardemaker attempted to overcome this problem by setting up a further soundproof room inside the already existing room. This room consisted of a movable box with walls made of peat that were covered both inside and out by heavy layers of horsehair. Only once the experimental subject was inside this *camera silentissima* was the Dutch physiologist able to achieve satisfying results. Yet even then one could not talk of absolute silence:

> Immediately upon entrance to the room a weak tinnitus in the ears begins. It resembles the wind in the tops of the trees in the woods. Besides this, a high tone appears which is very near the upper limit of hearing along the tone-scale; also more complex noises occur, which take on, in different degrees, the character of hallucinations, resembling the singing of birds, the crowing of cocks, melodies, etc.[70]

In this case as well, the isolated inner world connected up with an extended outer world and, as with Proust, it was a specific form of time that created the connection. Zwaardemaker was not certain whether the sounds in silence were caused by the circulation of the blood or whether they were acoustic aftereffects of noises that he had heard before entering the soundproof room. But regardless of whether they were due to the presence of the body or to bodily *après coup* effects, within the soundproof double room it was a kind of living time that made itself known, a mobile silence that did not deviate from the zero state being sought but was an exact characterization of it.

---

[69] Zwaardemaker, "An Intellectual History of a Physiologist with Psychological Aspirations," in *A History of Psychology in Autobiography*, ed. Carl Murchison (Worcester, 1930), 491–516, on 502.
[70] Ibid., 506.

Scripture as well discovered unexpected noises, which he related to time, on entering the quiet and darkness of the isolated room in his laboratory at Yale University:

> My clothes creak, scrape and rustle with every breath. The muscles of the cheeks and eyelids rumble; if I happen to move my teeth, the noise seems terrific. I hear a loud and terrible roaring in my head; of course, I know it is merely the noise of the blood rushing through the arteries in my ears . . . , but I can readily imagine that I possess an antiquated clockwork and that, when I think, I can hear the wheels go 'round.[71]

Remote from every apparent source of noise, in this case as well, there was an encounter with the internal connection between bodily depths and temporality. In the solitude of the isolation room the experimental subject was confronted with an inner clock that apparently could not be escaped. As a consequence, the experimental subject himself emerged as the ultimate disruptive factor in experimental psychology involved with the measurement of time: "All the sights and sounds can be shut out, all disturbances of touch can be made small by comfortable chairs, but, alas! we have let in a sad source of disturbance, namely, the person himself!"[72]

In other words, the time experiments in the psychological laboratory created a line of flight drawn by the very bearer of the characteristics that were supposed to be investigated. If Scripture had truly been able to exclude all possible disturbances, all access to the epistemic object would have been blocked as well: at least for followers of Wundt, there could be no psychological experiments without experimental subjects.

CONCLUSION: VARIABLES AND VARIETIES

Deleuze and Guattari credited Antonin Artaud with the discovery of the body without organs: "The full body without organs is the unproductive, the sterile, the unengendered, the unconsumable. Antonin Artaud discovered it wherever it was present, without form or shape."[73] But Artaud is not the only or even the first person who by means of a complex of practices experienced the body beneath the skin as an "overheated factory." In one of the anechoic rooms at Harvard University John Cage discovered access to this fluid *fabrica* as the zero point of unintended music: "No one *means* to circulate his blood." Proust attained the organless body in his combined bedroom and study, which was lined with cork to protect it from noise: the "insensibility" of this room was the precondition for his "brain" opening itself to a music that restructured the relationship between interior and exterior. Finally, in their soundproof rooms Zwaardemaker and Scripture not only encountered movements in their bodies that were removed from every hierarchy of the senses (respiration, circulation, etc.), they also experienced continuations of bodily movements that were initially triggered outside the soundproof rooms, by the tops of trees moving in the wind, the singing of birds, melodies, and so on.

The assemblages within which these bodies without organs were created resemble each other. Whether in Paris, Utrecht, or elsewhere the time experimenters were

---

[71] Scripture, *Thinking, Feeling, Doing* (cit. n. 40), 42.
[72] Ibid., 41–2.
[73] Gilles Deleuze and Félix Guattari, *Anti-Oedipus: Capitalism and Schizophrenia*, trans. Robert Hurley, Mark Seem, and Helen R. Lane (Minneapolis, 1992), 8; translation modified.

always sitting in their research sites like spiders in their webs: without eyes, nose, or mouth, they were waiting at the edge of an extended web of simultaneities for the slightest vibrations that their bodies could receive. Despite these similarities the bodies of the experimenters were not exactly the same. More precisely, the sought-for border, the intensity equal to zero, that Proust, Zwaardemaker, and Scripture were moving toward was dealt with in different ways. For the scientists it was a difficult-to-control *variable* in conventional simultaneity—that is, a disruption of the sought-for form of precise measurement of time; for the writer it was a kind of threshold that had to be crossed in order to create, on the other side, a new time regime that would be capable of cultivating *varieties* of certain experiences.[74] Only in the rarest instances was the border itself a subject of investigation, as it probably was for Cage, who—with Thoreau and Meister Eckart in mind—first created a concept for this *degré zéro* experience: the roaring silence of organlessness is a duration.

The camera silenta is, therefore, not primarily an acoustic space but a temporal one, a "time-space." In contrast with the camera obscura the silent room is not focused on a sense organ; and if the ear plays a special role in it, then it is only as a portal to the dynamic forms that characterize the body without organs: vectors, gradients, migrations.

The relationships of technology and knowledge that characterize the camera silenta are correspondingly different. As Jonathan Crary has shown, "The camera obscura allows the subject to guarantee and police the correspondence between exterior world and interior representation and to exclude anything disorderly or unruly."[75] The hope of discovering such correspondences certainly motivated the construction of soundproof rooms too. But the connection between reflective introspection and self-discipline that the darkroom seems to have so strikingly facilitated encountered clear limits in the camera silenta—not least because the ear stands open and as such is directed toward both outer space and inner time.

The silent room does not accommodate, therefore, an autonomous, individual self that has acquired the ability to comprehend intellectually the infinite existence of bodies in space. It accommodates a many-headed subject that is in the process of dissolution and that, by means of accelerations and decelerations, by movements in place, continually recreates anew the borders between the inside and the outside.

---

[74] On variables in science and varieties in the arts, see Gilles Deleuze and Félix Guattari, *What Is Philosophy?* trans. Hugh Tomlinson and Graham Burchell (New York, 1994), 202–3.

[75] Crary, *Techniques of the Observer* (cit. n. 7), 43.

CIRCULATION OF SOUND OBJECTS

Experiments on Tone Color
in Music and Acoustics:
Helmholtz, Schoenberg, and *Klangfarbenmelodie*

by Julia Kursell*

ABSTRACT

In the mid-nineteenth century, Hermann von Helmholtz developed a new, mathematically formalized representation of the quality of tones, which he termed *musikalische Klangfarbe*. He did so at the price of excluding change from this representation and from the sounds he experimented with. Later researchers and composers discovered the cognitive and aesthetic side effects of this new concept. Experimental psychologist Carl Stumpf found that stable tones veil their source; their recognition strongly depends on their characteristic beginnings and endings. Arnold Schoenberg in turn used this effect to merge the sounds of musical instruments into new orchestral colors. On the basis of a three-part case study, I argue that nineteenth-century research in perception has deeply affected twentieth-century concepts of music, bringing to the fore the aesthetic quality of experimental situations.

Every tone starts with a noise. Sometimes, however, this noise is barely audible. Hermann von Helmholtz reported such a situation in his book *On the Sensations of Tone as a Physiological Basis for the Theory of Music*:

> It is interesting in calm weather to listen from a high hill to the voices of people in the plain. Words can no longer be recognized, or at most only such as are composed of *M*, *N*, and vowels, as *Mamma, No, Noon*. But the vowels contained in the spoken words are easily distinguished. Wanting the thread which connects them into words and sentences, they form a strange series of alternations of quality and singular inflections of tone.[1]

Helmholtz recreated these strange inflections of tone in his famous experiment on sound synthesis with his "apparatus for the artificial construction of vowels."[2]

---

* Capaciteitsgroep muziekwetenschap, Universiteit van Amsterdam, Nieuwe Doelenstraat 16, 1012 CP Amsterdam, Netherlands; j.j.e.kursell@uva.nl.

Work on this article was supported by VolkswagenStiftung and the Max Planck Institute for the History of Science, Berlin. For helpful comments and support at various stages of this text the author wishes to thank Alexandra Hui and Myles W. Jackson, Tessa Dunkel, Matthias Flaig, Andrea Rusnock, and Rachel Kamins, the participants of the preparatory Berlin and Starkville conferences, as well as two anonymous reviewers. All translations are by the author unless otherwise credited.

[1] Helmholtz, *On the Sensations of Tone as a Physiological Basis for the Theory of Music*, 2nd ed., trans. from the 4th German ed. by Alexander Ellis (1885; repr., New York, 1954), 68; translation modified.

[2] Ibid., vi.

© 2013 by The History of Science Society. All rights reserved. 0369-7827/11/2013-0010$10.00

In this article, this experiment will be reconsidered as creating a new sound. I argue that for Helmholtz, the steady, internal repetition of periodic sound waves was the point of departure for an entire "experimental system."[3] Before a technological means of reproducing sound events was available, this particular sound allowed the experimenter to gain repeated access to one experimental object. The periodic sound wave proved successful in granting the tools of formal description that acousticians were so eager to find. The advantages of periodic sounds came with restrictions, as well. As long as the acousticians relied on stable, periodic sounds, they would not acquire the means to grasp, describe, and investigate those parts in a sound that change quickly.

More specifically, I will juxtapose Helmholtz's concept of the steady sound with a concept Arnold Schoenberg termed "tone-color melody." In one of his compositions, the orchestral piece *Farben: Ein Sommermorgen am See* (Colors: a summer morning at the lake; op. 16, no. 3), Schoenberg transforms Helmholtz's experimental setup for the production of vowels into an orchestral setting. Looking back at the physiological experiment from the vantage point of this composition, I argue that already within Helmholtz's research on hearing, experimentation transgressed the boundaries of physiology toward an aesthetics of hearing.

Helmholtz invented a new type of sound generation that avoided any characteristic beginnings and endings. For his purpose of adjusting the conditions of hearing to the physical knowledge of sound, the sounds he was thus able to generate served particularly well. No better physical approximation of the infinite waves required in the mathematical formulae that Helmholtz used to represent sound could be given. While Helmholtz himself did not acknowledge the compromise he made between mathematical waves and audible sounds, later experimental researchers, such as the psychologist Carl Stumpf, emphasized the time-critical aspect of the new definition of sound. Reducing sounds to the middle section of their envelope created unexpected effects, as Stumpf found in his own experiments on this matter. When the characteristic beginnings of sounds are suppressed, listeners cannot easily trace the sounds to their sources.

While scholars today are almost certain that Schoenberg had not read Helmholtz's famous book, it is well established that he was aware of Stumpf's work, though not in detail.[4] In a footnote that Schoenberg added at the last minute to the proofs of his no less famed manual *Theory of Harmony* (1911), he complained—tongue in cheek—that he had only now discovered that such a great scholar as "Professor Stumpf" had come to the same conclusions as he himself had: "At such moments I regret that I know so little. I have to guess at all of it. If I had only an inkling that a scholar with the reputation of Stumpf represents the same view as I! I am ignorant of all these sources and have to depend on a single source: on thinking."[5] Parallel to Stumpf's new reading of sound color, Schoenberg embraced the effect of the missing beginnings and

[3] Hans-Jörg Rheinberger transformed this term from everyday laboratory talk into an epistemological notion. See Rheinberger, *Toward a History of Epistemic Things: Synthesizing Proteins in a Test Tube* (Stanford, Calif., 1997).

[4] See Albrecht Schneider, "Akustische und psychoakustische Anmerkungen zu Arnold Schönbergs Emanzipation der Dissonanz und zu seiner Idee der Klangfarbenmelodie," in *Komposition als Kommunikation: Zur Musik des 20. Jahrhunderts*, ed. Constantin Floros, Friedrich Geiger, and Thomas Schäfer (Frankfurt am Main, 2000), 35–55.

[5] Schoenberg, *Theory of Harmony*, trans. Roy E. Carter (Berkeley, Calif., 1983), 431.

endings of sounds for his music. This compositional device enabled him to create colorings that were unheard of in orchestral music.

A sound's individuality lies to a great extent precisely in those noises that mark its beginning. The longer a sound continues, the more difficult it is for the ear to recognize it as specific. It is equally difficult for the ear to identify the sound if the beginning is not heard.[6] Researchers before Helmholtz who tried to emulate speech sounds had reported on these phenomena. There was one way out of this dilemma, which both Helmholtz and Schoenberg used, in their respective ways. If the quality of a sound changed stepwise at an appropriate pace, the listener could trace the change in quality at each step. Both Helmholtz and Schoenberg understood, however, that through such changes the sound would gravitate toward music and, more specifically, toward a melodic and/or harmonic progression. The changing qualities would then be recognized as one element of such a progression—one musical note or chord. This recognition, in turn, would distract attention from the shading of sound quality, or tonal color. Both Helmholtz and Schoenberg took precautionary measures against this distraction. In this, Helmholtz's concept of experimental sound synthesis and Schoenberg's concept of tone-color melody were similar: they both used a device to change the habits of perception.

This article shows how such experimental or quasi-experimental constraints can serve as conditions of possibility, from which an experimental system emerges that eventually produces new insights. Further, it is the transformation of such a system, the ways in which it turns into something else, that characterize the system—to the extent that its beginnings may disappear and its endings remain open.

TONE AND *KLANGFARBE*

The notion of "tone color"—or "timbre" or "sound quality"—poses a linguistic problem. As can be expected, the terms to describe sound and its properties are different in English and German. Both the translator of Helmholtz's seminal book *On the Sensations of Tone*, the phoneticist Alexander Ellis, and the editor of a later reprint edition, the historian of science Henry Margenau, were well aware of this. The latter wrote in the introduction to the reprint about Ellis's choice of terms: "The reader owes Ellis a debt of gratitude for sparing him such literal translations as 'clang tint,' proposed by Tyndall for the German 'Klangfarbe,' and rendering it as 'quality.' This term has now been generally adopted."[7] The reader also wishes that Margenau were right, but unfortunately the use of "quality" is the exception in Ellis's translation. In most cases he instead used "tone color," which is a very insightful choice but has not come into use.

With "tone color," Ellis chose a term that provided an interpretation rather than a translation: Helmholtz in fact created a strict definition of *Klangfarbe* that applied to the case of periodic sound phenomena only and designated the term *musikalische Klangfarbe* to refer to this case. In addition to this, he gave descriptions of what he meant by the term *Klangfarbe* in which the term gained a wider meaning. This wider meaning is encompassed by the German word *Klang* but not the English word "tone."

---

[6] This finding is still upheld today; see W. Jay Dowling, "Music Perception," in *Hearing*, ed. Christopher J. Plack, vol. 3 of *The Oxford Handbook of Auditory Science*, ed. David R. Moore (Oxford, 2010), 231–48, on 248.

[7] Helmholtz, *On the Sensations of Tone* (cit. n. 1), intro. [unpaginated].

Calling a sound a tone implies that the sound is heard as a unity, which is not necessarily the case for *Klang*. As these descriptions will be the focus of this article, I have chosen to use the German term *Klangfarbe* and explain the shadings of its meaning where necessary.

While the term appears to have been coined in the German-speaking countries, the content it refers to points to the French *timbre*, which arose in the mid-eighteenth century and prominently figures in Jean-Jacques Rousseau's dictionary as one of three properties (*objets*) one can distinguish in a sound or tone (*son*).[8] In Germany, a notion of *Klangfarbe* that referred to the characteristic sound of musical instruments emerged during the nineteenth century. Whereas Heinrich Christoph Koch's *Music Lexicon* of 1802 had no entry for it,[9] Gustav Schilling's *Encyclopedia of Musical Sciences* of 1838 identified the German *Klangfarbe* as the translation of the French *timbre* and provided the following definition: "K. [*Klangfarbe*] denotes mostly the accidental properties of a voice."[10]

For Schilling, however, Klangfarbe was not central to the notion of "musical sciences." The quality of sounds remained external to the relations of musical notes. Whether a musician played a good or a bad instrument or a singer was indisposed at a concert could not be considered a scientific issue. Klangfarbe or timbre thus epitomized the empirical and contingent aspects of music, as opposed to the rules of musical harmony that were thought to root music in the laws of nature.

The question of the quality of sounds took another direction when experimental physicists started to use instruments for sound generation that were meant for laboratory purposes exclusively. In the first half of the nineteenth century, such experimental physicists as Charles Cagniard de Latour, Félix Savart, Robert Willis, and August Seebeck began to employ such instruments.[11] The question of a sound's specific quality did not initially apply to these instruments. The focus of the experimental physicists' interest was the periodic sound wave, the tone in its relation to the physical concept of frequency. With the help of the new instruments experimental physicists attempted to control the number of vibrations propagated to the air.

If these experiments involved hearing at all it was in order to find out what caused the impression of a tone. Georg Simon Ohm was among the first to ask how Fourier's theorem (discussed in more detail below) could help in this. His main goal, however, was to formalize physicists' knowledge about the movement of sound waves. In his article "On the Definition of the Tone, in Connection with a Theory of the Siren and Other Tone-Producing Apparatuses," he replied to Seebeck's report on experiments

[8] See Rousseau, *Dictionnaire de musique* (Paris, 1768), s.v. "son," on 444–5; on this see Daniel Muzzulini, *Genealogie der Klangfarbe* (Bern, 2005), 248–56.

[9] See Koch, *Musikalisches Lexikon*, ed. Nicole Schwindt et al. (1802; repr., Kassel, 2001); Koch, *Kurzgefaßtes Handwörterbuch der Musik für praktische Tonkünstler und für Dilettanten* (Leipzig, 1807).

[10] *Encyclopädie der gesammten musikalischen Wissenschaften oder Universal-Lexikon der Tonkunst*, ed. Schilling (Stuttgart, 1838), vol. 6, s.v. "timbre."

[11] See Myles W. Jackson, "From Scientific Instruments to Musical Instruments: The Tuning Fork, the Metronome, and the Siren," in *The Oxford Handbook of Sound Studies*, ed. Karin Bijsterveld and Trevor Pinch (New York, 2011), 201–23; on the siren, see also Philipp von Hilgers, "Sirenen: Lösungen des Klangs vom Körper," *Philosophia Scientiae* 7, no. 1 (2003): 85–114; Caroline Welsh, "Die Sirene und das Klavier: Vom Mythos der Sphärenharmonie zur experimentellen Sinnesphysiologie," in *Parasiten und Sirenen: Zwischenräume als Orte der materiellen Wissensproduktion*, ed. Bernhard J. Dotzler and Henning Schmidgen (Bielefeld, 2008), 143–77.

with a siren.[12] While Seebeck had reported the sounds he had heard using a siren, Ohm aimed at supplementing a mathematical theory for the periodic sound of the siren. This instrument was an invention of Cagniard de Latour, consisting of a perforated disc that began to rotate when air was supplied to its openings.[13] When the disc was turning, the air passed through the openings, which cut the airstream into single puffs. Beyond a certain speed of rotation, these air puffs would result in a continuous tone whose pitch increased with the rotation speed.

Among the researchers of his time, Seebeck excelled in observing the audible sounds that resulted from his experiments. He even noted that he sometimes heard tones that he could not explain on the basis of his experimental setups. For example, he heard a tone whose pitch was lower than the number of perforations in his siren's disc suggested. In his reply to Ohm's article, Seebeck's primary interest was not the mathematical formalization of his observations. He reproached Ohm for a miscalculation, but his main charge was that Ohm disregarded the audible phenomena.[14] Ohm's calculus concerned a possible rather than a real form of vibration. In the cases where he mentioned the audibility of the tones and their components, he presupposed this audibility without proving it experimentally.

Mathematics even replaced the ear for Ohm, and here he relied on the most recent knowledge of periodic phenomena. In 1822, French mathematician Jean Baptiste Joseph Fourier published a theorem according to which every periodic vibration is composed of simple, sinusoidal vibrations in integer ratios.[15] Although this theorem promised new insights into all kinds of periodic phenomena through their decomposition into simple components, it remained unclear how such analyses would refer to audible phenomena. For Ohm, Fourier analysis guaranteed the existence of components in sound: "As a means to judge whether or not in a given impression the form $a \cdot \sin 2\pi \, (mt + p)$ exists as a real component, I use the theorem of Fourier that has become famous for its multiple and important applications."[16]

The mathematical analysis that Ohm carried out always stipulated a counterpart in hearing: "If we now imagine some impression $F(t)$ made in the time t upon our ear, then the equation that was mentioned above and that contains Fourier's theorem makes us understand that this impression can be analyzed into the partial impressions $A_0, A_1 \cdot \cos \pi \, \frac{t}{l} + B_1 \cdot \sin \pi \, \frac{t}{l}, A_2 \cdot \cos \pi \, \frac{2t}{l} + B_2 \cdot \sin \pi \, \frac{2t}{l}, A_3 \cdot \cos \pi \, \frac{3t}{l} + B_3 \cdot \sin \pi \, \frac{3t}{l}$, and so forth."[17]

Ohm's notion of a tone built upon this understanding of an "impression." He called

---

[12] August Seebeck, "Beobachtungen über einige Bedingungen der Entstehung von Tönen," *Annalen der Physik und Chemie* 53 (1841): 417–36; Ohm, "Ueber die Definition des Tones, nebst daran geknüpfter Theorie der Sirene und ähnlicher tonbildender Vorrichtungen," *Annalen der Physik und Chemie* 59 (1843): 513–65. Seebeck in turn replied in his article "Ueber die Sirene," *Annalen der Physik und Chemie* 60 (1843): 449–84.

[13] On the siren see Jackson, "From Scientific Instruments to Musical Instruments" (cit. n. 11).

[14] See Seebeck, "Ueber die Sirene" (cit. n. 12). On the debate between Ohm and Seebeck, see R. Steven Turner, "The Ohm-Seebeck Dispute, Hermann von Helmholtz, and the Origins of Physiological Acoustics," *British Journal for the History of Science* 10 (1977): 1–24.

[15] See, e.g., Olivier Darrigol, "The Acoustic Origins of Harmonic Analysis," *Archive for History of Exact Sciences* 64 (2007): 343–424; Johannes Barkowsky, *Das Fourier-Theorem in musikalischer Akustik und Tonpsychologie* (Frankfurt am Main, 1996).

[16] Ohm, "Ueber die Definition des Tones" (cit. n. 12), 519.

[17] Ibid., 520.

a tone an impression when it was caused by a uniform vibration of a certain minimal
duration:

> Impressions that return in a length of $2l$ generate a tone of the amount of vibration $\frac{1}{2l}$,
> if the impressions occupy the same positions in the consecutive intervals in which they
> lie, and if they remain of the same kind for at least the time our organ of hearing requires
> to grasp consecutive waves of tone.[18]

Based on his calculations, Ohm eventually inferred that the fundamental vibration
must be louder than all the others, thus explaining why a tone is heard as possessing
one single pitch. For Ohm, the question of why one single note is heard when many
vibrations occur was solved by way of calculation. In contrast to Seebeck, who did
not presuppose the dominating pitch but reexamined the concept of it through hearing, Ohm sought to illuminate physical phenomena with the help of their mathematical formalization. And while Seebeck relied on hearing to question the acquired
knowledge of physics, Ohm challenged hearing by adding a new formal expression
for sound to this knowledge. Neither of them, however, entrusted their analyses to
those phenomena that music had provided for such a long time.

Long before the publication of Fourier's theorem, acoustics and music had gained
some knowledge about the components of sound. Even the ratios between these components were known by 1700. Marin Mersenne had been among the first to recognize
additional vibrations in a string's movement. As he reported in his *Harmonia Univer-
salis*, he had been able to hear such components:

> When struck or bowed, an open string makes no less than five different sounds at the
> same time, the first of which is the string's natural sound that serves as the fundament for
> the others and that is the unique reference for the singing and instrumental parts in music,
> even more so as the others are so weak that only the best ears can hear them easily.[19]

Mersenne knew the pitches he could expect from a string, as he could resort to the
knowledge on the proportions of strings that had been passed down from antiquity.
Being not only a sensitive but also a learned listener, he could detect the notes produced by a string vibrating in two, three, or four equal parts. Other listeners would
not be disturbed by these additional sounds.

A century later, Jean-Philippe Rameau took advantage of these additional sounds
to provide musical harmony with a new foundation. He found in the sequence of
the fundamental and its harmonics a natural order of sounds that was replicated in
the triadic chords of music. He adopted the term "harmonics" (*sons harmoniques*)
for the additional vibrations from French physicist Joseph Sauveur, who had corroborated their existence and investigated their ratios.[20] It posed no problem for either
Sauveur or Rameau that music notated only one sound where there were many. On

---

[18] Ibid.

[19] Marin Mersenne, *Harmonie universelle, contenant la théorie et la pratique de la musique*, 3 vols.,
ed. François Lesure (1636–7; repr., Paris, 1986), 3:208.

[20] Sauveur, *Principes d'acoustique et de musique; ou, Système général des intervalles des sons, et de
son application à tous les systêmes et à tous les instruments de musique* (1701; repr., Geneva, 1973),
51. For early modern acoustics, see Sigalia Dostrovsky and John T. Cannon, "Entstehung der musikalischen Akustik (1600–1750)," in *Geschichte der Musiktheorie*, vol. 6, *Hören, Messen und Rechnen
in der frühen Neuzeit*, ed. Frieder Zaminer (Darmstadt, 1987), 7–79.

the contrary, the concept of harmonics seamlessly merged into harmonic theory. Rameau explicitly remarked that he heard the harmonics:

> The first note that hit my ear enlightened my understanding. All of a sudden, I was aware that it was not one, or, that the impression it made on my ear was composed: Lo and behold, I said to myself, this is the difference between noise [*bruit*] and sound [*son*]. Any cause that makes an impression on my ear that is one and simple, makes me hear a noise; any cause that makes on my ear a composed impression, makes me hear sound.[21]

A single note thus anticipated polyphonic music and its harmony. Being itself a complex sensory impression, it provoked combination with other notes that corresponded to its composition. A noise, in contrast, was not suitable for music because it did not contain the ordered multitude of tones contained in each musical note. In these deliberations on sounds, tones, and noises, no mention was made of Klangfarbe. The single musical tone was relevant for Rameau in terms of its relations to other musical tones. The musical harmony that resided in every note left no room for Klangfarbe.

Early nineteenth-century debates on tone, such as the dispute between Ohm and Seebeck, did not speak of Klangfarbe, either.[22] In an earlier essay, however, Ohm had already mentioned the possibility that vibrations might differ with regard to their composition. As he noted in his *Remarks on Combination Tones and Beats* (1839), the current hypotheses on the phenomena resulting from the superimposition of sound waves had disregarded the likelihood that instruments differed with regard to their sound components. Because their respective overtones were governed by different laws, the sounds of strings or air columns, for instance, would certainly yield different results when superimposed as compared to those of rods. Ohm's deliberations, however, do not refer to an auditory experience but to Ernst Florens Friedrich Chladni's experiments on the visualization of the complex vibration patterns of sounding plates and membranes. Chladni would disperse sand on the plates, which he then caused to vibrate by bowing them. As the sand remained still on the nodal lines, their patterns became visible. As Chladni explained in his manual on acoustics, these so-called sound figures had often been misunderstood as visualizing different sounds and, more specifically, different pitches of sounds. This was not at all the case, he emphasized. The great superiority of his visualization device over the ear was that the sound figures showed shapes that the ear was unable to detect. His device did not illustrate what was known from music. It made new phenomena accessible.

Because Chladni was his point of departure, Ohm would not integrate hearing into his research. Like Chladni, who had attempted to liberate acoustic research from dependence on the ear as its only observational tool, Ohm did not refer to it in his various essays. Thus, even when the form of sound vibrations came into play, their respective sounds would not contribute to the argument. Sound was discussed independent of its audibility. For the time being, Fourier analysis generated neither a notion of Klangfarbe nor a differentiation of sounds. The perception of a tone seemed to mask the tone's quality, instead focusing the attention on the tone's pitch. At the same

---

[21] Rameau, *Démonstration du principe de l'harmonie, servant de base à tout l'art musical théorique et pratique* (Paris, 1750); reprinted in Rameau, *Complete Theoretical Writings*, Miscellanea 3 (n.p. [American Institute of Musicology], 1968), 1–112, on 12.

[22] Among the first acousticians to use the term was the little-studied Friedrich Zamminer; see his *Die Musik und die musikalischen Instrumente in ihrer Beziehung zu den Gesetzen der Akustik* (Gießen, 1855).

time, the mathematical formalization and visualization of sounds, even though they disregarded hearing, cast into question the habits of hearing. Klangfarbe, however, was mostly absent from the discussion. It only appeared as a negative entity, Helmholtz explained in *On the Sensations of Tone*. As a tone's pitch and intensity referred to a wave's frequency and amplitude, the shape of the sound waves was considered to be the factor that might determine the characteristics of the sound.

SOUNDS FROM AFAR

The status of Klangfarbe changed when sensory physiology called for a redefinition of physical acoustics. The fact that some vibrating movements happen to be audible provided—at least for Helmholtz—a rather weak reason for singling out the study of such vibrations as a separate branch of physics. Acoustics, he argued, had to be based on knowledge of hearing. In *On the Sensations of Tone*, he wrote:

> It is physically indifferent whether observations are made on stretched strings, by means of spirals of brass wire (which vibrate so slowly that the eye can easily follow their motions, and, consequently, do not excite any sensation of sound) or by means of a violin string (where the eye can scarcely perceive the vibrations which the ear readily appreciates).[23]

Switching from eyes to ears was contingent on the observed physical phenomena, Helmholtz explained. The human mode of observation did not change the physical phenomena in any respect.

> The only justification for devoting a separate chapter to acoustics in the theory of the motions of elastic bodies, to which it essentially belongs, is, that the application of the ear as an instrument of research influenced the nature of the experiments and the methods of observation.[24]

Beyond the audibility of sound, no delineation of physical acoustics was available. Therefore, the science of sound had become a science of hearing.

In addition to physiology, mathematics challenged the established notion of sound. Fourier's theorem provided a tool for which no acoustic object had yet been found. Music did not help with this problem. As long as sounds could be subsumed under simple ratios, music theory felt no need for additional mathematical description. Rather than explaining some well-known problem, Fourier analysis solved problems in the study of vibrations, whose application to audible phenomena remained uncertain.

Helmholtz turned this upside down. He took up problems in the study of hearing and related the mathematical theorem to the faculty of the ear to distinguish sounds. The sounds of a flute and an oboe could have the same pitch and still differ in terms of the periodic waves that represented them for the physicist. This was not a vain attempt to formalize a vibration, but rather a mobilization of the very faculty of the ear to distinguish between the two instruments. This, however, required a new definition of Klangfarbe. In this, Helmholtz followed a double strategy. On the one hand, he

---

[23] Helmholtz, *On the Sensations of Tone* (cit. n. 1), 3.
[24] Ibid., 4.

restricted Klangfarbe to a very narrow definition whose purpose was to make the defined objects accessible to mathematical formalization:

> We shall at first disregard all irregular portions of the motion of the air, and the mode in which sounds commence or terminate, directing our attention solely to the musical part of the tone, properly so called, which corresponds to a uniformly sustained and regularly periodic motion of the air, and we shall endeavour to discover the relations between the quality of the sound and its composition out of individual simple tones. The peculiarities of quality of sound belonging to this division, we shall briefly call its *musical quality* [*musikalische Klangfarbe*].[25]

On the other hand, he embedded his notion into a narrative discussion of all those properties of sounds he did not have the means to investigate. Thus, the newly defined term on the one hand denoted those vibrations that could be formally described according to Fourier's theorem, and on the other hand referred to those properties that were perceived as distinguishing sounds.

Helmholtz discussed at length the beginnings and endings of sounds as well as the accompanying noises, even though his strict definition excluded them. For instance, he wrote that "it is very characteristic of brass instruments, as trumpets and trombones, that their tones commence abruptly and sluggishly. . . . It always requires a certain amount of effort to excite the new condition of vibration in place of the old, but when once established it is maintained with less exertion."[26] For woodwind instruments, in contrast, the tone is as easily changed as the length of the air column, by application of the fingers to the side holes, while the style of blowing remains the same. Such properties of sound generation, particularly important for differentiating the sound of instruments, were also significant in the sounds of speech. With the letters of the alphabet, the characteristics of sound could even be written down, the consonants denoting the beginnings and endings and the vowels the color of the sounds in the narrower sense of the notion.

More specifically, Helmholtz termed his narrow definition *musikalische Klangfarbe*, as it was meant to denote only those properties of a sound that defined it as a musical one. Musikalische Klangfarbe was restricted to the stationary sound, as it occurs in bowed string and woodwind instruments. It is in this part of the sound that its pitch can best be detected. The noises of the attack or those accompanying the movement of a bow or the air in an organ pipe were excluded from the definition. Helmholtz admitted that the definition required abstracting from many of the phenomena that constitute an instrument's sound, but he also pointed to specific situations when this definition came close to the actual impression on hearing:

> Such accompanying noises and little inequalities in the motion of the air, furnish much that is characteristic in the tones of musical instruments, and in the vocal tones of speech which correspond to the different positions of the mouth; but besides these there are numerous peculiarities of quality belonging to the musical tone proper, that is, to the perfectly regular portion of the motion of the air. The importance of these can be better appreciated by listening to musical instruments or human voices, from such a distance that the comparatively weaker noises are no longer audible. Notwithstanding the absence of these noises, it is generally possible to discriminate the different musical instruments,

---

[25] Ibid., 68–9.
[26] Ibid., 67.

although it must be acknowledged that under such circumstances the tone of a French horn may be occasionally mistaken for that of the singing voice or a violoncello may be confused with an harmonium.[27]

This passage describes a situation in which the faculty of discrimination is confronted with a minimal difference between two sounds. Yet, even when a sound could no longer be traced back to its source because this source was hidden from sight, and even though noises that could be heard at a short distance got lost in the course of the sound's propagation, some discrimination was still possible. Such sounds that were remote from their sources still possessed their own traits, although under these unfamiliar circumstances sounds became related to each other whose sources would not suggest the relation. This was, for instance, the case with the cello, a bowed string instrument, and the pipes of the reed organ. The distant sounds remained sufficiently different to be distinguishable. The discrimination was given over completely to hearing. Sound and its carrier medium, thereby, obtained their own materiality.

SYNTHESIS OF SOUNDS

Although physical acoustics focused on the periodic sound wave in the first half of the nineteenth century, little was known about the connection between the movements of sound sources and the shape of the propagating waves.[28] If, however, the theorem of Fourier applied, at least one sound source was well known, and this was sound itself. When a sound wave hits a body that is capable of vibration, it can excite that body to vibrate if the body's resonant frequency is present in the wave. When this is the case, even a minute repeated impact can cause a vibration. To illustrate this, Helmholtz, like other researchers before him, evoked the example of ringing a bell: "It is known that the largest church-bells may be set in motion by a man, or even a boy, who pulls the ropes attached to them at proper and regular intervals, even when their weight of metal is so great that the strongest man could scarcely move them sensibly, if he did not apply his strength in determinate periodical intervals."[29] Once such a bell is set in motion it continues to swing back and forth for some time, as would any physical system capable of oscillation. Whoever rings the bell therefore can use the system's capacity to oscillate to accumulate small amounts of force. One only has to keep pulling the rope downward a little further when the lever that holds the bell moves the rope downward anyway. The bell will then start ringing once the movement's amplitude has grown to the point where the clapper hits the sound rim. In contrast, if one kept pulling on the rope at a moment when it was being lifted by the lever, the bell would eventually stop ringing. What matters in this case is only the moment at which the rope is pulled. Similarly, Helmholtz explains, any system capable of vibration will be set into motion when struck in the appropriate frequency.

Understood in this way, resonance turned the relation between sound and carrier medium upside down: a sound resulting from resonance has no source in the usual sense, but at the same time it has two sources. It has no source because the resulting

[27] Ibid., 68.

[28] Robert T. Beyer, *Sounds of Our Times: Two Hundred Years of Acoustics* (New York, 1999), 50.

[29] Helmholtz, *On the Sensations of Tone* (cit. n. 1), 36; this example occurs also in Ernst Wilhelm Weber, *De Pulsu, Resorptione, Auditu et Tactu: Annotationes Physiologicae et Anatomicae* (Leipzig, 1834), 29.

sound has only those properties that the two vibrating systems share. It also has two because the exciting sound wave hits upon a new sounding body. Using this effect, Helmholtz created simple tones from the resonance between tuning forks and strings or hollow bodies. Starting from his first publication in the physiology of hearing, he introduced this new type of sound and explained how it could be produced. Even though the audible sounds only approximated their mathematical definition, which presupposed infinite duration, they came as close to a sinusoidal wave as possible. This sound did not depend on a specific materiality, as its production could involve several instruments of different materials that had nothing in common except resonant frequency.

This sound questioned the separation between pitch and Klangfarbe. It made audible the fact that frequency did not correspond to pitch as neatly as was assumed, nor could tone color be identified with a sound's characteristics. On the one hand, Helmholtz ascribed certain characteristics to simple tones, relating them to the high or low register:

> These tones are uncommonly soft and free from all shrillness and roughness . . . they appear to lie comparatively deep, so that such as correspond to the deep tones of a bass voice produce the impression of a most remarkable and unusual depth. The musical quality of such deep simple tones is also rather dull. The simple tones of the soprano pitch sound bright, but even those corresponding to the highest tones of a soprano voice are very soft, without a trace of that cutting, rasping shrillness which is displayed by most instruments at such pitches.[30]

On the other hand, the simple tones demonstrated that the very idea of pitch had to be related to Klangfarbe. This could be seen from the fact that famous musicians had erroneously estimated their pitch too high or too low.[31] The pitch of simple tones was in fact difficult to determine by the ear because pitch was usually recognized on the basis of more than one datum.

Yet the strict notion of Klangfarbe had to be denied to the simple tones. A simple tone could not have its own color, although Helmholtz characterized it as dull and soft. It was the elementary level of Klangfarbe because its sound could not be further differentiated. Since the form of simple waves of known periodic time was completely given when their amplitude was given, simple tones of the same pitch could only differ in force and not in musical quality.[32] As the basic element of all sound, the simple tone's color had to be neutral.

The neutral components of sound usually passed unnoticed. They did not occur in nature. Therefore, the ear had no practice in discriminating them, as Helmholtz repeatedly remarked.

> The ultimate simple elements of the sensation of tone, simple tones themselves, are rarely heard alone. Even those instruments by which they can be produced, as tuning-forks before resonance chambers, when strongly excited, give rise to weak harmonic upper partials, partly within and partly without the ear. . . . Hence in this case also, the opportunities are very scanty for impressing on our memory an exact and sure image of these simple elementary tones. . . . We are even without the ease that can be obtained by

---

[30] Helmholtz, *On the Sensations of Tone* (cit. n. 1), 69.
[31] Ibid.
[32] Ibid.

frequent repetitions of the experiment, such as we possess in the analysis of musical chords into their individual tones. In that case, we hear the individual tones sufficiently often by themselves, whereas we rarely hear simple tones and may almost be said never to hear the building up of a compound from its simple tones.[33]

The reason why the ear had no practice in recognizing the simple components was that the usual task for the ear was to discriminate sounds, rather than to analyze them. As long as sounds always occurred in the same compound way, there was no need to determine the partials: "We practise observation on sensation only to the extent necessary for clearly apprehending the external world."[34]

THE VOWELS AS A SET OF KLANGFARBEN

Helmholtz understood hearing as a means of access to the representations of external objects only insofar as it allowed for the identification of sound sources. As in the other senses, the data merged into an outer object when repeated experience corroborated the relations between these data.[35] This experience made listeners then hear a clarinet, or a violin, for instance. Music, however, did not employ such representations, Helmholtz explained: "When in hearing a concert we recognise one tone as due to a violin and another to a clarinet, our artistic enjoyment does not depend upon our conception of a violin or clarinet, but solely on our hearing of the tones they produce."[36]

Though it may be easy to discriminate a violin from a clarinet, it was not this capacity that was essential in music listening. It was instead the ability to abstract a single pitch from the mass of data that constituted one compound sound. And it was this way of hearing that was practiced most extensively in music. Here, Klangfarbe was ignored, and so was the fact that pitch depended on the presence of several components, since it had not yet been probed by such sounds as the simple tones. The systemic coherence of music seemingly did not need real sounds at all. The music theory of that time omitted Klangfarbe, claiming, in all events, priority for the tonal relationships.

For his sound synthesis, Helmholtz chose vowels as the object of recognition. Thus, he did not choose a material musical instrument with one specific sound character, but the product of the mouth's malleability. In addition, his term *musikalische Klangfarbe* applied to these sounds. Among the sounds of language, the vowels were the most appropriate for singing. Through their double function in language and singing, the vowels smuggled musical elements into language and speech elements into music. In both cases, these intruding elements merged into the other system while retaining some of their properties. In music, the sounds of vowels were exempt from the rules of composition. They did not have the status of musical instruments either, as shown by orchestration manuals. Although the human voice was counted among musical instruments and, further, composers were sensitive to text setting, the use of vowels was secondary. The human voice was characterized by the individual timbre

[33] Ibid., 65.
[34] Ibid., x.
[35] On "structural objectivity," see Lorraine Daston and Peter Galison, *Objectivity* (New York, 2007), 253–307.
[36] Helmholtz, *On the Sensations of Tone* (cit. n. 1), 3.

of each singer. Composers busied themselves with how they could prescribe the use of chest or head voice, and the preference for certain vowels was considered a consequence of this problem. Hector Berlioz, for instance, in his 1841 treatise on instrumentation, wrote, "This one will sing with great ease sustained phrases in the high register, in all nuances and any kind of movement; in the lyrics he will like the *i*'s and *e*'s." He continued: "This other one has a painful head voice and prefers to sing vibrating sounds of the chest all the time; this one will excel in passionate roles—he will prefer open syllables and sonorous vowels such as *a*."[37]

The voice singing vowels belonged to two regimes of sound organization at once. This was because two independent mechanisms of articulation take part in the singing of different vowels. Helmholtz described these mechanisms in his discussion of the sounds of different instruments. He explained that the vocal cords resembled a musical instrument in which the sound was produced by "membranous tongues."[38] These tongues opened and closed in rapid alternation, thereby interrupting a stream of air and thus producing a sound. The same process applied in brass instruments, where the lips caused the interruption of the air stream. The characteristic sound of the vowels was obtained when the vibrating airstream passed through the mouth cavity. This was the second mechanism. For each vowel, the mouth cavity took on a different shape and thus acted as a resonator for different components in the sound of the vocal cords. This implied, however, that resonance also took part in shaping the vowel sound. To some extent, the vowel sound was, therefore, independent from its material conditions, as were the simple tones. Any means of reinforcing the simple components in the same way achieved through the resonance of the mouth cavity would, therefore, produce a sound that resembled the vowels.

A number of experimenters had explored this before Helmholtz. Starting with Johann Caspar Kratzenstein's attempt in the eighteenth century to imitate the mouth cavity's various shapes in a series of organ pipes, these experiments had little by little given up the idea that an imitation of the vowel sounds needed a sound source that resembled the human vocal apparatus.[39] In Wolfgang von Kempelen's speaking machine, the resemblance to the human apparatus was equally present in a soft funnel whose shape was manually molded in order to obtain the appropriate sound. While

[37] Berlioz, *De l'instrumentation*, ed. Joel-Marie Fauquet (Bordeaux, 1994), 109.

[38] Helmholtz, *On the Sensations of Tone* (cit. n. 1), 97.

[39] The story of speaking devices has often been told, starting from the accounts by Sir Charles Wheatstone, through Giulio Panconcelli-Calzia's overview of research in experimental phonetics, and up to its integration into Homer Dudley's and Gerold Ungeheuer's reports on their empirical research; also, Friedrich Kittler, Joachim Gessinger, Thomas L. Hankins and Robert J. Silverman, Brigitte Felderer, and Mara Mills have investigated the devices from the viewpoint of the history of science and the history and theory of media. See Wheatstone, "Willis on Reed Organ Pipes, Speaking Machines etc.," *London and Westminster Review* 6 (1838): 27–41; Panconcelli-Calzia, *Geschichtszahlen der Phonetik: Quellenatlas der Phonetik*, ed. Konrad Koerner (1940–1; repr., Philadelphia, 1994); Dudley and T. H. Tarnoczy, "The Speaking Machine of Wolfgang von Kempelen," *Journal of the Acoustic Society of America* 22 (1950): 151–66; Ungeheuer, *Elemente einer akustischen Theorie der Vokalartikulation* (Berlin, 1962); Ungeheuer, "Über die Akustik des Vokalschalls im 18. Jahrhundert: Der Euler-Lambert-Briefwechsel und Kratzenstein," *Phonetica* 40 (1983): 145–71; Kittler, *Gramophone, Film, Typewriter*, trans. Geoffrey Winthrop-Young and Michael Wutz (Stanford, Calif., 1999) (original German ed. published Berlin, 1986); Gessinger, *Auge und Ohr: Studien zur Erforschung der Sprache am Menschen, 1700–1850* (Berlin, 1994); Hankins and Silverman, *Instruments and the Imagination* (Princeton, N.J., 1995), 179–220; Felderer, "Stimm-Maschinen: Zur Konstruktion und Sichtbarmachung menschlicher Sprache im 18. Jahrhundert," in *Zwischen Rauschen und Offenbarung: Zur Kultur- und Mediengeschichte der Stimme*, ed. Kittler, Thomas Macho, and Sigrid Weigel (Berlin, 2002), 257–78; Felderer and Ernst Strouhal, *Kempelen—Zwei Maschinen: Texte, Bilder und Modelle*

the speaking machine on the whole had a strong effect of disconnecting the acoustic phenomenon of the voice from the human body, the machine itself was based on the imitation of the organs of speech.[40] In a long article, "On Vowel Sounds, and on Reed-Organ Pipes," British physicist Robert Willis criticized his forerunners for this: "Kempelen's mistake, like that of every other writer on this subject, appears to lie in the tacit assumption, that every illustration is to be sought for in the form and action of the organs of speech themselves, which, however paradoxical the assertion may appear, can never, I contend, lead to any accurate knowledge of the subject."[41]

Willis emphasized that the means to produce the sounds did not have to resemble the organs of speech. He described an experiment that broke completely with the idea of such similarity. When a piece of watch spring was held against a revolving toothed wheel, an alternation of sound qualities was produced that depended on the length of the vibrating portion of the spring: "In effect the sound produced retains the same pitch as long as the wheel revolves uniformly, but puts on in succession all the vowel qualities, as the effective length of the spring is altered, and that with considerable distinctness, when due allowance is made for the harsh and disagreeable quality of the sound itself."[42]

Not only did these tools have nothing in common with the human organ of speech, but the resulting sounds also had little in common with the human voice. Willis insisted that the vowels were characterized by a quality other than the timbre of musical instruments. "Vowels are quite a different affection of sound from both pitch and quality," he explained.

> By quality, I mean that property of sound, by which we know the tone of a violin from that of a flute or of a trumpet. Thus we say, a man has a clear voice, a nasal voice, a thick voice, and yet his vowels are quite distinct from each other. Even a parrot, or Mr. Punch, in speaking, will produce A's, and O's, and E's, which are quite different in their quality from human vowels, but which are nevertheless distinctly A's, and O's, and E's.[43]

To create the voice of "Mr. Punch," puppet players held a piece of metal in their mouths in order to give their voices a distorted sound. Each vowel produced in this kind of speech would hardly be recognized in isolation, yet Mr. Punch's utterances could be understood.

While criticizing Willis for continuing to emulate the dual nature of vocal sound production in his combination of the toothed wheel and the watch spring, Helmholtz grasped the advantage of vowels for experiment. Willis had discovered a systemic effect in language. He attributed this to a specific quality, and yet all his experiments and explanations pointed to the importance of the quick succession between the

---

zur Sprechmaschine und zum schachspielenden Androiden Wolfgang von Kempelens (Vienna, 2003); Mills, "Medien und Prothesen: Über den künstlichen Kehlkopf und den Vocoder," in *Klangmaschinen zwischen Experiment und Medientechnik*, ed. Daniel Gethmann (Bielefeld, 2010), 127–52.

[40] Kempelen, *Mechanismus der menschlichen Sprache nebst der Beschreibung seiner sprechenden Maschine* (1791; repr., Stuttgart, 1970). French and Italian could be produced more easily than German. Also, the sentences could not be too long, as Kempelen admitted: "Ganze Redensarten kann ich nur wenige und kurze sagen, weil der Blasebalg nicht groß genug ist, den erforderlichen Wind dazu herzugeben. Z. B. vous etes mon ami" (455–6).

[41] Willis, "On Vowel Sounds, and on Reed-Organ Pipes," *Transactions of the Cambridge Philosophical Society* 3 (1830): 231–68, on 233.

[42] Ibid., 249–50.

[43] Ibid., 234.

vowels. Willis explained: "In repeating experiments of this kind, it must always be kept in mind that the difference between the vowels depends entirely upon contrast, and that they are therefore best distinguished in quick transitions from one to the other, and by not dwelling for any length of time upon any of them."[44] Vowels would readily be discriminated whenever a quick and consistent change in sound quality could be effected.

Helmholtz chose vowels as his object of experimentation, and in 1858 he had an apparatus for the synthesis of vowels constructed. The Bavarian king sponsored his research, hoping to encourage an invention of similar impact to Helmholtz's ophthal-moscope. The "apparatus for the artificial creation of vowels" consisted of a set of eight tuning forks and resonators that were tuned to a Fourier series on the note b_1. The eight simple tones that resulted from this arrangement were combined with an electromagnet and another tuning fork tuned to b_1 that served as an interrupter. This device set the resonating forks in motion until the electromagnet was turned off. The beginnings of the sounds were almost unnoticeable, as this device did not require touching the tuning forks, and the duration of the sounds was almost infinite. This was the closest reenactment of a Fourier series possible. In addition, a keyboard allowed Helmholtz to diminish or shut off single forks, which enabled him to move quickly from one combination of simple tones to another and thus profit from the systemic effect of the vowels.

With this complex instrument Helmholtz synthesized the vowels of the German language. While the vowels *a*, *o*, and *u* turned out well, *e* and *i* proved more difficult. Helmholtz admitted that he had to make the other vowels duller than necessary to produce a recognizable *e* (in a later version of the experiment he received better results with some additional forks). In this experiment, a minimal discrimination of sounds was approached from a different angle than in the example of sounds heard from afar. In that example, all the components of language that would transform into musikalische Klangfarbe had disappeared, yet the remaining sounds could be distinguished. In the experiments with tuning forks, however, these components were suppressed from the very beginning. The resulting sound was the best acoustic embodiment of the new definition of Klangfarbe. The sound synthesis systematically reified the mathematical description of a periodic wave as closely as possible. Yet again, the resulting sounds could be discriminated. Later researchers, however, reported that when trying to reenact this experiment they were unable to recognize anything.[45]

A few more aspects of this sound synthesis were significant. One was that each of the single components could be made to sound alone and thus be understood as an independent sound coming from its own source. Another was that the systemic effect of producing the complete range of vowels, rather than isolated sounds, enhanced

[44] Ibid., 234.

[45] See, e.g., the critical remarks by Alexander Ellis, phonetician and translator of *On the Sensations of Tone*, in that work (cit. n. 1), 543; see also David Pantalony, *Altered Sensations: Rudolph Koenig's Acoustical Workshop in 19th-Century Paris* (New York, 2009). Helmholtz's experimental synthesis has been the object of numerous excellent studies, which describe it from different angles; see, e.g., Stephan Vogel, "Sensation of Tone, Perception of Sound, and Empiricism: Helmholtz's Physiological Acoustics," in *Hermann von Helmholtz and the Foundations of Nineteenth-Century Science*, ed. David Cahan (Berkeley, Calif., 1993), 259–87; Timothy Lenoir, "Helmholtz and the Materialities of Communication," *Osiris*, 2nd ser., 9 (1994): 184–207; Patrick J. McDonald, "Demonstration by Simulation: The Philosophical Significance of Experiment in Helmholtz's Theory of Perception," *Perspectives on Science* 11 (2003): 170–207.

the recognition of language sounds and thereby helped mask the provenance from separate sources. But most important, the unmarked beginnings and complete lack of noise helped merge the components into one sound. These aspects became crucial for the acoustic technology that emerged in the wake of Helmholtz's study of the ear. The fact that the ear can be easily deceived about the origin of a sound was the basic insight that enabled transmitting and synthesizing sound. In fact, to ask about a sound's location was already to pose the wrong question, because this location was the ear, in the first instance. This implied a model of sensory perception that relied on reenactment, rather than representation. Helmholtz even went so far as to completely deny a faculty of sound localization to the sense of hearing.

At the same time, the properties of sounds that facilitated the proof of this hypothesis created a new and unheard sound, which continued to be explored by researchers as well as composers. These explorations will be discussed in the remaining two sections of this article. Beginning in 1910, the experimental psychologist Carl Stumpf submitted Helmholtz's work on hearing to a reexamination that included testing the importance of the beginnings and endings of sounds. At about the same time, the composer Arnold Schoenberg developed his concept of *Klangfarbenmelodie*, which extended sound synthesis into aesthetic experimentation.

A PSYCHOACOUSTICS APPROACH

If resonance created a sound that had no source and thus was spatially underdetermined, Fourier synthesis created a sound that had no temporal determination. Helmholtz's approximation of a mathematical model that implied the infinite duration of its elements could only succeed because the vibrations of the electromagnetically driven tuning forks reduced the habitual components of the beginnings and endings of sound. The synthesized sounds had no specific envelope—that is, they almost did not change in time—and the same held for their compounds.

Such temporal aspects had to be ignored in Helmholtz's approach. It had seemed that the only way to investigate a sound was to ensure identical reproduction, arguably impossible before sound recording. Helmholtz, however, preferred another way: he prolonged the sounds until he had examined them. This ensured the identity of the investigated sounds but it also created artifacts. Most important, it created a new relation between acoustic research and music. Many musical instruments proved too unstable in their sound production to convey proper research objects. Although he famously used the piano as a model for the physiological process of hearing, Helmholtz preferred the reed organ to the piano for experiments on musical objects. The sounds created by a piano were too short and contained too much change for his investigations of combination tones, beats, and tuning as well as the musical notions of consonance and dissonance, major and minor mode, or harmonic affinity.[46]

In experimental sound synthesis, however, the aspect of duration was congruent with another aspect; namely, the lack of a marked beginning and ending. It was this

[46] On Helmholtz and music, see Benjamin Steege, *Helmholtz and the Modern Listener* (Cambridge, 2012); Alexandra Hui, *The Psychophysical Ear: Musical Experiments, Experimental Sounds, 1840–1910* (Cambridge, Mass., 2012); Veit Erlmann, *Reason and Resonance: A History of Modern Aurality* (New York, 2010), 217–306; Julia Kursell, "Wohlklang im Körper: Kombinationstöne in der experimentellen Hörphysiologie von Hermann v. Helmholtz," in *Resonanz: Potentiale einer akustischen Figur*, ed. Karsten Lichau, Viktoria Tkaczyk, and Rebecca Wolf (Munich, 2009), 55–74.

latter aspect that Stumpf singled out to reinvestigate Helmholtz's claim about the discrimination of sounds. Stumpf started to experiment on the Klangfarbe of musical instruments in 1910. His experiments were part of a larger project of examining Helmholtz's work on hearing from the perspective of psychology. In 1886, Stumpf published the first of two volumes of his *Tone Psychology*; the second volume followed in 1890. In this work, he studied the perception of the basic elements of music. As he explained in the foreword to the first volume, he planned to move from single notes to their simultaneous combinations—that is, to musical chords—and, eventually, to the rules of harmony and composition. The project was not realized as planned, however. Stumpf wrote a number of articles concerning the concepts of consonance and dissonance, such as their history and their application to musical chords, but he did not venture to merge them into his *Tone Psychology*. Instead, he examined the sounds of speech and of musical instruments in a third major publication. This was published in 1926, under the title *Die Sprachlaute* (The sounds of language). His experiments on the discrimination of Klangfarbe were integrated into this book as an appendix.[47]

Stumpf began his experiments by collecting actual descriptions of sounds heard from afar. He placed musicians and experimental subjects in two adjacent rooms in the spacious bourgeois apartment that hosted his laboratory in Berlin. These rooms communicated through an opening the size of the palm of a hand. This aperture could be opened and closed, thereby enabling control of the beginning and ending of the sound the subjects would hear. The listeners were then confronted with a short section of a tone in such a way as to conceal its characteristic beginning and ending from them and to present them only with the middle part, which allegedly corresponded best to the representation of musikalische Klangfarbe.

The experimental subjects were clearly experts, as Stumpf reported. Among them were musicians, acousticians, psychologists who were familiar with acoustic experimentation, and, last but not least, a musical instrument maker. All of them were experienced in judging the sound of musical instruments. The sounds they were supposed to discriminate came from various string, woodwind, and brass instruments, such as a violin and a cello; a flute, a clarinet, and a bassoon; a French horn, a trumpet, and a cornet; and, finally, a tuning fork with resonator, which generated a simple tone. When the sounds were separated from their beginnings and endings, the listeners could identify only about 50 percent of the instruments. The instrument maker had significantly better results, as he had manufactured a few of the tested instruments. Nevertheless, Stumpf asserted that his results corroborated Helmholtz's claim that the stationary part of a sound could suffice to describe Klangfarbe.

While Helmholtz saw no way to experimentally verify his narrative description of sounds heard from afar, Stumpf saw in Helmholtz's description a temporal problem, rather than a spatial one. Consequently, he replaced the effect of distance with a modification of the sounds' development in time. In this process, Stumpf transformed the trajectory of the sounds into the transmission of signals. While Helmholtz could predict the loss of energy that occurred in the transmission of a sound from its source to the ear, he did not conceive of it in terms of a signal.

It was crucial for Helmholtz's model of hearing to describe the ear as the place

---

[47] Stumpf, *Die Sprachlaute: Experimentell-phonetische Untersuchungen nebst einem Anhang über Instrumentalklänge* (Berlin, 1926).

where sound emerged. He therefore would not admit a process of transmission that exposed the activity of the ear as deficient. Being familiar with a model of representation in which sound was assumed to exist eminently outside the ear, his readers might not have distinguished such a concept from the contemporary idea of the propagation of sound.[48] Only after the ear had been established as the place where sound emerged could the distinction between source, sound, and ear be reformulated in terms of sender, message, and receiver. And only after distortion had been located within the ear, as Helmholtz did in his book, could the concept of distortion be discriminated from the alleged deficiency of the ear in contrast to the "true" sound outside of it.

Helmholtz's description of remote sounds, together with the definition of musikalische Klangfarbe that it was supposed to exemplify, now turned into Stumpf's description of an experiment in which real sounds were modified in order to defamiliarize them. The experts who listened to these sounds had no access to the sources. Just as in Helmholtz's description, they had to rely on what they heard and thus found themselves in the position that Helmholtz claimed for the ear. The controlled distortion of the transmission made the listeners into receivers and turned the sounds into signals.

ORCHESTRATING SYNTHESIS

This article started from the idea that *Klangfarbe* was a term that acoustics borrowed from music and redefined for acoustic purposes. In this last section, I will return to music. The relation between experimental and musical experience can be discussed in the terms that philosopher Gilles Deleuze has suggested to describe the position of aesthetics between individual experience and sensory perception in general: "Aesthetics suffers from a wrenching duality. On the one hand, it designates the theory of sensibility as the form of possible experience; on the other hand, it designates the theory of art as the reflection of real experience. For these two meanings to be tied together, the conditions of experience in general must become conditions of real experience; in this case, the work of art would really appear as experimentation."[49]

By the early nineteenth century, experimental acoustics ceased to be dominated by the sounds of musical instruments. The experiments carried out with alternative devices, most famously the siren, were directed toward understanding the phenomenon of frequency, as mentioned above. The synthesis of vowels, however, suspended the question of tone. This was achieved by fixing the synthesized sounds to one single pitch. Helmholtz's apparatus of tuning forks and resonators built a Fourier series over b_1. This complex instrument that used frequency in the functions of sound production and electromagnetic interruption forbade transposing this series. A second frequency would have required a second apparatus.

The idea of synthesizing different Klangfarben on one pitch prominently featured the systemic effect of the vowels. A change of pitch in parallel to the change of color would have masked the latter. If the apparatus had played a melody, this would certainly have absorbed the attention of observers. Equally, the pitches that constituted the Fourier series disappeared once the sound merged into one vowel. In this respect,

---

[48] See Muzzulini, *Genealogie der Klangfarbe* (cit. n. 8), 367–75.
[49] Gilles Deleuze, *The Logic of Sense*, trans. Mark Lester with Charles Stivale, ed. Constantin V. Boundas (New York, 2001), 288–9.

the choice of the vowels turned out to be helpful. The nonmusical object helped to draw attention away from the tonal relations present in the Fourier series. In short, Helmholtz invented a conceptual device, not a musical instrument. As such, it shed new light on the concept of tone, which was in turn taken up by composers in the early twentieth century.

Arnold Schoenberg ends his manual of tonal harmony, written in 1911, with a utopian vision of a music in which the tonal relations would not dominate but would instead dwell in a realm of Klangfarbe: "The distinction between tone color and pitch, as it is usually expressed, I cannot accept without reservations. I think the tone becomes perceptible by virtue of tone color, of which one dimension is pitch. Tone color is, thus, the main topic, pitch a subdivision. Pitch is nothing but tone color measured in one direction."[50]

With this, Schoenberg proposed not only a new understanding of Klangfarbe, but also a new understanding of tone. In the German original, the usual expression for pitch, which contains the word "tone," was replaced by a neologism built from components that attribute pitch to a compound: *Tonhöhe* became *Klanghöhe*. This gave precedence to Helmholtz's new terminology for timbre and tone, creating equivalence between the objects and their color.[51] Schoenberg thus reversed the hierarchy of tone and color. In the traditional concept, color could be added to a note, but it remained accidental. For Schoenberg, no tone without color was possible; a color could, but did not necessarily have to, become a tone. The distinction between essence and accident was obsolete. If, in the traditional concept, pitch had been arbitrarily isolated as a parameter, this should be possible for other aspects of sound as well.

Schoenberg therefore postulated:

> Now, if it is possible to create patterns out of tone colors that are differentiated according to pitch, patterns we call "melodies," progressions, whose coherence (*Zusammenhang*) evokes an effect analogous to thought processes, then it must be possible to make such progressions out of the tone colors of the other dimension, out of that which we call simply "tone color," progressions whose relations with one another work with a kind of logic entirely equivalent to that logic which satisfies us in the melody of pitches.[52]

Schoenberg called these progressions "tone-color melodies" (*Klangfarbenmelodien*).[53] The third orchestral piece in his opus 16 shows how differentiations of Klangfarbe emerge. Giving his piece a title—by request of the editor—Schoenberg called it *Farben*, and in a later edition of 1922, he added the subtitle *Ein Sommermorgen am See*.[54] In it, all the musicians play notes of the same duration and pause before they

---

[50] Schoenberg, *Theory of Harmony*, trans. Roy E. Carter (Berkeley, Calif., 1978), 421.

[51] The English translation, which captures very well Schoenberg's style of English, also very correctly uses the term "tone color," which had been introduced by Helmholtz's translator Ellis but never came into use in English. In German, Helmholtz's term *Klangfarbe* had a rather successful career, as it still denotes sound color in scientific and—mainly classical—musical contexts.

[52] Schoenberg, *Theory of Harmony* (cit. n. 50), 421.

[53] Ibid., 422. See also Rainer Schmusch, "Klangfarbenmelodie," in *Terminologie der Musik im 20. Jahrhundert*, ed. Hans Heinrich Eggebrecht, special vol. 1 of *Handwörterbuch der musikalischen Terminologie* (Stuttgart, 1995), 221–34; Alfred Cramer, "Schoenberg's Klangfarbenmelodie: A Principle of Early Atonal Harmony," *Music Theory Spectrum* 24, no. 1 (2002): 1–34.

[54] In his diary for 1912 Schoenberg noted as the title of the piece the word *Akkordfärbungen*; in the fourth revised edition of op. 16 from 1922, the piece has the title *Farben* (*Sommermorgen am See*); see

begin another note, again of the same duration, but the points at which the different instruments begin each note are concealed in various ways. The notes overlap, creating a smooth transition from one stable sound to the next. The instruments taking part in one sound unnoticeably fade and are replaced—equally unnoticeably—by new instruments. At some point, a new sound is present, composed of new pitches and shades, whose starting point cannot be denoted because the process of replacement happens so subtly, and neither the temporal components nor the components of the sound can be clearly distinguished at any moment. Schoenberg requests this smooth transition explicitly in the score of the piece: "The change of the chords must happen so gently that no accent can be noticed on the beginning instruments, in order to make [the change of chords] noticeable only through the change of color."[55]

This impedes an individuation of the instruments in the traditional sense of the instrumental "voice" or part. None of the instruments is supposed to play a leading melody. Rather, the musicians are asked to reduce their playing to the stationary part of their instruments' sounds. This means that they must try to suppress the noises correlated with the beginning of a sound, maintain a steady intensity while they play, and fade out softly at the end of the note so that its disappearance is unnoticeable. Any two notes are separated by a break, so the musicians can give the maximum attention to the production of their note.

The overall sound of this piece is that of a constant flow, in which no individual movement occurs and which is only slowly hovering. The slow pace of movement in the piece adds another component to this effect. Any instrument can play at the tempo Schoenberg chose for his piece. This is all the more remarkable as most of them could not necessarily play any more slowly or more quickly. The clarinetists, for instance, would have difficulty if asked to play longer notes, whereas the bass is not appropriate for faster movements. Although bass players are able to play fast notes, a listener would have trouble grasping them because the low frequencies of this instrument take more time to be recognized by the ear than the high notes of a clarinet. In the resulting sound, a listener cannot pick out the colors of individual instruments. Instead they are merged into a new, previously unheard color. The harmonic progression is in the service of this sound creation. It does not follow the rules of harmonic progression, but freely adds sound to sound. This happens in a fashion that juxtaposes and overlaps the tones so as to create dissonant and unruly chords that do not immediately evoke the demand of harmonic resolution in expert listeners. The harmonic effect that typically makes chords gravitate toward a continuation is suspended. Rather than traditional harmony, the listener hears a harmonious flow of shimmering sound shades. Here and there, short motives spring out of the stillness. Schoenberg jokingly called these little melodic figures "jumping fish motives."[56]

The music of Schoenberg's op. 16, no. 3, is "atonal" in the strict sense of the word because it abolishes the unit of the tone. The sound changes without taking a detour through the domain of tonal relations. The role of the instruments is similar to that of the simple tones in Helmholtz's experimental synthesis. Their sounds merge into

Nikos Kokkinis, "Zur Werkgeschichte," in Schönberg, *Sämtliche Werke*, pt. 4, ser. B, vol. 12, *Orchesterwerke I: Kritischer Bericht*, ed. Kokkinis (Mainz, 1984), xiii–xv.

[55] Schönberg, op. 16, in *Sämtliche Werke*, pt. 4, ser. A, vol. 12, *Orchesterwerke I: Fünf Orchesterstücke*, ed. Nikos Kokkinis (Mainz, 1980), 1–69.

[56] Quoted after Jonathan Cross, "Fünf Orchesterstücke op. 16," in *Arnold Schönberg: Interpretationen seiner Werke*, ed. Gerold W. Gruber (Laaber, Germany, 2002), 216–28, on 219–20.

one, such that the components can no longer be recognized. This does not happen according to the traditional rules of musical performance. Schoenberg therefore advises the conductor not to try to profile individual instrumental parts that seem important or to modify sound mixtures that seem unbalanced. If a voice has to be accented, Schoenberg emphasizes, this is foreseen in the instrumentation. The conductor's task is thus to supervise every player in the reproduction of the intensity given in the score, "exactly (subjectively) corresponding to the musician's instrument and not (objectively) submitted to some idea of the overall sound."[57]

The conductor is not asked to anticipate the sound. Rather, he is invited to take part in an experiment in the differentiation of colors within sound. Thus, with his instructions, Schoenberg set conditions for an experience that was supposed to both be valuable for the specific case and enable generalization. The relation to the sound resulting from the score that Schoenberg designed for the conductor as well as the musicians is telling. No single musician was supposed to make judgments about the whole of the sound. In fact, it was not possible for a musician to do so, because it was not possible for Schoenberg to encode the intended sound in musical notation. Schoenberg's approach was a break from the tradition of music making that had assigned to each instrument its predefined role in order to recreate a well-known sound. His music, instead, created a situation from which a sound was to emerge that was unheard of. Any preconceived knowledge would have distorted the emergence of this sound.

This way of setting conditions for an unpredictable experience can be related to experimental research in hearing. Once experimental physiology introduced experimentation into the life sciences in the mid-nineteenth century, this soon became the model for other disciplines as well as the arts in the most general sense. The case of Schoenberg illustrates how both a methodology and a specific understanding of sound spread into the realm of music. Although Schoenberg does not explicitly refer either to Helmholtz or to experimentation, both the concept of Klangfarbe and the modes of defining a set of conditions were vital in the culture and, more specifically, in the musical avant-garde of his time. Musical composition had become an experiment whose outcome was open.

[57] Schönberg, op. 16 (cit. n. 55), 1. Note that Schoenberg uses "subjectively" here to indicate that each musician shall follow exactly his own measure of the intensity of his instrument. By "objectively" he refers to the intensity as judged by the conductor—i.e., an authority who disregards the absolute intensity but considers the relative weight of each instrument's input.

Craftsmen-Turned-Scientists?

The Circulation of Explicit and Working Knowledge in Musical-Instrument Making, 1880–1960

*by Sonja Petersen**

ABSTRACT

During the nineteenth century a major change took place in musical-instrument making. In the course of industrialization many crafts transformed into industries, and production methods changed significantly. Prompted by this development, many trained piano makers, like Siegfried Hansing, published "educational books," in hopes of establishing a fixed canon of explicit knowledge. This canon, elucidating physics and acoustics to an audience of craftsmen (and, to a lesser extent, lay theoreticians), was intended to replace the craftsmen's working knowledge. Despite efforts to bring explicit knowledge to industrialized musical-instrument making, the working knowledge of the craftsmen remained irreplaceable.

BETWEEN EXPLICIT AND WORKING KNOWLEDGE

Siegfried Hansing (1842–1913) at first glance appears to have been a typical craftsman of his time. He was born in Germany and educated as a cabinetmaker. He eventually worked for several piano-making companies in Germany and the United States. Despite his lack of formal education in physics or math, Hansing would become one of the best-known researchers in the field of the acoustics of piano making. He died in 1913 as the author of over forty journal articles and an innovative book on the acoustics of pianos. His critical role is made clear in a statement by Alfred Dolge, a felt producer from the United States:

> The science of acoustics as developed by Chladni, Tyndall, Helmholtz, and in its direct relation to the piano, especially by Siegfried Hansing, has given us much enlightenment as to the proper and correct laying out of a scale, also the laws controlling the production of sound by percussion and otherwise . . .[1]

* Universität Stuttgart, Historisches Institut, Wirkungsgeschichte der Technik, Keplerstraße 17, D-70174 Stuttgart, Germany; sonja.petersen@hi.uni-stuttgart.de.

This paper is based on my research about the history of piano making for my PhD thesis: "Vom 'Schwachstarktastenkasten' und seinen Fabrikanten: Wissensräume im Klavierbau, 1830–1930" (Münster, 2011). All translations in the article are my own.

[1] Dolge, *Pianos and Their Makers*, vol. 1, *A Comprehensive History of the Development of the Piano* (1911; repr., New York, 1972), 106.

© 2013 by The History of Science Society. All rights reserved. 0369-7827/11/2013-0011$10.00

Figure 1. *Siegfried Hansing. Pfeiffer, "Siegfried Hansing" (cit. n. 2), 981.*

With his 1888 book *Das Pianoforte in seinen akustischen Anlagen* (The pianoforte and its acoustical properties; hereafter referred to as *Das Pianoforte*), which was also translated into English, Hansing created one of the first comprehensive acoustical studies of the piano. The text was based on the common literature of acoustics and physics, combining the work of Hermann von Helmholtz and John Tyndall with his own research, observations, and experience. On the basis of his writings, then, Hansing would appear to have been a recognized theoretician in his field and not an ordinary craftsman. But he was not an ordinary scientist either—not in the classic sense. He never studied at a university, nor was he affiliated with an accepted research institution. Before the publication of *Das Pianoforte*, Hansing had written several articles on acoustics in the *Zeitschrift für Instrumentenbau* (Journal for instrument making), one of the most influential journals for the musical-instrument-making field at that time. His aim was not only to apply scientific acoustics to piano making but also to present his results in a way that was understandable to trained craftsmen. His transition from solidly trained craftsman to the theoretician his contemporaries saw occurred in incremental steps, as he progressed through several positions as technical leader in piano-making factories.[2] This development culminated in his book.

A case study of Hansing illustrates the important role that the circulation of explicit and working knowledge played during the industrialization of musical-instrument

[2] Walter Pfeiffer, "Siegfried Hansing: Zu seinem siebzigsten Geburtstag," *Zeitschrift für Instrumentenbau* 32 (1911–2): 981–2; Alfred Dolge, "Dem Andenken Siegfried Hansing's," *Zeitschrift für Instrumentenbau* 34 (1913–4): 220; Emma Hansing, "Siegfried Hansing's Grabmal," *Zeitschrift für Instrumentenbau* 34 (1913–4): 1129–30; Hansing, *Das Pianoforte in seinen akustischen Anlagen* (Breslau, 1888).

making. This case study gives insights into the interplay of research, technology, music, and acoustic experimentation carried out not by a scientist, but by a trained craftsman. Further, this study shows that scientific methods and knowledge circulated beyond the scientific community and made their way down to the technicians and craftsmen, the "invisible technicians," as Steven Shapin calls them, who ultimately built the instruments.[3]

Hansing's efforts to standardize the knowledge of piano makers can be seen in the context of, on the one hand, a process of professionalization and diversification of music, and, on the other, an increasing scientific interest in acoustics, which took place side-by-side during the nineteenth century. In this period, several types of piano players developed: virtuosos, female dilettantes of the bourgeoisie, and professional musicians. These different typologies of piano players correlated to instruments of different quality. The virtuosos, pianists who gained attention for their unique technique, marked a new level of professional piano playing. Often they served as an advertising medium for piano makers and were in a unique position to consult on the technical development of the instrument. They advanced this development by demonstrating the technical advantages of particular pianos and companies to the public.

Learning to play the piano was also part of the bourgeois education of young ladies. In the absence of an overt economically productive role, they were expected to be capable of playing the pieces that were considered classic by the increasingly sophisticated middle and upper classes. They therefore needed to learn to play the piano semiprofessionally in order to entertain at private concert evenings in the houses of the bourgeoisie. This generated a demand for pianos similar in quality to those played by the virtuosos in the great concert halls.

The third group of piano players earned their living by playing in *variétés*, music halls, and, later, cinemas. These players mostly played on second- or thirdhand instruments. Of the three typologies, Dorothea Schmidt identifies virtuosos as key drivers for innovations in piano construction.[4] They influenced the way pianos were designed and built. The innovations initiated by the virtuosos trickled down to the other groups of users over time.

During the nineteenth century, the second half especially, the standardization of the knowledge of piano making was related to the increasing interest of scientists in acoustics. Helmholtz's groundbreaking work *Die Lehre von den Tonempfindungen* (On the sensations of tone) was published in 1863,[5] and in 1869, Tyndall's *Sound* was

[3] "In my working definition, technicians are persons in a setting dedicated to the production of scientific knowledge who are remuneratively engaged to deploy their labor or skill at an employer's behest." Shapin, *A Social History of Truth: Civility and Science in Seventeenth-Century England* (Chicago, 1995), 353–407, on 361.

[4] Schmidt, " 'Das Klavier kann alles'—Klavierbau und Klavierspiel im 19. Jahrhundert," in *Homo Faber Ludens: Geschichten zu Wechselbeziehungen von Technik und Spiel*, ed. Stefan Poser and Karin Zachmann (Frankfurt am Main, 2003), 135–54, on 140–54.

[5] Hermann von Helmholtz, *Die Lehre von den Tonempfindungen als physiologische Grundlage für die Theorie der Musik* (Brunswick, 1863). See also David Cahan, *Hermann von Helmholtz and the Foundations of Nineteenth-Century Science* (Berkeley, Calif., 1993); "Helmholtz, Hermann Ludwig Ferdinand von," on the website "Virtual Laboratory: Essays and Resources on the Experimentalization of Life," Max Planck Institute for the History of Science, Berlin, http://vlp.mpiwg-berlin.mpg .de/people/data?id=per87 (accessed 26 Feb. 2010); Julia Kursell, "Wohlklang im Körper: Kombinationstöne in der experimentellen Hörphysiologie von Hermann v. Helmholtz," in *Resonanz: Potential einer akustischen Figur*, ed. Karsten Lichau, Viktoria Tkaczyk, and Rebecca Wolf (Munich, 2009), 55–74, on 65–70.

translated into German (*Der Schall*).[6] In less than twenty years, the work of Helmholtz, Tyndall, and many other physicists was being carefully studied by craftsmen, Hansing included. For example, Helmholtz and Tyndall's books, as well as several editions of Hansing's *Das Pianoforte in seinen akustischen Anlagen*, all thoroughly well used, can be found in the library of the Grotrian-Steinweg Piano Company. Evidently, piano makers were well aware of the current scientific research in acoustics and endeavored to implement its results in their work. The diversification and professionalization of piano playing, the increasing scientific interest in acoustics, and the standardization and scientification of the knowledge of piano making were all linked during the nineteenth century.

The case study of Hansing and his work also illustrates that working knowledge, as the sociologist Douglas Harper calls this specific kind of craft knowledge, is still needed, even in industrialized production.[7] Harper's working knowledge can be distinguished from Michael Polanyi's definition of tacit knowledge as embodied, informal knowledge, the concept "that we can know more than we can tell."[8] A myriad of studies in the fields of history of technology, history of science, science and technology studies, sociology of science, and philosophy of technology employ Polanyi's concepts to examine the origin of experience and tacit knowledge. Eugene S. Ferguson, for example, emphasizes the importance of intuitive action and nonlinguistic thinking in the engineering process, especially for design drawings.[9] Other scholars emphasize the importance of tacit knowledge for scientific work, as Harry M. Collins did in his well-known studies on the reproduction of a TEA laser, or as Trevor Pinch, Collins, and Larry Carbone did in their examination of the skills of veterinary surgeons.[10] Myles W. Jackson, in his study on Joseph von Fraunhofer, emphasizes

[6] John Tyndall, *Der Schall: Acht Vorlesungen, gehalten in der Royal Institution von Großbritannien*, ed. H. Helmholtz und G. Wiedmann (Brunswick, 1869). See also William Hodson Brock, Norman MacMillan, and R. Charles Mollan, eds., *John Tyndall: Essays on a Natural Philosopher* (Dublin, 1981).

[7] Harper, *Working Knowledge: Skill and Community in a Small Shop* (Chicago, 1987), 31–7, 117–33; Georg Mildenberger, *Wissen und Können im Spiegel gegenwärtiger Technikforschung* (Berlin, 2006).

[8] Polanyi, *The Tacit Dimension* (New York, 1966), 4. See also Polanyi, *Implizites Wissen* (Frankfurt am Main, 1985), 14–5, 21–3; Polanyi, "Knowing and Being," in *Knowing and Being: Essays by Michael Polanyi*, ed. Marjorie Grene (Chicago, 1969), 123–37; Mildenberger, *Wissen und Können* (cit. n. 7), 102.

[9] Ferguson, *Das innere Auge: Von der Kunst des Ingenieurs* (Basel, 1993), 9, 18–22. For Matthias Heymann and Ulrich Wengenroth, tacit knowledge also plays an important part in the construction work of engineers. And Mikael Hård stresses that technical work is mostly local and practical, as opposed to its common characterization as universal and cognitive. See Heymann and Wengenroth, "Die Bedeutung von 'Tacit Knowledge' bei der Gestaltung von Technik," in *Die Modernisierung der Moderne*, ed. Ulrich Beck and Wolfgang Bonß (Frankfurt am Main, 2001), 106–21; Hård, "'Die Praxis der Forschung': Zur Alltäglichkeit der Technikwissenschaften am Beispiel einer britischen Ingenieurfirma," *Dresdener Beiträge zur Geschichte der Technikwissenschaften* 27 (2001): 1–17; Hård, "Technology as Practice: Local and Global Closure Processes in Diesel-Engine Design," *Social Studies of Science* 24 (1994): 549–85.

[10] Pinch, Collins, and Carbone, "Inside Knowledge: Second Order Measures of Skill," *Sociological Review* 44 (1996): 163–86; Collins, *Changing Order: Replication and Induction in Scientific Practice* (London, 1985); Collins, "The TEA Set: Tacit Knowledge and Scientific Networks," *Science Studies* 4 (1974): 165–85. Collins still works on tacit knowledge, which he divides into three parts—relational, somatic, and collective—in order to analyze the relation between explicit and tacit knowledge not only in science or the use of technology but also in daily routines. See Collins, *Tacit and Explicit Knowledge* (Chicago, 2010). Hans-Jörg Rheinberger emphasizes the importance of tacit knowledge for scientific research; see Rheinberger, *Experimentalsysteme und epistemische Dinge: Eine Geschichte der Proteinsynthese im Reagenzglas* (Frankfurt am Main, 2006), 92–4.

that tacit knowledge can be related to skilled actions through nonarticulated and hidden knowledge. Polanyi's concept of tacit knowledge not only helps us to understand practical actions, it also helps us to analyze natural philosophy or mathematics.[11]

An important advantage of Harper's conception of working knowledge over Polanyi's understanding of tacit knowledge is that Harper makes it possible to ask about the concrete operations of skilled workers. He defines working knowledge as consisting of (1) learning by doing, mostly informally during childhood—for example, when playing with materials; (2) knowledge of materials, a sensory skill in understanding the characteristics of materials and how they react; and finally (3) kinesthetic sense, a specific ability to control the body while working with materials and tools. Each of these features remains unconscious and only reveals itself during the actions of the worker.[12] Harper's three categories enable us to draw a detailed picture of working knowledge, even if we are unable to decode it entirely, which would be contradictory to Polanyi's concept of tacit knowledge.

Working knowledge can be found in musical-instrument making, even though the nineteenth-century development of this field into a mechanized industry was characterized by an extensive division of labor. In this setting, furthermore, the working knowledge was combined with explicit knowledge.

A CRAFT TURNED INTO AN INDUSTRY

If you ask someone nowadays how he or she pictures the production of a concert grand, you usually get an answer along these lines: It's made by hand, in a small workshop. The master applies his superior skills, and years later a masterpiece is complete. This picture of piano making as an artisanal craft is still cultivated today, not only by society, but also by institutions like museums as well as the producers themselves.[13] During the second half of the nineteenth century, however, musical-instrument making was largely transformed from a craft, performed in small workshops, into an industry with large mechanized factories. Even in the small workshops at the beginning of the century, a piano or a concert grand was not built by a single person. Besides the master, many journeymen, craftsmen, and helpmates worked on the instruments.

In Germany, England, and the United States, the development of piano making ran parallel to industrialization and the development of modern transportation and communication systems. As Christoph Buchheim points out, piano making is one example of the transformation of Germany from a developing into an advanced economy during the process of industrialization.[14] The extensive mechanization of German piano making mirrors the country's *Hochindustrialisierung* (extensive industrialization)

---

[11] Jackson, *Fraunhofers Spektren: Die Präzisionsoptik als Handwerkskunst* (Göttingen, 2009). Originally published as *Spectrum of Belief: Joseph von Fraunhofer and the Craft of Precision Optics* (Cambridge, Mass., 2000).

[12] Harper, *Working Knowledge*, 31–7, 117–33; Mildenberger, *Wissen und Können*, 114–5 (Both cit. n. 7).

[13] Gabriele Zuna-Kratky, ed., *Technisches Museum Wien* (Munich, 2005).

[14] Buchheim, "Grundlagen des deutschen Klavierexports vom letzten Viertel des 19. Jahrhunderts bis zum Ersten Weltkrieg," *Technikgeschichte* 53 (1987): 231–40, on 239. For a general overview of how processes of industrialization overlapped throughout the European continent, see Buchheim, *Industrielle Revolution: Langfristige Wirtschaftsentwicklung in Großbritannien, Europa und in Übersee* (Munich, 1995), 69–104.

from 1870 onward, although it was achieved relatively late (around 1900) compared
to Germany's other industrial sectors or the American piano-making industry. In
contrast, the American piano-making industry was characterized by extensive mechanization as early as 1870, at the company Steinway & Sons, for example.[15] Steinway
& Sons is also an example of the vertical integration of all of the manufacturing sectors that contribute to a finished product that characterized American industry after
1840. For pianos this required the integration of, among other processes, iron and
copper smelting, lumber milling, and metal processing. In the case of Steinway &
Sons, the factory was relocated in 1870 to the East River in Queens, New York, in
order to integrate the transportation of raw materials on the river into the production
system; modern machinery and an extensive division of labor were also utilized.[16]
Such comprehensive vertical integration was not to be found anywhere in German
industry at that time.[17]

Instead, in Germany small workshops continued to dominate piano making from
the end of the eighteenth through the middle of the nineteenth century. At that time
pianos were luxury goods, made to order.[18] During the nineteenth century this production method changed, and the small workshops became less important. New production units took over: mechanized factories in which pianos were mass-produced.
This change began slowly around 1830, accelerated around 1850, and was complete by 1870. Production units were enlarged, more workers were employed, and
an extensive division of labor was introduced. Some companies also ran huge raw-
material inventories, and single components were obtained from specialist suppliers.
A geographically distributed network of producers arose. The specialist supply industry produced components using specialized machine tools. This allowed the large
factories to use components of good quality at a much lower price than if they manufactured the components themselves. Machine-tool use and additional purchase of
components increased. Consequently, around 1850 a rationalization and standardization of piano models became necessary to produce instruments using interchangeable precision parts.[19] All of this enabled pianos, which had previously been made as

[15] Flurin Condrau, *Die Industrialisierung in Deutschland* (Darmstadt, 2005), 8–9; Richard K. Lieberman, *Steinway & Sons: Eine Familiengeschichte um Macht und Musik* (Munich, 1996), 43–4,
125–8.
[16] Alfred D. Chandler, *The Visible Hand: The Managerial Revolution in American Business* (Cambridge, Mass., 1980), 72; Lieberman, *Steinway & Sons* (cit. n. 15), 43–4, 125–8; Petersen, "Schwachstarktastenkasten" (cit. n. *), 60–1.
[17] Chandler, *Visible Hand* (cit. n. 16), 77.
[18] Christina Meglitsch, *Wiens vergessene Konzertsäle: Der Mythos der Säle Bösendorfer, Ehrbar
und Streicher* (Frankfurt am Main, 2005), 17; Edwin M. Good, *Giraffes, Black Dragons, and Other
Pianos: A Technological History from Cristofori to the Modern Concert Grand* (Stanford, Calif.,
1982), 58; Cyril Ehrlich, *The Piano: A History* (Oxford, 1990), 9.
[19] Buchheim, "Grundlagen des deutschen Klavierexports" (cit. n. 14), 231–40; Georg Pfeiffer, "Die
Entwicklung der deutschen Pianoforteindustrie" (PhD diss., Wien Wirtschaftsuniv., 1989), 29, 71;
The Piano: An Encyclopedia, ed. Robert Palmieri (New York, 2003), s.v. "Germany—Piano Industry," by Carsten Dürer; Zuna-Kratky, *Technisches Museum Wien* (cit. n. 13), 142; Ehrlich, *Piano* (cit.
n. 18), 19; Norbert Ely, "Pianofortebau in Deutschland," in *Faszination Klavier: 300 Jahre Pianofortebau in Deutschland*, ed. Konstantin Restle (Munich, 2000), 163–226; Hubert Henkel, *Besaitete
Tasteninstrumente: Deutsches Museum—Kataloge und Sammlungen; Musikinstrumenten-Sammlung*
(Frankfurt am Main, 1994), 9; Irmgard Bontinck, "Das Klavier im 19. Jahrhundert: Technologie,
künstlerische Nutzung und gesellschaftliche Resonanz," in *Das Klavier in Geschichte(n) und Gegen-
wart*, ed. Michael Huber et al. (Strasshof, 2001), 11–31, on 13; Schmidt, "Klavier kann alles" (cit.
n. 4), 135.

individual instruments for the former nobility and bourgeoisie, to be mass-produced at a reasonable price for families of the middle class.

By the end of the nineteenth century piano making had attained impressive production numbers, reaching a climax around 1913. In 1870, Britain produced 25,000 pianos, the United States produced 24,000, and France produced 21,000. Germany, however, only produced 15,000 pianos. As of 1913 German piano makers had increased their production to 160,000 pianos and 12,000 concert grands. In terms of percentages, by 1913 German piano makers had increased production by 1,147 percent over 1870, had commanded a 20 percent share of the world market, and had made Germany the second-largest producer of pianos after the United States and, hence, the leading exporter. It was the quick rationalization and mechanization of production in large factories, combined with the outsourcing of the production of special components, like keys, actions, and cast-iron frames, that led to their success. The Germans also rapidly adopted new technological developments concerning the instruments themselves, such as the so-called American system, consisting mainly of a cast-iron frame and cross stringing. Additionally, modern marketing techniques at such events as the Great Exhibitions and the expertise of famous pianists were well exploited for advertising.[20]

This fundamental change in German piano manufacturing went hand in hand with the development of formalized and standardized knowledge, which was necessary for these industrialized methods. At first glance, the several books written by trained craftsmen between 1880 and 1960 suggest that the working knowledge of craftsmen was replaced by explicit scientific knowledge. Hansing's work seems to be a prime example of this development. In his publications he attempted to standardize and formalize the acoustics of piano making and to translate the complicated physical and acoustical work of Helmholtz, for example, into the language of craftsmen. At second glance, however, Hansing's work makes it clear that even in industrialized piano making, standardized, formalized, explicit knowledge still needed to interact with working knowledge. Furthermore, explicit knowledge did not stand in opposition to working knowledge but was instead complementary to learning by doing, knowledge of materials, and a kinesthetic sense.

THE SCIENTIFICATION OF A CRAFT; OR, HOW TO TRAIN CRAFTSMEN IN ACOUSTICS

Hansing's book *Das Pianoforte* can be traced back to a series of more than forty articles he wrote for the *Zeitschrift für Instrumentenbau*. The journal was published between 1880 and 1943 in 63 volumes. The journal was unique in its scope: it covered all types of musical-instrument making. It attracted attention both inside Germany as well as abroad. To facilitate the exchange of ideas among colleagues and allow them to discuss scientific and technical questions, the *Zeitschrift für Instrumentenbau* offered a special section, the "Sprechsaal" (discussion room).[21] Until the publication

---

[20] Christoph Buchheim, *Deutsche Gewerbeexporte nach England in der zweiten Hälfte des 19. Jahrhunderts: Zur Wettbewerbsfähigkeit Deutschlands in seiner Industrialisierungsphase; Gleichzeitig eine Studie über die deutsche Seidenweberei und Spielzeugindustrie, sowie über Buntdruck und Klavierbau* (Ostfildern, 1983), 109–13; Buchheim, "Grundlagen des deutschen Klavierexportes" (cit. n. 14), 231–6; Ehrlich, *Piano*, 56–65; Good, *Black Dragons*, 176–87 (Both cit. n. 18).

[21] The "Sprechsaal" existed until the volume of 1935–6. See Verlag und Schriftleitung der *Zeitschrift für Instrumentenbau*, "Paul de Wit: 50 Jahre *Zeitschrift für Instrumentenbau* und Verlag Paul de Wit," *Zeitschrift für Instrumentenbau* 51 (1930–1): 9–12; Paul de Wit, "Zum zehnten Jahrgange,"

of his book, Hansing, for one, used the "Sprechsaal" extensively to publish his research results and to explain acoustics and physics to craftsmen. The "Sprechsaal" allowed Hansing both a platform for discussion and a venue in which to lay the foundations for his later book. For example, as will be discussed below, he debated in the "Sprechsaal" whether theoretical or practical knowledge was necessary for tuning a piano. Both Hansing's articles and his later book demonstrate the scientification of a craft by a craftsman.

Hansing drew on the work of several respected scientists. He referred to Helmholtz, who in the chapter of *Die Lehre von den Tonempfindungen* dedicated to harmonics identified harmonics as crucial for tone pitch.[22] Hansing also studied the work of Georg Simon Ohm and Tyndall.[23]

Hansing discussed the science of acoustics in the "Sprechsaal" from the very first volume of the *Zeitschrift für Instrumentenbau*.[24] One of his main interests was to elaborate the oscillations of piano strings. This led to a series of articles in which his special position as a theoretician of piano acoustics became clear. Hansing explored transverse, pendulum, side, and molecular oscillations of strings and their associated material properties.[25] He stressed that in order for a body to generate a tone, it had to be capable of molecular, transverse, and pendulum oscillations.[26] A body capable of transverse oscillations should be flexible enough to perform molecular oscillations, too. But beyond transverse and pendulum oscillations, both of which were, in addition to the fundamental tone, necessary for secondary tones, Hansing concentrated on the side-transverse oscillation. He postulated that if the secondary tones were higher than the fundamental tone, they turned into harmonics.[27]

Hansing did not formulate a new acoustic theory. His work is interesting, however, for his efforts to both integrate acoustic theory into his own field of activity and make it understandable for his craftsman colleagues.[28] He started his own experiments on oscillations, focusing on *stumme Schwingungen* (mute oscillations).[29] Hansing began

---

Zeitschrift für Instrumentenbau 10 (1889–90): 1; Die Redaction der *Zeitschrift für Instrumentenbau*, "Sprechsaal," *Zeitschrift für Instrumentenbau* 1 (1880–1): 236; Der Herausgeber, "Zur Einführung," *Zeitschrift für Instrumentenbau* 1 (1880–1): 1–2.

[22] Helmholtz deals with harmonics in several chapters; e.g., "Von der Zerlegung der Klänge durch das Ohr," in Helmholtz, *Lehre von den Tonempfindungen* (cit. n. 5), 84–112. Jackson summarizes: "Helmholtz limited his scientific study of music to the experience of musical notes in their most basic forms, since the comprehension and explanation of the extraordinarily complex effects of music on the listener required a detailed knowledge of historical circumstances and national characteristics." Myles W. Jackson, *Harmonious Triads: Physicists, Musicians, and Instrument Makers in Nineteenth-Century Germany* (Cambridge, Mass., 2006), 7.

[23] Hansing, "Ueber das Tragen und die Breite des Tones," *Zeitschrift für Instrumentenbau* 1 (1880–1): 305–6, on 306. For Helmholtz, see Kursell, "Wohlklang im Körper" (cit. n. 5). See also Jackson, *Harmonious Triads* (cit. n. 22), 178; "Ohm, Georg Simon," on the website "Virtual Laboratory: Essays and Resources on the Experimentalization of Life," Max Planck Institute for the History of Science, Berlin, http://vlp.mpiwg-berlin.mpg.de/people/data?id=per488 (accessed 26 Feb. 2010).

[24] Hansing, "Beobachtungen über Transversal-, Pendel- und Nebenschwingungen," *Zeitschrift für Instrumentenbau* 1 (1880–1): 236–8; Hansing, "Ueber das Tragen" (cit. n. 23).

[25] A string swings by transverse oscillations vertical to the propagation. In contrast, a body making pendulum oscillations swings from right to left and back again.

[26] Hansing, "Beobachtungen" (cit. n. 24), 238.

[27] Ibid.

[28] Ibid.

[29] Hansing, "Schwingungssystem der tönenden Claviersaite (Fortsetzung)," *Zeitschrift für Instrumentenbau* 2 (1881–2): 94–5; Hansing, "Ueber Obertöne, Schwingungssysteme, Tonfarbe und Anschlagslinie der Claviersaiten: Schwingungssystem der tönenden Claviersaite (Fortsetzung)," *Zeitschrift für Instrumentenbau* 2 (1881–2): 106–8.

his presentation of these experiments with a brief explanation of transverse oscillations. He gave the example of an elastic body, a string, that is fixed at its endpoints.
Due to its own weight it sags in the middle. If an oscillation is induced throughout the
entire mass of the string, it leaves its equilibrium position, swinging back and cross-
ing the equilibrium position again. If someone attempts to stop the string with his
hand in its upward movement, the hand is felt to be pulled downward by the force of
the string. This force comes from the endpoints, what Hansing called the *Ruheknoten*
(nodes) of transverse oscillations.

In his discussion, Hansing reduced the scientific complexity and explained this
phenomenon using real-life examples. For instance, to explain the characteristics
of transverse oscillations, he presented a scenario of two boys holding a rope and
then stretching it. At the beginning, due to its own weight, the rope is lying on the
ground. The rope is longer than the original distance between the boys, so to tighten
it, the boys have to move away from one another, pulling on each end.[30] In the case
of piano strings the increased tension is created by tuning. Returning the discussion
to the string of his earlier example, Hansing stressed that the string possesses its own
weight, which pulls it downward and consequently has an effect on the *Ruheknoten*
(represented in his present example by the two boys). The downward force of the
string is the strongest when the string reaches its maximum amplitude. Hansing illustrated with the example of the two boys again: if the boys move away from each
other to put the rope in tension, then the rope is lifted upward in the middle and the
two boys must move toward each other.[31]

Strings make twice as many of these longitudinal oscillations as transverse ones.
Hansing concluded that longitudinal oscillations influence transverse oscillations
and that they affect the tone produced—but only the volume, not the pitch.[32] A higher
amplitude of longitudinal oscillations corresponds to differences in pressure in the
transport medium (a gas or liquid), perceived as a higher volume. Longitudinal oscillations propagate through gases and liquids through these pressure changes. Transverse oscillations, on the other hand, spread via shearing forces, which cannot be
transmitted through gases or liquids.[33] Through his extensive analyses Hansing explained the theoretical and scientific basis of these phenomena. Piano makers may
have known of these phenomena from their daily working experiences, but would
have been unaware of their theoretical explanations.

In addition to providing easily comprehensible examples, Hansing combined his
theoretical explanations with drawings to illustrate the longitudinal and transverse
oscillations of a string.[34] He presented the image of a horizontal string (*cd*) to which
he fixed a vertical string (*ab*) (see fig. 2). Inducing the horizontal string to oscillate

---

[30] Hansing, "Schwingungssystem" (cit. n. 29), 94.

[31] In his example Hansing chose to change the equilibrium position upward, as opposed to downward, as is the case with piano strings. See ibid., 94–5; Hansing, "Ueber Obertöne" (cit. n. 29), 108.

[32] Hansing, "Ueber Obertöne" (cit. n. 29), 108.

[33] Transverse waves expand in solid material, which transfer shear forces. The term *wave* refers
to both chronological and spatial description of vibrations, whereas oscillations can be described
only chronologically. See Christian Gerthsen, Hans O. Kneser, and Helmut Vogel, *Physik: Ein Lehrbuch zum Gebrauch neben Vorlesungen* (Berlin, 1989), 733–6; Gerhard Müller and Michael Möser,
Taschenbuch der Technischen Akustik (Berlin, 2004), 11–3, 753. For longitudinal oscillations, see
Jackson, *Harmonious Triads* (cit. n. 22), 35–44.

[34] Molecules of a string that is activated to oscillate expand and contract, thus causing the string to
oscillate both longitudinally and transversely.

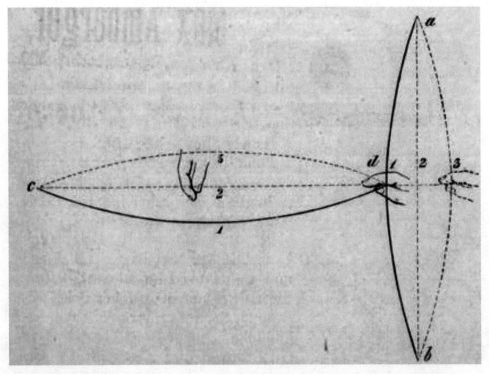

Figure 2. Visualization of Hansing's oscillation experiments. Hansing, "Ueber Obertöne" (cit. n. 29), 106.

would induce an oscillation in the vertical string. He demonstrated his conclusion—that the string would contract in length as a result of this oscillation—in relation to piano strings in a second drawing.[35]

Hansing's article about the oscillation of strings illustrates his efforts to explain established physical and acoustic knowledge to craftsmen. Interestingly, he never used equations to explain his research results. Instead he chose practical examples, closely related to real life, to simplify the complicated work of Helmholtz, Ohm, Tyndall, and others. His efforts brought the work of these important physicists to piano makers in a way they could understand and apply to their own work. Hansing never cited specific books or even chapters of Helmholtz or other physicists, but his knowledge of acoustics suggests that he studied their work intensively. His "Sprechsaal" articles also show that science in general and acoustics in particular found their way into the fixed canon of explicit knowledge of piano making. A process of scientification becomes evident. The practice of writing and drawing for the "Sprechsaal" allowed Hansing to develop his ideas. The "Sprechsaal" format also allowed him to present his knowledge to a broad audience and start a discussion about the acoustics of the piano. Hansing therefore inhabited the important position of a mediator between science and craft, between physicists and piano makers.

[35] Hansing, "Ueber Obertöne" (cit. n. 29), 106.

THE PIANOFORTE AND ITS ACOUSTIC PROPERTIES

Hansing's efforts to explain acoustics to craftsmen culminated in the publication of his book *Das Pianoforte* in 1888. He was, however, not the only craftsman to write about piano making. One of the best-known books on piano construction was written by the piano maker Julius Blüthner and his coauthor Heinrich Gretschel in 1872. It is still in print in updated editions today and contains a chapter about acoustics, albeit written by an independent expert.[36] The formalization of craft knowledge was also proceeding in the United States. A famous example is the work of Dolge, *Pianos and Their Makers*, published in two volumes. In addition to an extensive history of the piano and its components, Dolge also offered an overview of "Men Who Have Made Piano History" and described the American piano industry, paying special attention to single companies around 1910.[37] Other craftsmen also published books on component parts, such as Walter Pfeiffer, who wrote a work on piano action.[38]

The books mentioned above are part of the extensive literature concerning the construction of pianos. But only a few books about the acoustics of pianos were published by craftsmen. Hansing's book appears to have been well received. It was published in a second edition in 1909, and a reprint followed in 1950. In 1904 he also produced an English translation.[39] As mentioned previously, Hansing established the foundations for his book with his earlier "Sprechsaal" articles, and he directly referred to them in the preface. He understood his book to be a continuation of these first deliberations.

Dolge described Hansing's work as pioneering and groundbreaking: "His studies in the realm of acoustics disclose a most penetrating mind capable of exact logical reasoning. He bases his conclusions on exhaustive studies, without regard to the accepted theories of earlier scientists. As a thoroughly practical piano maker and master of his art, Hansing studied cause and effect in its application to the piano."[40] Hansing's combination of his own acoustic knowledge, his experience as a piano maker, and knowledge gleaned from other sources, such as scientific works, impressed Dolge the most.

But how can we know that Hansing's book was noticed and studied by piano makers? During my research I found two independent sources attesting to this fact, written by piano makers: one from 1906, by Willi Grotrian, owner of the Grotrian-Steinweg Piano Company, and one written around 1950 by Carl Georg Berger, the technical director of the Viennese company Bösendorfer between 1901 and 1928. Both refer directly to Hansing and express their admiration for him and his work. Grotrian in particular studied Hansing's work extensively. He owned the first and second editions of Hansing's book and paid particular attention to the first edition.

[36] Blüthner and Gretschel, *Lehrbuch des Pianofortebaus in seiner Geschichte, Theorie und Technik* (Weimar, 1872), v–vi.
[37] Dolge, *Pianos and Their Makers*, vol. 1 (cit. n. 1); Dolge, *Pianos and Their Makers*, vol. 2, *Development of the Piano Industry in America since the Centennial Exhibition at Philadelphia, 1876* (1913; repr., New York, 1979).
[38] Pfeiffer, *Vom Hammer Untersuchungen aus einem Teilgebiet des Flügel- u. Klavierbaus* (Stuttgart, 1948); Dolge, *Pianos and Their Makers*, vol. 1 (cit. n. 1); Dolge, *Pianos and Their Makers*, vol. 2 (cit. n. 37).
[39] Hansing, *Das Pianoforte in seinen akustischen Anlagen*, 2nd ed. (Schwerin, 1909); Hansing, *Das Pianoforte in seinen akustischen Anlagen* (1888; repr., Schwerin, 1950); Hansing and Emmy Hansing-Perzina, *The Pianoforte and Its Acoustic Properties* (Schwerin, 1904).
[40] Dolge, *Pianos and Their Makers*, vol. 1 (cit. n. 1), 426.

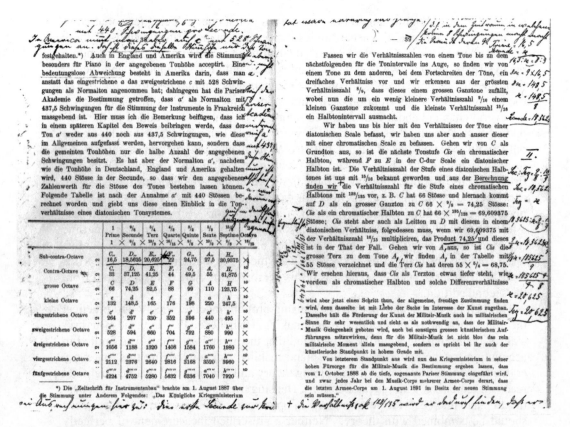

Figure 3. *Notes by Willi Grotrian in Hansing's book* Das Pianoforte *(cit. n. 2), 38–9. Reproduced courtesy of Grotrian-Steinweg Pianofortefabrikanten, Brunswick.*

A short look inside Grotrian's copy shows that he valued Hansing's work greatly. A plethora of underlinings and—sadly, mostly unreadable—notes by him are evidence of his detailed studies (see fig. 3).[41]

The cryptic nature of Grotrian's notes could be an entirely innocent result of casual handwriting, since he knew he alone would have to decode them. But a more intriguing explanation for their unreadability also suggests itself: *Betriebsgeheimnisse* (company secrets). Obscurantism of this sort was (and still is) quite common in all types of business, not only in the craft of piano building, in an understandable attempt to retain competitive advantage and foil industrial espionage.

In *Das Pianoforte in seinen akustischen Anlagen*, Hansing offers fifteen chapters of detailed acoustic study of the piano. He starts with the question of why objects get loud. A chapter concerning the characteristics of the piano's tone, one about its

---

[41] Hansing, *Pianoforte* (cit. n. 2), 176; Grotrian, "Aufgaben von Teilen und Arbeitsfolgen im Klavierbau" (manuscript, Brunswick, 1906), company archive of Grotrian-Steinweg Pianofortefabrikanten, Brunswick; Berger, "Aus der Werkstatt des Klaviermachers" (manuscript, Vienna, 1955), company archive of L. Bösendorfer Klavierfabrik GmbH, Wiener Neustadt.

timbre, and one about its sound ratio follow his initial query. A chapter about combination tones and one about harmonics concerning piano strings round off these general considerations. Chapters 7 and 8 deal with the oscillation system of sounding piano strings, and the two following chapters deal with the soundboard and the action. The first edition concludes with three chapters concerning *Luftkörper* (air mass), room acoustics, and interruption of rhythmical movements of sounding objects.

Several chapters about individual components of pianos were added to the second edition. Hansing thus expanded his work to be a comprehensive handbook not only about the acoustics of pianos but also about piano construction in general. As in his articles for the "Sprechsaal," he drew upon Helmholtz, Tyndall, Ohm, and Ernst Florens Friedrich Chladni.[42] In the first edition Hansing had drawn on his own experience; for example, he described his examination of several instruments to explain the construction of his *Mensurtheorie* (scale theory).[43] For him the text was a result not only of his physical and acoustical knowledge but also of his longtime experience as a piano maker and his studies on several instruments: "To publish these conclusions, gained from my practical studies, is the purpose of writing this book."[44] Curiously, such descriptions of his personal experiences disappeared from the second edition.

Through his practical experiments Hansing brought together theory and practice: theoretical acoustics and practical piano making. He believed that acoustics was crucial for piano making,[45] but he recognized that theoretical acoustics was limited in accounting for the real conditions of piano making. Acknowledging that not every craftsman would be capable of combining theory and practice, and that he might not reach all craftsmen with his publications, Hansing tried to explain theoretical acoustics to piano makers.[46] He never missed an opportunity to emphasize that experience should be combined with theory: "Neither a prescribed measurement of perfectly curved surfaces nor the slavish imitation of a model guarantees a perfect instrument. Rather the feeling, the tactile sense, must also be brought to bear. This, in combination with a scientific knowledge of the essence of the instrument, will result in the building of something extraordinary."[47]

He acknowledged that a piano maker could indeed build a fine instrument without any knowledge of theoretical acoustics, drawing only on his experience. He believed that experience could and indeed should supplement theory only up to a certain point. The knowledge required beyond this point could only be described by a craftsman for craftsmen. Hansing understood his role as that of a mediator between physicists and craftsmen, between science and a nonscientific public.

Hansing's work was but one example of explicit knowledge and working knowledge coming together in piano making. He was certainly not the only craftsman gaining an impressive knowledge of physics and acoustics. There was also the famous scientific-instrument maker Rudolph Koenig. From his beginnings as a violin maker, he developed into the leading scientific-instrument maker in Paris. He had craft-based experience in addition to well-founded scientific knowledge, gained through

[42] Hansing, *Das Pianoforte* (cit. n. 2), 6–7, 25–6, 70.
[43] Ibid., 72.
[44] Ibid., 1.
[45] Ibid.
[46] Ibid., 2.
[47] Ibid., 3.

private studies and public lectures. He also performed experiments in his workshop that have since been recognized by science. Koenig published on acoustics and gave at least one lecture in Toronto. Like Hansing, he combined these different kinds of knowledge in his work to develop and improve scientific instruments. Koenig described his workshop as a space where different kinds of knowledge flowed together and were inscribed in his instruments.[48]

In addition to published books, craftsmen also wrote several unpublished manuscripts in which their working knowledge is evident. Grotrian's 1906 text, *Aufgaben*, was intended for the staff and executives of the company. It included descriptions of production and construction processes in the Grotrian-Steinweg company.

Berger, the Bösendorfer technical director, finished his manuscript of a book for piano makers, *Aus der Werkstatt des Klaviermachers* (From the piano maker's workshop) in 1950. It was rejected by the publisher Österreichischer Bundesverlag in 1956 and was never published.[49] Berger's manuscript contains nineteen chapters, covering topics such as construction drawings, strings, scales, frames, models, soundboard, action, tuning, and voicing. Berger's manuscript was based on his own experience, gained during his career as a piano maker and technical director. Berger's father Franz was also employed by Bösendorfer as technical director, so Berger was raised in the household of a piano and organ builder and started his apprenticeship at Bösendorfer at a young age.[50] He had already learned much about the materials and procedures of piano making during his childhood and early teens (reflecting the category of learning by doing in Harper's concept of working knowledge). *Aus der Werkstatt des Klaviermachers* formalized and systematized not only his own experiences but also the Bösendorfer company's internal knowledge. This distinguishes Berger's book from Hansing's, even though Berger wanted to write for a wider public. He made it very clear that his aim was to make his experiences, or working knowledge, as Harper would call it, available for others: "During the writing of this book, my intention was not to publish another book about piano making. Rather I wanted to share my experiences, collected over the course of sixty years as a piano maker."[51]

Berger was familiar with the technical literature on piano making. He referred to Hansing's book (illustrating the extent to which that work had become and remained part of the fixed canon of explicit knowledge of piano making), as well as the tuning textbook *Theorie und Praxis des Klavierstimmen* (Theory and practice of piano tuning) by Otto Funke, and Karl Jung's *Dehnbuch Klavierbau* (Expandable book of piano making).[52] Jung, who had been a teacher at the technical Fachschule für Orgel- und Klavierbau Ludwigsburg (College for Organ and Piano Making in Ludwigsburg) since 1924, developed his *Dehnbuch* based on his teaching and practical experience. He intended it to be a kind of handbook for his pupils. The title *Dehnbuch* was chosen with care to describe its format of notes on loose sheets, consisting of text, drawings, tables, and diagrams concerning, for example, the soundboard, tuning, the calculation of string scales, construction work, and so on. Although it was initially written

[48] David Pantalony, *Altered Sensations: Rudolph Koenig's Acoustical Workshop in Nineteenth-Century Paris* (Dordrecht, 2009), 1–17, 42–56, 83–7, 91–110, 145–50.
[49] Österreichischer Bundesverlag Wien to Carl Georg Berger, 14 Mar. 1956, company archive of L. Bösendorfer Klavierfabrik.
[50] Carl Hutterstrasser, *Hundert Jahre Bösendorfer* (Vienna, 1928), 24.
[51] Berger, "Aus der Werkstatt" (cit. n. 41), 3.
[52] Ibid., 121, 76.

only for the students of the technical college, Jung made it available to a wider circle around 1949 by publishing it himself.[53] Berger was also well informed of the contemporary theoretical and scientific knowledge pertaining to piano making.[54] In addition to this formalized knowledge, working knowledge was also crucial, and Berger's manuscript was an attempt to make his own experience accessible to others. Building a superb instrument necessitated more than detailed technical drawings and exact measurements. Berger's working knowledge and its indispensability are particularly apparent in the chapter about tuning and voicing: the skills necessary could only have been gained after years of training. Berger compares tuning and voicing to an "art"; this is the first time he uses the term, and in doing so he intentionally differentiates tuning and voicing from construction work. Critical for the process of tuning and voicing was the ability to listen for, though not necessarily count, the *Schwebungen* (beats).[55] Berger formalized the *Schwebungen* into a chart, but he also explained that "to hear the beats is often difficult for beginners. The beginner should not be dissuaded, however, because the ear will soon become familiar with the beats . . . and in time will develop a feeling for their tempo."[56] The piano tuner must be able to handle the tuning hammer sensitively but also feel the tempo of the beats.[57]

One could only gain the working knowledge that was crucial for tuning and voicing through experience and the development of knowledge of materials and kinesthetic sense. Piano tuning and voicing, in addition to establishing the relative pitch of each individual key, gave each piano its unique timbre and secured a consistent stroke for each key. The felt tips of the hammers that strike the strings hardened over time, altering the instrument's timbre, and could be softened using a voicing needle.[58] This tool consists of several small needles positioned in a wooden grip. The piano tuner had to gauge, with both his fingers and his ears, which hammers were too hard and how often to insert the voicing needle into them to maintain the desired timbre. The felt must also not be too soft—hence there was danger in inserting the voicing needle too often. The tuner had to recognize when the perfect timbre was reached. He needed a knowledge of materials and a kinesthetic sense that could be acquired only from experience.

Both Berger and Hansing emphasized the equal importance of theory and practice. But Hansing was unique in his attempt to develop a detailed acoustic theory of piano making and in his role as mediator between theory and practice, science and craft. On the basis of the sources analyzed for this article, it is impossible to say what kind of knowledge flowed from craft into science and how. What is certain is that Hansing

[53] The school was originally named the Gewerbeschule Ludwigsburg, and today has been renamed the Oscar-Walcker Schule. Since 1924 it has offered evening courses for organ and harmonium makers. Already in 1907 organ makers were receiving lessons in drawing on a voluntary basis. In 1935 a preparation course for masters was created, and since 1938 courses for piano makers have been offered. The school provides a systematic, formalized education in organ, harmonium, and piano making. Karl Jung, "Dehnbuch Klavierbau" (manuscript, Ludwigsburg, 1949), archive of Oscar-Walcker-Schule, Ludwigsburg. Jung also wrote a "Dehnbuch Orgelbau" [Expandable book of organ making] (manuscript, Ludwigsburg, 1959), archive of Oscar-Walcker-Schule. For more on Jung, see Hans-Christoph Knippel's self-published book, *Chronik der Bundesfachschule für Musikinstrumentenbau an der Oscar-Walcker-Schule* (Ludwigsburg, 2008), 37, 107–8.

[54] Berger, "Aus der Werkstatt" (cit. n. 41).

[55] Ibid., 114–5.

[56] Ibid., 120–1.

[57] Ibid.

[58] Ibid., 121.

had contact with other piano makers. In his memoirs Grotrian describes a meeting with Hansing in 1890, when he passed through Schwerin during his Wanderjahre: "After an exchange of opinions I left . . . the beautiful seaside town."[59] There are two additional examples of direct exchange between craftsmen and scientists: First is the contact between the company Steinway & Sons and Helmholtz. Theodore Steinway studied Helmholtz's *Lehre von den Tonempfindungen* and corresponded with him, making several instruments available to Helmholtz for his experiments. The second example can be found in the Grotrian-Steinweg Piano Company, where Grotrian established an acoustic laboratory in 1927 for the physicist Heinrich Hörig to perform acoustic experiments according to his own interests, the interests of the technical manager, and the construction work of the company. There was direct, face-to-face communication between Hörig, Grotrian, and the factory staff and indirect, written communication between the factory owner and the scientist through the series of *Mitteilungen aus dem Laboratorium* (Reports from the laboratory). Although not all of Hörig's developments were accepted and implemented by the craftsmen, the internal company laboratory is a nice example of the flow of knowledge from science to craft and vice versa.[60]

BETWEEN THEORY AND PRACTICE

Hansing's role as mediator became clear in a theory-practice dispute that unfolded in the *Zeitschrift für Instrumentenbau*, in a particularly well-known "Sprechsaal." Hansing regularly reacted to articles by other authors in the "Sprechsaal," and he usually discussed their theses intensively. The dispute erupted after Victor Hainisch of Vienna published an article titled "Querschnittmensur des Klavierbezuges: Differentialtöne" (Cross-sectional scaling of piano stringing and differential tones) in the main part of the *Zeitschrift für Instrumentenbau*.[61] Hainisch wrote about the use of *Saitenchöre* (string sets). These contain two or three strings for each bass or discant tone, respectively. This redundancy both reinforces the tone and provides backup for the contingency of string breakage during playing. A differential string set is characterized by a small difference in pitch between the individual strings in the set. The effect is a richer and fuller tone. In a series of three articles Hainisch offered the pros and cons of employing differential string sets.

Hainisch was critical of the usual method of strengthening the tone through equal-pitched string sets. He explained that pianos were the only keyboard instruments in which a minimal pitch differentiation within string sets was not used to strengthen the tone.[62] This differentiation, Hainisch argued, was responsible for harmonious sound. For him minimal pitch differentiation followed a kind of natural law of variation. Like Hansing, Hainisch used real-life examples to present his thesis, illustrating

[59] Willi Grotrian, "Curricula Vitae" (manuscript, Brunswick, 1929), 41, company archive of Grotrian-Steinweg Pianofortefabrikanten.

[60] Hörig, "Mitteilungen aus dem Laboratorium Nr. 11: Stimmhaltungsprobleme und die damit zusammenhängenden konstruktiven Fragen, vom 13. Oktober 1926" (unpublished report, Brunswick, 1926), company archive of Grotrian-Steinweg Pianofortefabrikanten. For more on the Grotrian-Steinweg laboratory, see Petersen, "Schwachstarktastenkasten" (cit. n. *), 165–97.

[61] Hainisch, "Über die Querschnittmensur des Klavierbezuges: Differentialtöne," *Zeitschrift für Instrumentenbau* 29 (1908–9): 4–5.

[62] Ibid., 4.

this natural law with examples such as the variation among individual voices in a choir or the interplay of different musical instruments in an orchestra.

However, Hainisch seems to have put more faith in technical writings on acoustics—specifically, Helmholtz's work on interference.[63] He cited Helmholtz's experiments with organ pipes and related them to the sets of piano strings. To make his case that equal-pitched string sets were inappropriate for strengthening notes and in fact weakened them, Hainisch needed to show that equality of the *Mensur* (scale) of single string sets resulted in interference. If the scale was exactly equal, the strings could cancel one another out, which would result in a tone that lacked strength and would only sound for a short period of time.[64] Such interference had to be prevented.

Differentiated strings were critical for the instrument's timbre and harmonic overtones. With equal-pitched strings, a threat of interference occurred here as well. When all strings in a set were of equal pitch, Hainisch argued, only identical harmonics occurred, which, as in the case of identical fundamental tones, weakened the timbre. Small differentiations of pitch were the only way in which the strings in a set could, aided by the soundboard, complement and strengthen one another.

For Hainisch, the need for a differentiated string set was obvious. The appropriate differentiation was characterized by a marginal variable amplitude. To differentiate a set of strings, the middle one was tuned to the exact frequency of the note, and the tuning of the two strings on either side was then slightly altered. The string to the right of the middle tone was shortened by about two to three millimeters, while the string to the left was lengthened by the same amount. This method of differentiation preserved the average pitch of the set and kept the composite tension the same as before, while preventing an *Einklang* (unison sound). Crucially, Hainisch pointed out, the construction of pianos was unaffected by differentiation of the string sets, aside from some minor changes to the bridge, damping, and stringing. Differential sets would not change the timbre, but perfect it.[65]

Enter Hansing to mediate between practice and theory.[66] Two individuals named Eugen Espert and Georg Brauer also participated in the discussion, but, sadly, no further details on Espert and Brauer are available. The engagement at first circled around Hainisch's technical statements. It then turned into a discussion about the relationship between practice and theory in piano making. A close reading of the dispute reveals that Hansing was at pains to demonstrate his roots in craftsmanship to Hainisch, even though he was far more than a simple piano maker. He argued that Hainisch's preferred method of differentiation did not, in fact, result in positive changes to the timbre. Then, invoking his craft origins, Hansing asserted that there was no reason to differentiate among the strings within a set, because a differentiation in piano making already took place as a result of the length of the scale. In addition, he presented the common scenario of a string breaking during a piano tuning

[63] Interference occurs when two or more waves are out of phase and the overall amplitude is subject to the addition and subtraction of the amplitudes of the peaks and troughs of the individual waveforms. This most commonly can be heard as a "beating" of the note. See Gerthsen, Kneser, and Vogel, *Physik* (cit. n. 33), 135–6, 152–3. For the significance of beats for tuning, see Jackson, *Harmonious Triads* (cit. n. 22), 151–82.

[64] Hainisch, "Über die Querschnittmensur" (cit. n. 61), 4–5.

[65] Victor Hainisch, "Über die Querschnittmensur des Klavierbezuges: Differentialchöre; Schluß," *Zeitschrift für Instrumentenbau* 29 (1908–9): 77–80.

[66] Hansing, "Erwiderung auf den Artikel: 'Über die Querschnittmensuren des Klavierbezuges: Differentialchöre,'" *Zeitschrift für Instrumentenbau* 29 (1908–9): 115–6.

at a customer's house. A replacement string with the correct *Drahtnummer* (wire gauge) would often be lacking. In this case the tuner had to use the strings he had at hand, overlooking the incorrect wire gauge. In doing so, the tuner would, due to the different mass of the replacement string, introduce a tuning differential to the string set. Lastly, differences in manufacturing tolerances and the specifications of strings from various manufacturers also introduced an element of unpredictability to the tuning of string sets. Hansing concluded that the piano maker had sufficient opportunities to establish a differential tuning of strings within equal-pitched sets.

This statement of Hansing's prompted an extended dispute, pitting theory against practice as the question of differential sets faded in significance. Instead the argument turned on whether a piano maker needed explicit or working knowledge to correctly tune a piano. On the one hand, piano makers in the early twentieth century were well aware of the contemporary scientific knowledge in the field of acoustics. On the other, there remained a discrepancy between this scientific explicit knowledge and their working knowledge, gained through a lifetime of experience. Science had clearly found its way into craft, of course, but it could not replace working knowledge entirely. As Hansing summed up: "We might best modify Dr. Hainisch's concluding sentence as follows: 'To be sure, the way to the optimum timbre of a string was learned long ago not from theory but from experience.'"[67]

Hansing's critique of Hainsich indicates that while he was aware of Helmholtz's findings and valued them greatly, he would not slavishly adopt theoretical knowledge if it contradicted his own experience. He remained skeptical of the infallibility of science and emphasized the individual experience of piano makers. Not every application of the dictates of physics led to the desired effect—in this case, to a full and rich timbre.

Hainisch's response to Hansing's critique followed directly in the next issue of *Zeitschrift für Instrumentenbau*.[68] He immediately rejected Hansing's arguments and again invoked scientific knowledge to argue for differential string sets. Furthermore, he referred to the interplay of the tension, length, and cross section (*Querschnitt*) of strings. To bring these components together it was essential, in his opinion, to act in accordance with the laws of acoustics. For him, science was irrefutable. Hainisch was particularly perturbed by Hansing's statements concerning Helmholtz. It was unimaginable for Hainisch that a practical piano maker would, given his exposure to the behavior of many pianos, consider Helmholtz's theory to be untrue.[69] Hainisch was not alone in defending the importance of scientific knowledge in piano making. Indeed, Hansing did not say that science was unimportant for piano making. He did, however, clearly state that in the case of the differentiation of string sets, he considered his own experience an indispensable basis for judging certain scientific statements to be inappropriate.

For the most part, Hainisch balanced personal experience and scientific knowledge and saw both as important parts of the fixed canon of explicit knowledge of

[67] Ibid., 116. Hainisch's original sentence was, "Practice learned from theory and experience how the optimum timbre of a string could be reached." Victor Hainisch, "Über die Querschnittmensur des Klavierbezuges: Differentialchöre; Fortsetzung," *Zeitschrift für Instrumentenbau* 29 (1908–9): 39–41, on 40.

[68] Victor Hainisch, "Zur Frage der Differentialchöre," *Zeitschrift für Instrumentenbau* 29 (1908–9): 150–2.

[69] Ibid.

early twentieth-century industrialized piano making: "It is an irrefutable fact that piano makers have gained their current success mostly by use of their ears. . . . But to deny theory its existence would not only contradict the facts, but would also voice a thought from which piano making would not benefit. The luminaries of the art of German piano making have never lost touch with theory. No industry could afford to turn its back wholly on theory as a matter of principle."[70] He believed that theory should be a piano maker's first basis for judgment, supplemented by experience. Hainisch asked Hansing to recognize that Hainisch's work was the result of a two-year study that was both theoretical and practical.[71] His absurd final riposte, in a reversal of his earlier attack, was to criticize Hansing, who had been publishing on science and acoustics in piano making since 1880 and had written a groundbreaking book about theoretical piano making in 1888, as being blind to theory.

Unsurprisingly, Hansing defended himself in the pages of the "Sprechsaal" against these accusations,[72] this time referring to several experiments to substantiate his criticisms of Hainisch. After this rejoinder, Hansing bowed out of the dispute. In his stead, Espert and Brauer took up the torch of arguing for the primacy of experience in piano making, with theory to be considered a supplement when useful.[73] Espert argued, like Hansing, that taking the experience of piano makers into account revealed that differentiated string sets were not fundamentally new in piano making and that they could be found in old upright grands or square pianos with straight stringing. Both theory and practice had always been important, Espert explained, but no one should blindly follow the one or the other. He argued that it was necessary instead to mediate between them, as Hansing did in his publications. Despite the advancements of science and industrialized production methods, the working knowledge of piano makers was still required. Espert explained: "It would take too long to describe all the circumstances and measures that have to work together, to be observed and taken from theory, to achieve a rich and beautiful timbre. One does not learn this from books. It is acquired only from practical work, from years of observation, slowly groping toward the solution, restlessly testing and carefully taking into account the crucial principles."[74] Explicit knowledge alone was insufficient for tuning a piano. Working knowledge was also crucial, especially for dealing with unforeseeable circumstances overlooked in theory; for example, the problem of not having the correct wire gauge on hand when a piano string broke.

Hainisch again raised his voice to staunchly defend himself against the accusation of prioritizing theory to the disadvantage of practice. He insisted that he also employed practical knowledge: "On closer study I have come to the conclusion that the question of differential string sets can only be answered by falling back on the principles of acoustic theory, once superficial attempts have failed. Once the theoretical basis had been specified . . . the practical experiments followed . . . to prove my

[70] Ibid., 151–2.
[71] Ibid.
[72] Hansing, "Zur Frage der Differentialchöre," *Zeitschrift für Instrumentenbau* 29 (1908–9): 186–7.
[73] Eugen Espert, "Kritische Betrachtungen über Differentialchöre," *Zeitschrift für Instrumentenbau* 29 (1908–9), 187–8; Espert, "Nochmals ein Wort über Differentialchöre," *Zeitschrift für Instrumentenbau* 29 (1908–9): 333–4; Georg Brauer, "Zur Frage der Querschnitt-Mensur, bzw. Differenzierung der Einklänge bei Pianos," *Zeitschrift für Instrumentenbau* 29 (1908–9): 408–9.
[74] Espert, "Kritische Betrachtungen" (cit. n. 73), 188.

thesis."[75] Hainisch emphasized that he was not an ivory-tower theorist and that he was very aware of his practical skills and experience.

We leave the dispute at this point, though it continued in subsequent issues of *Zeitschrift für Instrumentenbau*, without a concrete resolution. Hansing and Espert were clearly proponents of the experience of individual piano makers, but not to the exclusion of a theoretical approach, which they did recognize as an important part of piano making. It may be more instructive to think of their views as leaning toward the empirical, inasmuch as they considered experience to be necessary for the verification, completion, or possible rejection of the theoretical. If nothing else, the dispute shows that Hansing was not the only piano maker who gained theoretical knowledge, especially in the field of acoustics.

CONCLUSION

As a result of the increasing mechanization of production in piano making during the second half of the nineteenth century, explicit knowledge grew in importance. And yet the working knowledge of the craftsmen retained its status not only in instrument development but also in production. The example of Siegfried Hansing illustrates that the confluence of three inexorable nineteenth-century trends—the industrialization of piano making, the professionalization of piano playing, and the increasing scientific interest in acoustics—made the acquisition of theoretical knowledge positively advantageous, though as an enhancement to and not a replacement for individual experience. Hansing gained knowledge of physics and introduced theory not just to his own practical work, but also to his colleagues. For him it was indisputable that theory and practice should supplement each other in instrument development and production. In the dispute about differential string sets outlined above, we saw his attempts to clarify his views on the simultaneity of explicit and working knowledge. These efforts had been apparent earlier on in his writings in *Zeitschrift für Instrumentenbau* and his book *Das Pianoforte in seinen akustischen Anlagen*. In addition to scientific research, literature, and formalized and standardized production methods, working knowledge was irreplaceable in piano making. Other piano makers themselves were certainly well aware of the importance of their personal experience, and combined it with rather than replacing it by scientific knowledge.

To sum up, this case study of Hansing shows that both explicit and working knowledge circulated during the industrialization of musical-instrument making. Moreover, aesthetic criteria prominent among instrument makers circulated and mediated between explicit and working knowledge. It is also clear that scientific knowledge not only circulated in a scientific community but also made its way down to technicians and craftsmen. When Hansing died in 1911, his absence was deeply felt by his colleagues. Appreciation of his influence on piano making and the importance of his work as a mediator between science and practice, as well as personal admiration, were expressed by Pfeiffer in his obituary published in *Zeitschrift für Instrumentenbau*: "With him we lost a leader, in fact the only one manifested in the current literature."[76]

---

[75] Hainisch, "Zur Frage der Differentialchöre," *Zeitschrift für Instrumentenbau* 29 (1908–9): 259–61, on 260.

[76] Walter Pfeiffer, "Siegfried Hansing †," *Zeitschrift für Instrumentenbau* 33 (1912–3): 1082–3.

Listening to the Piano Pedal:

Acoustics and Pedagogy in Late Nineteenth-Century Contexts

by Elfrieda Hiebert*

ABSTRACT

Until the 1870s the practice of artistic pedaling in piano performance lacked guide-lines and was left largely to the whim of individual pianists. The kinship that de-veloped between science and music, drawn largely from Hermann von Helmholtz's experiments on musical sound, left a positive imprint on the art of pedaling. This article elucidates the contribution of Hans Schmitt by way of his acoustical analysis in *Das Pedal des Claviers* (1875). In conclusion the article probes Schmitt's genera-tive role in shaping the practice of piano pedaling. It demonstrates the potential for the collaboration of acoustics and music to enhance piano performance.

> "Musicians look as yet far too much on acoustics as something of no practical value, but this is a mistake."
> — Friedrich Niecks[1]

In the increasingly scientific culture of the late nineteenth century, scientists and mu-sicians brushed shoulders as they probed for answers to puzzling musical problems. As a result the various domains of music and the sciences became noticeably inter-woven; aesthetic concepts changed and new musical ideals were strengthened and validated. The connection between the analysis of musical acoustics at the intellec-tual and practical levels rendered new areas of expression possible.[2]

The interplay between the imagination of piano builders and the desires of musi-cians appears to have been reciprocal—a veritable dialogue between the material means and the longed-for sounds. Shortly before the mid-nineteenth century piano builders began to seek additional scientific analysis and assistance from physicists.

* This essay has grown out of my lectures and earlier papers on the links between acoustics and music featuring the piano and the performer. I am grateful to the Max Planck Institute for the History of Science for several residencies in Berlin, where much of the material was investigated. I wish to thank the editors of *Osiris* 28 for suggestions, especially Julia Kursell. I also wish to thank Erwin Hiebert for helpful comments on various versions of the paper, Lin Garber for research assistance, and Sara Wyse-Wenger for aiding in the final preparation of the paper. All translations are my own unless otherwise credited.

We editors report with great sadness that Elfrieda Hiebert passed away on 2 Sept. 2012. This vol-ume is dedicated to her memory.

[1] Niecks, "On the Use and Abuse of the Pedal, with Special Reference to a Book on That Subject," *Monthly Musical Record* 6 (1876): 179–83, on 181.

[2] See Georgina Born, *Rationalizing Culture: IRCAM, Boulez, and the Institutionalization of the Musical Avant-Garde* (Berkeley, Calif., 1993), 41, 346 nn. 4, 6.

© 2013 by The History of Science Society. All rights reserved. 0369-7827/11/2013-0012$10.00

For example, the piano builder Heinrich Welcher von Gontershausen suggested that what was needed most urgently was information on acoustics that would not simply repeat the outdated material of Marpurg, Chladni, and other earlier writers, but would set forth clear procedures for the application of theory to practice in piano building based on recent scientific advances.[3]

The piano, immensely popular during the nineteenth century, played a dual role in society: as an instrument for practical music making and for scientific research in the laboratory.[4] Not until the mid-1870s is there significant published evidence that piano performers and pedagogues reached out to the scientific community for new insights. The state of piano pedaling had been in confusion for decades, and efforts to clarify and to reform practices were long overdue. Without defined guidelines students of the piano used the damper pedal according to whim and fancy. When pedagogues finally began to produce guidelines for tasteful and efficient pedaling practices, including insights drawn from acoustical analyses, particularly the work of Helmholtz (see below), in the 1870s, it was clear that some musicians were beginning to seek illumination and validation from the sciences for their professional interests.

The widespread practice (amateur and professional) of playing the piano, and the great variety of piano designs originating in the vast number of piano-building workshops, contributed to the indiscriminate use of the pedal. How to time the use of the pedal and how to acquire techniques to produce intended color effects imagined by individual composers remained unclear. Also left undiscussed was how pedaling effects needed to be adjusted to specific venues. In addition, appropriate pedaling required an understanding of historical practices as well as knowledge of an individual composer's preferences. Until late in the nineteenth century these issues had been merely touched upon.

The aim of this article is to examine selected historically contingent associations between scientific information and pedagogical thinkers who were intent on improving piano performances by means of informed pedaling. How was the experimental information from musical acoustics disseminated, received, stabilized, and promoted among piano pedagogues? More specifically, how was acoustical information assembled, assimilated, and applied to the practice of piano pedaling? How did it generate new insights for piano performance? To what extent did attention to acoustical analysis lead pianists to increase and refine their listening to the sounds they produced? The effects of Helmholtz's acoustical investigations, the ideas of the innovative piano pedagogue Hans Schmitt, and the community of scientists and musicians drawn from many countries enabled and fostered an interdisciplinary exchange of information dealing with pedaling practices. I show how the merging of scientific thought (objective, rigorous analysis) and a musical aesthetic based on subjective elements (with little patience for acoustical analysis) sometimes led to controversy and ambiguity on

---

[3] Welcher, *Der Flügel oder die Beschaffenheit des Pianos in allen Formen*, 2nd ed. (Frankfurt am Main, 1853), 88. Whereas Welcher does not refer to the acoustical work of Hermann von Helmholtz even in the latest editions of his own writings in 1870, other builders did rely on scientific information from Helmholtz's work (see below): e.g., Julius Blüthner and Heinrich Gretschel, *Lehrbuch des Pianofortebaues* (1872; reprinted facsimile, Frankfurt am Main, 1992); Siegfried Hansing, *Das Pianoforte in seinen akustischen Anlagen*, 2nd ed. (Schwerin, 1909).

[4] The terms "piano" and "fortepiano" were used interchangeably during the late eighteenth and early nineteenth centuries; the term "piano" is used in this article to designate the instrument.

the music scene. But I also argue that the search for evidence from acoustics for the adequate employment of the pedal emerged as a dynamic element in improving piano performance; it led to useful and consequential results.

THE PIANO IN TRANSITION AND THE ABUSE OF PEDALING

By the middle of the nineteenth century thoughtful musicians were becoming increasingly concerned about mediocre piano performances due to poor and uninformed pedaling. Although it was commonly observed that the artistic pedaling of Franz Liszt and Anton Rubinstein should be emulated, there was little guidance for students of the piano from pedagogues and master performers or teachers.

In *Music Study in Europe* the pianist Amy Fay (1844–1928) pointed to the laxness with which piano teachers were dealing with the damper pedal (the right pedal, often called the "loud" pedal because it opens all the strings of the piano to vibrate, creating louder sounds). Fay declared that her own teachers, Carl Tausig, Adolph Kullak, and Liszt, did not guide her in the use of the pedal except to admonish her to avoid pedaling while playing scales.[5] The artistic effects Liszt achieved in what Fay called "peculiar" pedaling (which probably was syncopated pedaling) were a puzzle to her.[6] Syncopated pedaling is defined as lowering the pedal (i.e., raising the dampers) after the sound of the tone and lifting it (i.e., lowering the dampers) with the sound of the following tone; this connects the tones. It appears that Liszt was aware of the benefits of pedaling in this manner. In a letter dated 27 July 1875, he wrote to Louis Köhler, "The entrance of the pedal after the striking of the chords . . . is very much to be recommended . . . especially in slow tempi."[7]

During her last year of piano study in Europe (1874–5), Fay finally met a teacher in Berlin, Ludwig Deppe, who showed her how to achieve some of the sound effects from pedaling that she admired in Liszt. Deppe's concern about the importance of pedaling techniques was exceptional among piano pedagogues. His instruction was based on treading the pedal "after the chord" (syncopated pedaling).[8] When Deppe wanted a brilliant sound, the treading of the pedal was made to coincide with the tone or chord (often designated "rhythmic pedaling"), so that the strings of the piano would begin to vibrate immediately with the sound. Treading the pedal simultaneously with the sounding note had been the prevailing mode of pedaling during the late eighteenth and early nineteenth centuries. It meant damping with the change of harmony or at the barline, and resulted in gapped or disconnected sounds. Syncopated pedaling, in contrast, led to connected sounds and supported the growing aesthetic ideal of continuity and connectedness during the nineteenth century; but systematic guidance in achieving this came slowly.[9] The first direction composers put in

[5] Fay, *Music Study in Europe* (1880; repr., New York, 1965), 297.
[6] Ibid., 213, 224.
[7] *Franz Liszt Briefe*, ed. La Mara [Marie Lipsius] (Leipzig, 1893), 2:223.
[8] Fay, *Music Study* (cit. n. 5), 298. Deppe, who was a conductor by profession and not a bona fide pianist, contributed to piano-pedaling reform as an outsider. He made an enormous impact on the field of piano pedagogy.
[9] Friedrich Kalkbrenner is sometimes credited for being the first to hint, in the 1830s, at syncopated pedaling. See Adolph Kullak, *Die Aesthetik des Klavierspiels*, 4th ed., revised and enlarged by Walter Niemann (Leipzig, 1905), 105. In *Méthode pour apprendre le pianoforte* (Paris, 1830), Kalkbrenner remarks, "The loud pedal may be . . . occasionally taken after a note has been struck." Cited in David Rowland, *History of Pianoforte Pedalling* (Cambridge, 1993), 115. As early as 1839 Carl Czerny, in his *Vollständige . . . Pianoforteschule*, op. 500 (Vienna, 1839), vols. 3–4, discussed ways of achiev-

their scores with the intention of connecting tones with the damper pedal may have been the sign "Ped.," without an indication for ending the pedal hold. The earliest clear description of syncopated pedaling is found in Louis Köhler's general text *Systematische Lehrmethode für Clavierspiel und Musik*. Köhler emphatically instructs the piano player to sound the note before lifting the dampers—"Thus: do not strike and tread at the same time!"[10] Syncopated pedaling was slow to capture the imagination of students and performers. As late as 1913, Tobias Matthay observed, "Even the most primitive and antediluvian of teachers have now at least some hazy sort of notion as to the nature and importance of 'syncopated pedaling.'"[11]

With changes in sound ideals came changes in the piano, as builders sought experimentally to meet the imagination of musicians. Innovation in the mechanism of the piano during the nineteenth century allowed for the production of a fuller, more intense, connected, and sustained piano sound. This was achieved at first, in the early part of the century, largely by strengthening the wooden piano frames with metal crossbars. After midcentury partial or full iron frames took the place of the wooden frames on most pianos. This allowed for increased string tension and added to the sound possibilities by allowing the use of thicker and longer strings, cross stringing, and larger and heavier hammers. Builders paid attention to the aesthetics of connectedness by making the damping system more responsive and more effective.

The mechanics of modifying and extending the sounding tone went through different stages in the history of the pedal, beginning with the hand stop (mutation) system, which was replaced on Viennese pianos in the late eighteenth century by knee levers under the keyboard. During the first decade of the nineteenth century the foot pedal controlling the dampers became the common device to help sustain tones, create fuller sounds, extinguish sounds, and color piano music. English pianos were outfitted with foot pedals earlier, during the late eighteenth century. Development of the technique of pedaling took place during the period of the Viennese classical composers, especially with Beethoven. He personally lived through the changes from knee lever to foot pedal but limited his indications mainly to highlighting special effects. Until op. 53 (published in 1805) he used the terms *con Sordino* (with lifted dampers) and *senza Sordino* (without dampers). Thereafter his intended pedal usage was indicated by "Ped." and "*".[12]

Early nineteenth-century composers were slow to notate pedaling in their scores, and when they attempted to do so the details were ambiguous. As early as Schumann (1810–56), the aesthetics of pedal usage were noticeably changing. He often wrote *mit Pedal* or *con Pedale* at the beginning of a piece, leaving the actual pedaling prac-

---

ing more sensitive results in pedaling and hinted at syncopated technique. A rather vague reference to syncopated pedaling is located in Charles Chaulieu's article "Des Pedales," *Le Pianiste* 9 (1833–4): 131–2. It is generally understood that syncopated pedaling was practiced intuitively at least to some degree after the 1830s, long before it was first briefly described. I thank Sandra Rosenblum for discussions about syncopated pedaling and for bringing Chaulieu's work to my attention.

[10] Köhler, *Systematische Lehrmethode für Clavierspiel und Musik* (Leipzig, 1857), 187. This brief description is often misdated in references, to 1862, 1860, etc. Köhler does not refer to acoustical factors.

[11] Matthay, *Musical Interpretation* (London, 1913), 131.

[12] For more details, see Sandra P. Rosenblum, *Performance Practices in Classic Piano Music: Their Principles and Applications* (Bloomington, Ind., 1988), 102–41. Also see Elfrieda F. Hiebert, "Reflections on the Piano, Pedagogical Thought and the Practice of Pedaling during the Late Nineteenth Century," in "Physiologie des Klaviers" (Preprint 366, Max Planck Institut für Wissenschaftsgeschichte, Berlin, 2009), 71–100, on 82 n. 46, 85 n. 63.

tice to the discretion of the individual performer. Mendelssohn's instructions were sparse, sporadic, and indefinite. Chopin was exceptional in this regard, since he notated his scores in great detail.[13] Whether or not the composers mentioned here intended syncopated pedaling at various times cannot be discerned from their markings, although, as Sandra Rosenblum has observed, Chopin "may well have used syncopated pedaling, intuitively, especially in cantabile textures."[14]

Independent piano builders adapted damping systems to their individual and distinctive piano designs. Variations from workshop to workshop and country to country challenged the pianist's expertise and added to the complexities of choosing and annotating sensitive pedaling. The differences between the English/French and Viennese/German actions and damper pads complicated the process of making decisions about pedaling. In this context the practice of pedaling also was intertwined with trends in the changing properties of the piano (such as increased sustaining power and strengthened sound). Not until the beginning of the twentieth century could standardization in the construction of the piano even be considered. Builders retained and protected their individual styles and materials and the details of their construction methods throughout most of the nineteenth century, although some were more closely linked than others—Steinway, Bechstein, and Blüthner, for example.[15]

The respective mechanisms involved in the English/French and the German/Viennese damping systems proved crucial in determining pedal usage. The different shapes and weight of the dampers defined their effects, especially the length of time required to extinguish the sound of the strings. Since the dampers on the Viennese pianos were denser and heavier than those on the English, the sounds were clearer and decayed faster. The Viennese dampers extinguished sound more efficiently. The English dampers were lighter and less concentrated (more diffuse) and tended to allow the strings to vibrate with an after-ring that could last several seconds, often resulting in blurry sounds.[16] In the view of Bart van Oort, since English builders could easily have "adopted a more effective system, based on the Viennese [damping system] . . . we must conclude that this particular sound effect was based on a different musical aesthetic."[17]

Reliance on the damper pedal in order to connect tones in skips beyond the compass of the hand liberated the pianist and helped to sustain a more widely flung range

---

[13] See Jean-Jacques Eigeldinger, *Chopin: Pianist and Teacher, as Seen by His Pupils*, ed. Roy Howat, trans. Naomi Shohet with Krysia Osostowicz and Roy Howat (Cambridge, 1986), 57–8.

[14] Rosenblum, "Pedaling the Piano: A Brief Survey from the Eighteenth Century to the Present," *Performance Practice Review* 6, no. 2 (1993): 158–78, on 167.

[15] Even today each piano has its own individual character. This is true whether one compares a Steinway with another Steinway or a Steinway with a Bechstein, etc. Pianists therefore are not content to have someone choose a piano for a concert; they often visit factories to try out instruments to satisfy their own individual needs and tastes.

[16] See Bart van Oort, "Haydn and the English Classical Piano Style," *Early Music* 28 (2000): 73–89, on 75. Apparently some pianos, such as those made by Collard and Collard (originally owned by Clementi), operated more efficiently—at least individual ones are reported to have extinguished sounds promptly. See Stephen Bicknell, "There to Be Played [Richard Burnett's Historic Keyboard Instrument Collection at Finchcocks in Kent]," *International Piano* 9 (Mar./Apr. 2005), 20–2, on 22, where he discusses instruments in Finchcock's collection of pianos.

[17] See Van Oort, "Haydn" (cit. n. 16), 75–6. Van Oort was referring to the pianos of the late eighteenth and early nineteenth centuries. Steinway, Bechstein, and Blüthner incorporated the English action during the 1850s. Writers on pedaling toward the end of the nineteenth century still differentiated between the English/French and German/Viennese damping systems.

of tones—a phenomenon of great assistance to the virtuoso. How and when to lift and lower the dampers in order to produce extended, clear, or enhanced sonority, support rhythm, and separate articulating groups were matters to be examined carefully by the pedagogues.

The struggle to overcome difficult issues in the development of piano technique, such as increased key resistance, greater key dip, and longer sustained sounds, is reflected in a lull in the publication of pedagogical texts between 1830 and the late 1850s. Thereafter, pedagogues attacked problems of performance with new strategic techniques that took advantage of the adaptive function of the keyboard and the use of the pedal. They provided guidelines for pedaling that prompted pianists to produce judicious pedal effects. This required a new kind of listening based on the visualization of acoustical analysis and the application of knowledge.[18]

At the same time, the burgeoning popularity of solo piano playing, in which the bourgeoisie deified the individual achiever, was matched by the consuming interest in amateur piano playing in the home—a phenomenon that expanded increasingly during the nineteenth century.[19] With every household in the market for an instrument by mid- to late century, the demand for pianos reached a feverish level. To accommodate this need, as the German piano builder Julius Blüthner remarked, piano factories "shot up like mushrooms from the ground" all over German-speaking lands.[20]

The abuse of the pedal in this setting is well documented. Critics and musicians, from Heinrich Heine in 1843 to Eduard Hanslick in 1884, expressed deep dissatisfaction with the state of piano playing.[21] Hanslick spoke of a "piano epidemic" or "fever" (Clavierseuche) that infiltrated every level of musical culture from the amateur at home to the professional on the concert stage. He deplored the level and extent of poor piano playing.[22] In his chapter "About the Sustaining Pedal" in Piano and Song (1853), Friedrich Wieck, Clara Schumann's father, expressed outrage at the "tonal cataclysm" he had just experienced at a concert by a virtuoso. He cried, "What a fateful, frightful invention! I mean the pedal that raises the dampers on the piano. . . . One of the many deaf modern virtuosos has just stormed through a bravura piece with continuously depressed sustaining pedal—but with rapture!"[23]

It was in this context of dissatisfaction that the developing scientific culture in nineteenth-century society made inroads into the world of music. After almost universally neglecting pedaling technique for the first half of the nineteenth century, pedagogues began by the 1870s to initiate meaningful changes and guidelines. Assistance came from scientists who had developed acoustical investigations related

[18] The changing practice of pedaling contributed to "newness" in compositions (a subject for development) and can be traced throughout the century. Mature piano students and professionals today spend much of their practice time listening to their own playing and considering imaginative, variable approaches to pedaling depending upon the composition (or composer), venue, and particular instrument.

[19] See Leon Plantinga, "The Piano and the Nineteenth Century," in Nineteenth-Century Piano Music, ed. Larry Todd (New York, 1990), 1–15, esp. 11–2.

[20] Blüthner and Gretschel, Lehrbuch (cit. n. 3), 240. See the hundreds of nineteenth-century builders listed in Herbert Henkel's Lexikon deutscher Klavierbauer (Frankfurt am Main, 2000).

[21] Heine, Sämtliche Werke (Leipzig, 1910), 9:271–9.

[22] Hanslick, "Ein Brief über die 'Clavierseuche,'" in Suite: Aufsätze über Musik und Musiker (Vienna, 1884), 164ff.

[23] Wieck, Piano and Song (Didactic and Polemical), trans. Henry Pleasants (Stuyvesant, N.Y., 1988), 53. Originally published as Clavier und Gesang (1853; repr., Peer, Belgium, 1986).

to musical perception. As pedagogues explored, absorbed, and visualized new ideas from the physical world, they were able to realize new imaginative effects in the practice of pedaling.

HELMHOLTZ AS DISSEMINATOR OF MUSICAL ACOUSTICS

By the 1860s and '70s attempts were being made to bridge the more purely intellectual considerations of early acoustical studies with an enlightened examination of sound applied to the practice of music. The foremost catalyst in awakening pedagogues to a new level of perception was Hermann von Helmholtz, who acted as a key "informant" through his far-reaching work *Die Lehre von den Tonempfindungen*—a study in which acoustics was related to the analysis of sound as music.[24] The rudiments of Helmholtz's work, begun already in the 1850s, led to the adaptation of knowledge from acoustics to the appraisal of the quality of tone in experiments with piano strings (as well as the violin, organ, and harmonium). This, in turn, was drawn on by piano pedagogues to assist in building a theory of acoustical perception of piano resonance that was brought to life through informed pedal technique.

Helmholtz's early education in medicine and physiology and his later work in physics combined to open new areas of thought, interdisciplinary connections between the physical and physiological conceptions of the material of music. From his youth through his student days and onward he studied music, became an accomplished amateur performer on the piano, participated in singing groups, and frequently attended concerts. Helmholtz was culturally embedded in a changing musical world that stimulated him to adopt a historical orientation in his work. He was thoroughly schooled in music theory and literature and had a special fondness for the music of Bach, Beethoven, Mendelssohn, and Wagner.[25] Indeed, he introduced his lecture "Physiological Causes of Harmony in Music" in Bonn at the end of September 1857 with a tribute to Beethoven as "the mightiest among heroes of harmony."[26]

Over the years a wide circle of friends—among them professional music scholars, composers, and performers, such as Otto Jahn in Bonn, best known for his biography of Mozart; the composer Heinrich von Herzogenberg and his pianist wife, Elizabeth; and the violinist Joseph Joachim—kept Helmholtz in touch with the world of music.[27] His students in Berlin, Shohei Tanaka and Paul von Jankó (both scientists and

[24] Helmholtz, *Die Lehre von den Tonempfindungen als physiologische Grundlage für die Theorie der Musik* (Brunswick, 1863). A French translation of the second German edition (1865) was made by Georges Guéroult, with the assistance of piano builder August Wolff, and with a preface by Helmholtz: *Théorie physiologique de la musique fondée sur l'étude des sensations auditives* (Paris, 1868). A second French edition appeared in 1874. An English translation of the third German edition (1870) was made by Alexander J. Ellis: *On the Sensations of Tone as a Physiological Basis for the Theory of Music* (London, 1875).
[25] John Gray McKendrick, *Hermann Ludwig Ferdinand von Helmholtz* (London, 1899), 58–9, 134; Leo Koenigsberger, *Hermann von Helmholtz* (Brunswick, 1902), 1:22–30. See David Cahan, ed., *Letters of Hermann von Helmholtz to His Parents, 1837–1846* (Stuttgart, 1993), 45, about the piano he took to his living quarters in Berlin during his student days beginning Sept. 1838. For more details, see Alexandra E. Hui, "Instruments of Music, Instruments of Science: Hermann von Helmholtz's Musical Practices, His Classicism, and His Beethoven Sonata," *Annals of Science* 68, no. 2 (2011): 149–77.
[26] Helmholtz, "Physiological Causes of Harmony in Music," trans. Alexander J. Ellis, in *Popular Lectures on Scientific Subjects* (New York, 1873), 61–106. Originally published in *Populäre wissenschaftliche Vorträge*, vol. 1 (Brunswick, 1865).
[27] Helmut Rechenberg, *Hermann von Helmholtz: Bilder seines Lebens und Werkens* (Weinheim, 1994), 66–7, 114, 143, 243–4; Koenigsberger, *Hermann von Helmholtz* (cit. n. 25), 1:186.

musicians), became well known in musical circles and extended the dissemination of his work.[28]

The Wagner and Helmholtz families occasionally met at social events in Berlin, but there seems to be little or no evidence that the two men interacted on scientific or musical problems. In 1876 Helmholtz and his wife, Anna, together with distinguished friends from Berlin, attended the first Bayreuther Festspiele. Extant letters from an ongoing friendship between Anna von Helmholtz and Wagner's wife, Cosima, date from 1888 to 1899.[29]

Already in Koenigsberg, Helmholtz began to explore problems connected with the mechanics of hearing and sound; but it was not until after his move to Bonn in September 1855 that he embarked on the expansive treatment of sensations of tones as he attempted to account for the perception of quality of tone. Early in 1856 he wrote his friend Emil du Bois-Reymond that he had begun examining the problem of combination tones (Tartini tones).[30] A year later, on 18 May 1857, he reported to du Bois-Reymond that his work was developing fast and that he now had a plan to view physical acoustics from the physiological standpoint.[31]

After his move to Heidelberg in 1858, Helmholtz brought together his experiments and studies in the exhaustive treatise *Tonempfindungen*, completed in 1862 and published in 1863. The book falls into three large parts: part 1, "On the Compositions of Vibrations (Upper Partial Tones, and Qualities of Tone)"; part 2, "On the Interruptions of Harmony (Combinational Tones and Beat, Consonance and Dissonance)"; part 3, "The Relationship of Musical Tones (Scales and Tonality, Esthetical Relations)." Helmholtz analyzes the sensations of a single tone as well as the relationship of tones. He examines how sound behaves in context, which leads to the extension of the perception of dissonance.[32] Essential to Helmholtz's orientation was the view that "music does not rest on immutable natural laws, but is the consequence of aesthetic principles and is . . . subject to change during the continuing march of human evolution."[33]

Drawing on others' previous studies, Helmholtz demonstrates that the quality of sound depends on the order, number, and intensity of the overtones or harmonics that enter into the makeup of a musical tone. Playing a tone therefore generates an entire family of other tones involving, in this case, the entire piano string.[34] It must be noted that a single damper on the piano is activated by a single stroke of a key. Different constellations of overtones account for diversity in quality of tone. The mixture of overtones modifying the quality of tone is altered by the strength and length

[28] See Hans Schmitt, *Das Pedal des Claviers* (1892; repr., Vienna, 1907), 37–40, for Helmholtz's piano student Von Jankó and the keyboard he developed.

[29] Petra Werner and Angelika Irmscher, *Kunst und Liebe müssen sein: Briefe von Anna von Helmholtz an Cosima Wagner, 1889 bis 1899* (Bayreuth, 1993).

[30] Koenigsberger, *Hermann von Helmholtz* (cit. n. 25), 1:267–71.

[31] Ibid., 1:281–3.

[32] See Erwin Hiebert and Elfrieda Hiebert, "Musical Thought and Practice: Links to Helmholtz's *Tonempfindungen*," in *Universalgenie Helmholtz: Rückblick nach 100 Jahren*, ed. Lorenz Krüger (Berlin, 1994), 295–314, on 302–3.

[33] Carl Dahlhaus, *Nineteenth-Century Music*, trans. J. Bradford Robinson (Berkeley, Calif., 1989), 322–3. Originally published as *Die Musik des 19. Jahrhunderts* (Wiesbaden, 1980).

[34] See also the discussion in Elfrieda F. Hiebert, "Helmholtz's Musical Acoustics: Incentive for Practical Techniques in Pedaling and Touch at the Piano," in *The Past in the Present: Papers Read at the IMS Intercongressional Symposium and the 10th Meeting of the Cantus Planus, Budapest and Visegrád, 2000* (Budapest, 2003), 1:427–30.

of the stroke of the piano hammer and by the sustaining power, attacks, endings, and incidental noises. This creates a plurality of overtones that is a veritable reservoir (*Speichertyp*) of sound possibilities.[35] Helmholtz further develops the theory of differential and summational tones and "singing" resonance.[36] He notes that sounding bodies have the capacity to set up sympathetic vibrations in other bodies whether they contact them directly or not. This has profound implications for making music at the piano. To demonstrate this, Helmholtz gives an example in which the damper is raised from a piano string and a tone similar to that of the string is produced by other means; for instance, by another instrument, such as the horn. This causes the piano string to vibrate. When the foreign tone ceases, the vibration of the string continues briefly as a type of after-sound.[37] The vibrating piano string is central to Helmholtz's acoustical explorations. He relates his findings to "hearing"; that is, to how the ear receives vibration and becomes a sounding body.[38] For much of Helmholtz's early research with piano strings a Kaim and Guenther piano was used.[39] After the publication of *Tonempfindungen*, from 1871 onward, American Steinway pianos were at his disposal; they were presented to Helmholtz as gifts in 1871, 1881, and 1893.[40]

Musical communities throughout the Western world were alerted to *Tonempfindungen* almost immediately; the publication of several comprehensive reviews generated animated discussions. Although it was agreed by some (such as Edward Krueger) that Helmholtz's studies in the physics and physiology of sound were novel and deserved high praise, there were critics (such as Moritz Hauptmann) who rejected the essential relevance of his findings for music.[41]

Tonempfindungen soon inspired the publication of texts on science and music, first in England and France, and later in German-speaking countries and the United States.[42] Helmholtz was recognized as one who gave new life to the study of sound. In the 1881 preface to his widely read book *Musical Acoustics*, often referred to as "The Student's Helmholtz," John Broadhouse pointed to the fact that Helmholtz's *Tonempfindungen* was required reading for music students at Oxford. He proclaimed

[35] See Wolfgang Scherer, *Klavier-Spiele: Die Psychotechnik der Klaviere im 18. und 19. Jahrhundert* (Munich, 1989), 174.

[36] See Helmholtz, "Physiological Causes" (cit. n. 26), 23–51.

[37] Ibid., 37–8.

[38] For the mechanization of measuring hearing, see David Pantalony, *Altered Sensations: Rudolph Koenig's Acoustical Workshop in Nineteenth-Century Paris* (Dordrecht, 2009), 19–63.

[39] Oscar Paul, *Geschichte des Claviers* (Leipzig, 1868), 41. This manufacturer had a piano factory in Kirchheim, Germany, that was in production from 1845 to 1882. Henkel, *Lexikon* (cit. n. 20), 297. In addition to the piano, the harmonium also was a vehicle for experimentation, since it allowed Helmholtz to investigate problems (such as temperament) peculiar to sustained sounds, unlike the rapidly decaying sounds of the piano. He also used the violin and the organ as tools for investigation.

[40] This information has been confirmed in personal communications courtesy of Robert Berger (Steinway & Sons, New York). William Steinway's diary, in which there is further documentation, has recently (Dec. 2010) gone online at the Smithsonian National Museum of American History's website, http://americanhistory.si.edu/steinwaydiary/ (accessed 21 Feb. 2013). For confirmation in letters that the Steinways consulted with Helmholtz, see Hiebert and Hiebert, "Musical Thought and Practice" (cit. n. 32), 308.

[41] See Krueger's lengthy positive review in the *Allgemeine musikalische Zeitung*, 1 July, 8 July, and 15 July 1863. Among Krueger's students at Göttingen was Hugo Riemann, who also became involved in the discussions concerning the effect of science on music. Alexander Rehding deals with some of these issues in *Hugo Riemann and the Birth of Modern Musical Thought* (Cambridge, 2003). See also Hauptmann, *Allgemeine musikalische Zeitung*, 30 Sept. 1863, for his critical stance, and Hiebert and Hiebert, "Musical Thought and Practice" (cit. n. 32), 299–301.

[42] For a selected listing of books appearing shortly after 1863 on scientific attitudes to music, mainly outgrowths of Helmholtz's work, see Hiebert, "Reflections on the Piano" (cit. n. 12), 79 n. 30.

that many universities would soon follow suit.[43] Over the years, as Helmholtz's ideas were absorbed into explanatory scientific thinking, direct credit receded. His work, it appears, became swallowed up in the musical acoustics of later scientists.

Historical and traditional in its analysis, Helmholtz's approach allowed others to extend his ideas with new, innovative concepts. A burst of awareness and understanding by pedagogues gave rise to an application of musical acoustics to piano pedaling that slowly gathered momentum. The effects were soon noticeable. New areas of insight toward the end of the nineteenth century clearly connected theory and practice, knowledge and doing.[44]

MUSICIANS ABSORB ACOUSTICS IN THOUGHT AND PRACTICE: LINKS TO THE SPIRIT AND RIGOR OF SCIENCE

As the intensity of the scientific enterprise in fields of learning slowly permeated late nineteenth-century European musical thought, theoretical and pedagogical concepts that largely had been generated from *Tonempfindungen* in Berlin migrated to centers of learning in Vienna and Prague. The University of Vienna as well as the Vienna Conservatory were open to charting new directions in both acoustical analysis and music teaching. Great diversity of opinion nevertheless was evident both in those who favored scientific inquiry and those who were not only uninformed about the issues but opposed to the so-called scientific direction. In some quarters there were concerns about whether the various categories of inquiry that constitute the musical disciplines could meet the demands of the scientific (empirical) approach. In this context musicians in the various branches of the field of music became actively involved in the ferment of thought.[45] Guido Adler embodied both enthusiasm and concerns about the scientific direction as he struggled to define the discipline of musicology; he wished to establish for the study of music a respected place in the scientific academy, while at the same time he was concerned about retaining the vitality of the arts in a scientific age. In 1885 Adler set out to define his position in the well-known essay "Umfang, Methode und Ziel der Musikwissenschaft" (Scope, method, and goal of musicology).[46] Adler proposed to transform art into a rigorous discipline that would express the spirit of the natural sciences.[47] As Kevin C.

---

[43] Broadhouse, *Musical Acoustics; or, The Phenomena of Sound as Connected with Music* (London, 1881). In America, at Harvard University, Daniel Gregory Mason was heavily influenced by Broadhouse's book.

[44] For a comprehensive background account of the historical place of *Tonempfindungen*, see Burdette Green and David Butler, "From Acoustics to Tonpsychologie," in *The Cambridge History of Western Music Theory*, ed. Thomas Christensen (Cambridge, 2002), 246–71.

[45] For example, Philipp Spitta in Berlin was a leading proponent of transforming the study of the arts into a discipline as rigorous as the natural sciences. Friedrich Chrysander in Vienna joined him in this view. See Kevin C. Karnes, *Music, Criticism, and the Challenge of History: Shaping Modern Thought in Late Nineteenth-Century Vienna* (Oxford, 2008), 10–1, 13, 25–7, 40–1.

[46] Adler, "Umfang, Methode und Ziel der Musikwissenschaft," *Vierteljahresschrift für Musikwissenschaft* 1 (1885): 5–20. Translated in Bojan Bujić, ed., *Music in European Thought, 1851–1912* (Cambridge, 1988), 348–53, with a slightly different title, "The Scope, Method and Aim of Musicology." This essay appeared in the first issue of the *Vierteljahresschrift*, in which the announced single purpose of the journal was "to serve science." Karnes, *Music, Criticism* (cit. n. 45), 41. The writings in the *Vierteljahresschrift* during the 1880s followed Helmholtz's idea of explaining musical conventions on the basis of the overtone series. Brahms was skeptical of this direction in musical analysis. Max Kalbeck, *Johannes Brahms* (Vienna, 1904), 1:289.

[47] Karnes, *Music, Criticism* (cit. n. 45), 9.

Karnes has observed, Adler aimed to develop a "perception of the particular," dividing the discipline of music history (or the infant musicology) into sub- and ancillary branches like the physical sciences.[48] Yet he feared the decline of the art in the intellectual interchange of the sciences and music. Added to these concerns, he wondered about the adequacy of objective or scientific study for covering the diversity of the discipline of music.[49]

Adler wrote his dissertation (in 1880) under the mentorship of Hanslick, who had been established as professor at the University of Vienna since 1856. During the early part of his career Hanslick favored an emulation of the methodological rigors of science in music study. When Anton Bruckner was denied a professorial position at the University of Vienna in 1874, it was Hanslick who seems to have sized up the situation with the cutting comment, "He probably has not even read Helmholtz."[50] It must be added that Hanslick later in life changed his views on the importance of objective, empirical means of studying music. He gradually lost interest in an empiricist approach because in his view it was inadequate.[51]

The rigors of science were not always understood or accepted by musicians. It is well to remember that while physicists were looking at the quality of sound objectively, with perspectives substantiated by technology, most musicians allowed for human elements to shape subjectively their perceptions, although the two views intersected profitably at times. The ferment among music scholars extended to other German-speaking areas and to England, France, and the United States. John McKendrick, in England, pointed out that most musicians expected little from the study of physics and physiology: "They were unacquainted with the methods of scientific analysis, and they rather dreaded investigations into the minute structure of the parts and also the torturing expedient of physical experiment."[52] McKendrick's view was a simple one that compartmentalized the objective scientist and the subjective musician.

At the same time, at the Vienna Conservatory the intellectual orientation of certain pedagogues was affected by aspects of the bold advances in scientific thought. For example, Hans Schmitt, in his search for more adequate guidelines in teaching piano, turned to acoustical analysis and evidence to bear out his views on sound with regard to the practice of piano pedaling. In his exploration the boundaries of the discipline were pushed in both constructive and controversial directions.[53] Concerned about the adequacy of teaching, the administration of the conservatory undertook a complete reorganization of the curriculum in the 1870s. At this time Schmitt, a well-known and highly esteemed piano professor, was requested to give a series of lectures on piano pedaling. He sought to explain and clarify the art of pedaling by means of the

[48] Ibid., 10.

[49] Ibid., 134ff. The term *Wissenschaft* in German thought is complex to define. See ibid., 6–7; Bujić, *Music in European Thought* (cit. n. 46), 142.

[50] Alfred Orel, *Anton Bruckner: Das Werk—Der Künstler—Die Zeit* (Vienna, 1925), 134.

[51] See Karnes, *Music, Criticism* (cit. n. 45), 11ff. After 1898, when Adler received tenure in Vienna, he too changed his attitude radically.

[52] McKendrick, *Hermann Ludwig Ferdinand von Helmholtz* (cit. n. 25), 137.

[53] In this context, Adler may have already been sensitized to the invasion of acoustical analysis and its linkage to performance before he attempted to adapt and extend scientific thought to the study of music criticism and history at the university. Instead of appealing directly to the seminal sources for acoustical studies (e.g., Helmholtz and Ernst Mach), Adler deferred to adaptations of scientific investigations in the humanities; for example, Professor Moritz Thausing's work in art history. See Karnes, *Music, Criticism* (cit. n. 45), 133–58. Karnes discusses the "turbulent political and social world" of the 1870s and 1880s in Vienna in which musicians were active.

rigor and perspective of science. He voiced the hope that he would please the scientific community, and perhaps give it something to think about.[54] In adapting guidelines from the study of acoustics he expounded a technique of objective, efficient, and effective pedaling. In essence Schmitt aimed to validate acoustically his explanations of sound phenomena in pedaling. By adding a host of scientific principles to the already existing subjective (intuitive) and complex phenomena related to pedaling, Schmitt became a pioneer in the field. He sought to apply the analyses of Helmholtz's resonance theories to piano pedaling and to the production of tone—quality of sound. Scant, if any, systematic attention had been given earlier to these aspects of piano playing.[55] However, influential pianists during the late nineteenth century demanded and supported innovative pianistic thinking. Anton Rubinstein, for example, regarded Schmitt's work highly.[56] In the case of Liszt, the absorption of new perspectives (e.g., acoustical analyses) in piano technique was not expressed directly, but was implied in his later years. This also applies to a certain extent to the pedagogue Ludwig Deppe, who was indirectly influenced by the impact of the sciences and was the first among teachers to cultivate thoughtful pedaling with his students.[57]

Although Schumann had already emphasized listening to oneself in performance in order to draw out the essence of a piece of music, it was only later in the century that this idea, supported by acoustical analysis, was promoted more specifically.[58] As piano pedagogues, who saw in acoustical science an aid to explaining judicious pedaling, dealt with the mechanism of the piano, which Helmholtz described as a machine in his *Tonempfindungen*, they were drawn to recover and reconsider historical attitudes and shifting aesthetics in sound and performance. The development of this sensibility required more careful listening.[59] Pianists listened with greater intensity to their own sounds, and this resulted in more judicious, effective, and imaginative pedaling.

Recent discussions by the philosopher Jean-Luc Nancy on "listening" are clearly applicable to the insights achieved by pedagogues during the latter part of the nineteenth century as they aimed to objectify pedaling technique. Similar to Nancy's analysis, they sought to create totality in the performance of a piece of music by putting the essential parts together—pedaling being one component whose own parts were in need of clarification and unification. In this process listening became a "straining" toward the whole.[60]

---

[54] Hans Schmitt, *Das Pedal des Claviers* (Vienna, 1875), foreword. (On the various editions of Schmitt's book and my use of them, see n. 63.) Schmitt may have been referring to Helmholtz's inability to "explain the wonders of great works of art." See Helmholtz, *On the Sensations of Tone* (cit. n. 24), 371.

[55] Perhaps the earliest pedagogue to give even cursory attention to the sustaining tone activated by the knee lever and the ideal of singing style was Louis Adam, who in his *Méthode de piano du Conservatoire* (Paris, 1804) related how overtones and sympathetic vibrations of strings had the effect of strengthening and lengthening total sound. He referred particularly to slow pieces because of the quick decay of sound on the piano of his time.

[56] Schmitt, *Das Pedal* (cit. n. 28), foreword, xi.

[57] George Kochevitsky, *The Art of Piano Playing: A Scientific Approach* (Evanston, Ill., 1967), 7–8; Fay, *Music Study* (cit. n. 5), 2–3.

[58] Schumann, *On Music and Musicians*, ed. Konrad Wolff, trans. Paul Rosenfeld (New York, 1946), 140. Schumann further admonished, "Do what the head wills" (36).

[59] See Hiebert, "Reflections on the Piano" (cit. n. 12), 78–85. For the piano as a machine, see, e.g., Helmholtz's description of piano action and its parts in *On the Sensations of Tone* (cit. n. 24), 74–80.

[60] Nancy, *Listening*, trans. Charlotte Mandell (New York, 2007), 6–7.

HANS SCHMITT: MOBILIZER OF ACOUSTICAL ANALYSIS IN PIANO PEDALING

> "I maintain that the discipline of how the pedal should be used is
> the most difficult challenge for advanced piano teaching; and if
> we have not yet heard the piano at its best, the fault perhaps lies
> in that we do not yet understand how to bring out the possibili-
> ties of the pedal."
>
> —Anton Rubenstein[61]

Schmitt's series of lectures on piano pedaling in the early 1870s were intended by
the administration of the Vienna Conservatory to invigorate its piano-teaching program.[62] The four lectures became the basis of a book in 1875, which was the first volume devoted exclusively to the art of piano pedaling—*Das Pedal des Claviers*.[63] In
this text Schmitt developed and exemplified the kinship between acoustical analysis
and piano pedaling.

Reviews—favorable, but not entirely uncritical—promptly followed the publication of *Das Pedal* in both German- and English-speaking lands. On 28 January 1876,
in the *Neue Zeitschrift für Musik*, the reviewer applauded Schmitt's work not only
for finally having clarified this most difficult aspect of piano teaching, but for having placed a shaky discipline on a solid foundation. "Good fruits from this book
are inevitable."[64] That same year, Friedrich Niecks, critic for the *Monthly Musical
Record* and the *Musical Times*, applauded Schmitt's book in a lengthy, energetically
phrased essay entitled "On the Use and Abuse of the Pedal."[65] Niecks praised Schmitt
for opening "new vistas" in piano pedaling and expressed enthusiasm for acoustical
insights that were applied to pedaling. In this connection he elaborated on the studies of Helmholtz. Niecks declared, "Musicians look as yet far too much on acoustics
as something of no practical value; but this is a mistake."[66] While somewhat critical
of the organization of Schmitt's text, Niecks proclaimed it to have "great merit"; and
from the point of view of the practical application of acoustics to pedaling, he wrote
that "it leaves hardly anything to be desired."[67]

---

[61] Quoted in Schmitt, *Das Pedal* (cit. n. 54), 1–2.

[62] Schmitt was born in Koben, Bohemia, in 1835 and died in Vienna in 1907. He began his career
as an oboist, became first chair in the opera in Bucharest, and later played in the orchestra of the Hof-
burgtheater and Hofkapelle in Vienna. Because of throat problems he later, in 1860, became a student
of the piano at the Vienna Conservatory, studying music theory with Simon Sechter and piano with
Joseph Dache. Two years later, in 1862, Schmitt won the silver medal in piano performance that led
to his appointment as piano teacher at the conservatory. He was head of the Piano Pedagogy (*Klavier Ausbildungsklasse*) Department from 1875 until 1900. Students of all ages came to study with
Schmitt, including the young Arthur Schnabel, from age 7 to 9, in 1889–91, and Ferruccio Busoni, at
8 years of age. Béla Bartók was brought to him as a boy.

[63] The original 1875 edition (cit. n. 54) was reprinted in 1889; an enlarged edition (cit. n. 28) was
reprinted in 1907 and 1919. An English translation by Frederick S. Law came out in 1893: *The Pedals
of the Piano-forte and Their Relation to Piano-forte Playing and the Teaching of Composition and
Acoustics* (Philadelphia, 1893). Law provides no explanation regarding the sources (German editions)
from which the English translation is drawn. His text is freely adapted and material is shifted around
and often condensed, with examples missing. The translation leaves much to be desired. Most comments in this paper are based on the 1907 reprint of the 1892 enlarged German edition. The translations are my own.

[64] "Zur Clavierpedalfrage," *Neue Zeitschrift für Musik* 72, no. 5 (1876): 41–3.

[65] Niecks, "On the Use and Abuse of the Pedal" (cit. n. 1), 179–83. In 1891 Niecks was appointed
Reid Professor of Music at the University of Edinburgh.

[66] Ibid., 181.

[67] Ibid.

In the introduction to the enlarged edition of 1892/1907 Schmitt expressed gratitude for the high respect that virtuosi and composers like Rubinstein and Liszt bestowed on his work. He also noted with appreciation a plaque presented to him by the Gesellschaft der Musikfreunde in Vienna in recognition of the outstanding contribution he was making to the teaching of piano. Schmitt was gratified by the positive reviews in journals and the approval of what he termed "important physicists"—but who were they? Schmitt does not name them.[68]

In *Das Pedal* Schmitt examined and illustrated pedaling problems by employing Helmholtz's thoughts in *Tonempfindungen* as a rich source of ideas. By means of experimentation, explanation, and demonstration Schmitt exhibited objectivity and reliability in his work. His orientation stood in contrast to the attitude of most musicians, who relied instead on tradition and intuition (i.e., subjectivity and validation) in the master-apprentice system in which they worked. Schmitt obviously aimed to fill a serious gap in the teaching of piano-pedaling technique; the positive response from the music community accounts for the wide circulation of the book.

Schmitt's suggestions were complicated by the evolving piano—an instrument in transition. In his text Schmitt does not specify the particular piano he had as a model, but since he refers numerous times to the Bösendorfer instrument, expressing a preference for the Viennese damping system over the English/French, and mentions contacts with the Bösendorfer firm, which had long been well established in Vienna, we have some idea of his predilection. Moreover, it is known that Bösendorfer supplied the pianos for the conservatory in Vienna.[69]

How influential the powerful-sounding Bechstein and Steinway pianos were toward the end of the century is a question Schmitt does not address. By 1868 both were being built with iron frames and heavier cross-strung strings, and had greater sustaining power than the Viennese instruments. The Viennese Streicher and Bösendorfer pianos were usually wooden framed (albeit with metal-strengthened crossbars), having thinner straight-strung strings with a faster-decaying sound. In Vienna it was possible to purchase a Steinway in 1875, but there was only one sales outlet.[70] In this context Schmitt laments the lack of time to include a history of the piano itself.[71] The attributes of the changing instrument would have added considerable depth to his discussion on pedaling.

Schmitt's orientation was based on the prevailing ideal of connected as opposed to detached tones or sound that developed during the nineteenth century. The full richness and subtlety of auditory experience is conveyed in *Das Pedal*, where Schmitt demonstrates the remarkable abilities of the ear-and-brain system to distinguish between sound qualities of the piano, dynamic levels and shading in between, and qualities of tones in various registers. In other words, Schmitt takes into account aspects of pitch, timbre, loudness, time, and space (register) in pedal technique. How does he accomplish this?

---

[68] One can speculate that they may have been Schmitt's students, Von Jankó and Tanaka, who studied physics with Helmholtz in Berlin. See Schmitt, *Das Pedal* (cit. n. 28), 37–40; Erwin N. Hiebert, "Science and Music in the Culture of Late 19th Century Physicists: The Role and the Limits of the Scientific Analysis of Music," plenary lecture, in *Science and Cultural Diversity: Proceedings of the XXIst International Congress of History of Science, July 2001*, vol. 1 (Mexico City, 2003), 107.

[69] Leon Botstein, *New Grove* (Oxford, 2000), 4:53. See also Carl Hutterstrasser, *Hundert Jahre Bösendorfer* (Vienna, 1928), 19.

[70] *Signale für die musikalische Welt* 33 (1875): 79.

[71] Schmitt, *Das Pedal* (cit. n. 28), xii.

Clusters of ideas give organization to the text, although occasionally there are distractions, especially in chapters 3 and 4, that were criticized by the reviewers. The chapters of *Das Pedal* are clearly related by the common theme of listening to sounds, bringing the parts of sounds together in order to make critical pedaling decisions. Schmitt requests that the student of piano listen to varying effects of raising the dampers at different times in the context of sounding tones. Moreover, Schmitt's discussion of acoustical matters treats them as permanent factors linking pedal technique to effective performance. How does Schmitt explain the importance of acoustics in relation to pedaling?

For Schmitt the object in writing a text on pedaling was twofold: to demonstrate the importance of the subject matter and to explain the basic reasoning that well-prepared pianists carry out instinctively. In both the 1875 and 1892/1907 editions he observed that the state of the art of pedaling, although a burning issue among musicians, had not advanced much beyond notions of intuition or feeling (*Gefühlstandpunkt*).[72]

Schmitt's book is organized into four chapters, following the plan of his lectures at the conservatory: (1) basic information on the general use of the pedals; (2) acoustical orientation on single tones and chords; (3) use of the pedal in succession of tones and chords; (4) pedal decisions and a variety of other matters such as pedal notation, damper pedal (half pedal), soft pedal, sostenuto pedal, and *Kuntspedal*.[73] Deeply conscious of his innovative objective approach to pedaling, Schmitt supports the newness of his ideas by viewing pedaling from a historical perspective. He suggests consulting early piano tutors—Johann Nepomuk Hummel, Carl Czerny, and Adolph Kullak—to get an overview of past instruction.[74]

Schmitt guides the student step-by-step (with many examples) in acquiring facility and in understanding the different types of pedaling appropriate to various contexts; he refers to Helmholtz for validation.[75] Listening to the effects created by raising the dampers at different times, Schmitt builds his guide on the following types of pedaling: (1) lifting the dampers before sounding the tone (anticipatory), (2) lifting the dampers while sounding the tone (rhythmic), (3) lifting the dampers after sounding the tone (syncopated), (4) pedaling in tremolo fashion, and (5) half-pedaling.

Schmitt initially reflects on the pedal as a sustaining mechanism through rests (*klingenden Pausen*) and for long-held tones and connecting tones. This allows the pianist to connect widely ranging tones and chords beyond the expanse of the hand. This is his initial step in analyzing the nature of pedaled sound. In syncopated pedaling, he explains, one treads the pedal after the note is sounded, but as soon as the pedal is felt (sensed) the finger is removed and goes to the next note. The moment the next note is sounded, the pedal is released, but quickly depressed again, thus binding

---

[72] E.g., Schmitt, *Das Pedal* (cit. n. 28), 1.

[73] Schmitt concentrates on the damper pedal; he speaks very little about the soft pedal. The sostenuto pedal has a history of its own and is almost completely dismissed by Schmitt (*Das Pedal* [cit. n. 28], 116). He discusses the short-lived *Kunstpedal* in the 1892/1907 edition (ibid., 113–5). Because the damper pedal increases the fullness of sound immediately after a tone (chord) is produced, it sometimes, especially during the nineteenth century, was designated the "loud pedal." See Niecks, "On the Use and Abuse of the Pedal" (cit. n. 1), 183.

[74] Schmitt, *Das Pedal* (cit. n. 28), 40–1. Referring to volumes 2 and 4 of Czerny's *Clavierschule*, op. 500 (Vienna, 1839), Schmitt cautions the reader that in requesting treading together with the tone, Czerny did not understand the developing pedal technique, even though he directed the pianist to change the pedal quickly as in syncopated pedaling. He gives Kullak's orientation on pedaling in *Die Kunst des Anschlages*, op. 17 (Leipzig, 1855), high marks, but it is too brief.

[75] Schmitt, *Das Pedal* (cit. n. 28), 28, 43. Commentary on Helmholtz is on 70–1.

Figure 1. An example from Schmitt's exercises demonstrating the timing of the pedal for syncopated pedaling, resulting in connecting and filling in sound. Schmitt, Das Pedal (cit. n. 28), 5.

the notes together. Finger first, then pedal. With exercises Schmitt leads the reader step-by-step in achieving precision in the technique (see, e.g., fig. 1).[76] Schmitt observes in the original edition of 1875 and again in the enlarged edition of 1892 that few players are aware of the benefits of syncopated technique and other types of pedaling techniques.[77]

The sustained pedal sound—syncopated pedaling—also serves to enhance the nature of touch (*Anschlag*). By allowing time for the fingers to position themselves between notes or chords during the sounding rest, a pianist can prepare a particular type of touch to execute the next sounding tone. In this way, Schmitt comments, the pianist also is allowed to rest and recover (*sich erholen*) from the demands of touch. Moreover, lifting the dampers reduces the resistance of the keys.[78]

Beyond the syncopated technique, Schmitt clarifies the method of lifting the dampers concurrently with the finger/tone (rhythmic pedaling), useful at the beginning of a composition or after a general pause or rest. Historically, rhythmic pedaling was usually timed at bar lines and resulted in gapped sounds. It was the normal, ongoing approach to pedaling from the late eighteenth century, with some exceptions, until the middle of the nineteenth century. Schmitt furthermore observes that staccato notes must be pedaled together with the sound (finger) since they imply pauses between notes. If tones are to be absolutely staccato (extremely short) no pedal should be used.[79] He could have added that short articulation groups are given significance by means of the rhythmic pedal.[80] This may be one aspect of the concept of bringing the musical "self" of the piano to life.[81]

[76] Schmitt notates length of pedal by means of note and rest values as in fig. 1 and explains the notation in lecture 4. Few pedagogues have accepted Schmitt's method of marking pedaling. Arthur Whiting is an exception. See his *Pianoforte Pedal Studies*, 2 bks. (New York, 1904), where he directs the application of the pedal by means of notes and rests of definite length. Today we have no agreement about notation for pedaling; anxieties about aesthetic ideals, habits of individual composers, and differences in pianos and room size all stand in the way of any agreed-upon system of guidance.

[77] Schmitt, *Das Pedal* (cit. n. 28), 1–2, 10.

[78] Ibid., 15–8, 117.

[79] Ibid., 6–11.

[80] See fig. 2 for Brahms's method of marking the pedal for short articulation groups to support metrical shift.

[81] See Scherer, *Klavier-Spiele* (cit. n. 35), 114–78. There are other aspects to this notion. For example, Julia Kursell recently has explored the complex implications of timbre in the character of sound sustained by the pedal in "Shaping Differences: Hermann von Helmholtz's Experiments on Tone Colour," in "Conference on the Shape of Experiment" (Preprint 318, Max-Planck-Institut für Wissenschaftsgeschichte, Berlin, 2006), 215–24.

Figure 2. *Brahms's method of marking the pedal for short articulation groups. Sonata for Clarinet, op. 121, no. 1 (London, 1895), 3rd movement, mm. 98–104.*

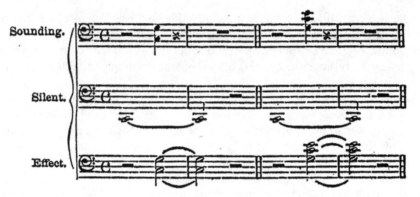

Figure 3. *Overtones are generated. The middle stave represents a note that is played silently. The notes of the top stave are then played loudly enough for the silently played string to generate overtones that sustain the sounded notes as heard in the bottom stave, even though they are no longer held.* Schmitt, Das Pedal *(cit. n. 28), 46.*

Having introduced the reader to syncopated pedaling and rhythmic pedaling, Schmitt proceeds to explain and exemplify his acoustical approach to decision making in pedaling. In the second chapter he illustrates various phenomena associated with the family of tones and their related overtones as they spread beyond the fundamental and initiate vibrations of interrelated strings according to Helmholtzian principles.[82] Schmitt thus demonstrates how the individual strings of the piano generate overtones and actually form families of chords. The dampers, when lifted from the strings by the pedal on the right, release sound from all the piano strings and allow the vibrating string to excite other strings that have common harmonics. The rising overtones can be detected from every tone in a definite order (see fig. 3).

Among the numerous illuminating examples, Schmitt shows how melodic lines (i.e., certain important tones) can be reinforced by strategic selective sounding (placement) of the overtones. The effect is illustrated in the three tests shown in figure 4.[83] After dealing with the acoustical effects of single tones, in lecture 3 Schmitt launches into topics such as pedaling a series of tones/chords, the tempered tuning

---

[82] Schmitt, *Das Pedal* (cit. n. 28), 42–7.

[83] In a brief *Anhang* to chap. 2 Schmitt speculates on the importance of "undertones," a topic much in the air at the end of the nineteenth century but outside the realm of this article.

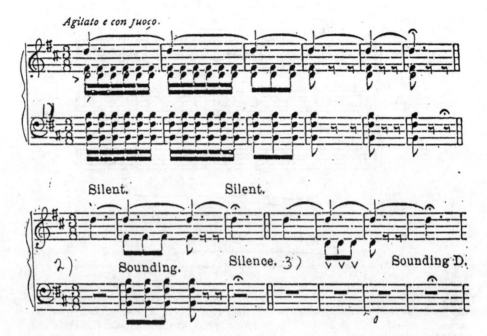

Figure 4. Reinforcement of melodic lines by means of strategic sounding of overtones. Three tests of this effect are marked by numbers within the figure. Drawn from Felix Mendelssohn's Song without Words, *no. 10 in B Minor, final cadence.* Schmitt, Das Pedal *(cit. n. 28), 66–7.

system, how the size of a hall affects pedaling, and aspects of the short-lived *Kunst-pedal* (promoted by Eduard Zachariae).[84] He deliberately spends considerable time with tones in the low registers of the piano and especially those in closed position (i.e., where the notes of a triad are as close as possible), warning the pianist that this area of the sound spectrum should receive very little pedal. Arpeggios based on the overtone series, especially in the low registers, provide maximum sonority and beauty (see fig. 5).

At the beginning of the passage, Schmitt suggests treading a full pedal and then, if the pianist wishes to tread again without losing the low tone, fluctuating the pedal very quickly with the foot up and immediately down again, as notated in example C in figure 5. Since the vibrational force of the higher strings is less than that of the lower ones, the damping effect in cutting off the tones is very brief. With each short, light tread the bass note is reinforced. The short contact is too slight to bring the lower strings to rest and silence, so they continue to sound and are repeatedly refreshed (*aufgefrischt*); by this method bass notes are sustained more successfully than notes in the higher registers. It also is possible to sustain the higher tones by playing them louder while the other tones are reduced to a softer level (see fig. 6). The resulting lightness of the keys when employing the anticipatory pedal technique aids in producing a more reliable touch, for fullness in both *forte* (*fortissimo*) and softness in *piano* (*pianissimo*) contexts.

Schmitt concludes his guide by summarizing his ideas for making knowledgeable

[84] Schmitt, *Das Pedal* (cit. n. 28), 103–8, 113–6.

Figure 5. *Half-pedaling—a tip provided by Brahms to Schmitt. Holding the pedal down as in A would create chaos; pedaling as in B would not help to sustain the pedal tone. By contrast, C would create an artistic interpretation of this passage. Schmitt,* Das Pedal *(cit. n. 28), 108–10.*

Figure 6. *Importance of anticipatory pedal. Technique 1 is less reliable. Technique 2 is softer and more reliable; all the dampers are lifted before the tone is sounded. Drawn from Beethoven, F Minor Sonata, op. 2, final cadence of the second movement. Schmitt,* Das Pedal *(cit. n. 28), 116–7.*

decisions on pedaling. He reminds students to listen to piano sounds carefully and to observe distinguished artists performing. He affirms the point that artistic pedaling can be learned; it is not simply a talent. Schmitt's ideas presented here only begin to reveal the rich source of information on pedaling that he provides for the pianist.

Das Pedal des Claviers made a deep impression on piano pedagogues and spawned reconsiderations of pedal technique. While advocating for the improvement of piano pedaling through objective means, Schmitt at the same time did not neglect its subjective elements, such as feeling expressed in tempo, dynamics, and so forth. He never lost sight of the importance of grounding decisions in knowledge of the individual composer and specific compositional contexts, including the size of the particular venue for which the piece was composed, the individual instrument on which it was composed, and the historical milieu of the composition.

AFTER SCHMITT

In his dynamic impact on the musical community Hans Schmitt heightened the significance of acoustical resources for piano pedaling. In 1894 he published a text on touch at the piano in which he provided detailed information and guidance for the pianist in the "spirit" of science.[85] A few years earlier his colleague, Professor of Harmony Leopold A. Zellner, presented a series of sixty-two lectures on acoustics and music at the conservatory; they were published in 1892.[86] Not long after, in 1896, Ludwig Riemann, a pedagogue from Essen, ventured into the field of acoustical analysis by condensing Helmholtz's *Tonempfindungen* into a concise guide for music students and for the general musical public entitled *Populäre Darstellung der Akustik in Beziehung zur Musik*.[87] Riemann aimed to bring the acoustical studies of Helmholtz to the conservatories in a short, understandable form by reducing the text to extracts that dealt with purely musical issues. His mission was to bring neglected theoretical knowledge to those who thought feeling in music was enough.[88]

Riemann followed the Helmholtz abridgement with a lengthy book on touch, *Das Wesen des Klavierklanges*, in which, similar to Schmitt's text on the subject, the approach is from the physical acoustical point of view, separating, defining, and yet relating various elements of tone production in musical contexts.[89] Determining factors for tone building include the strings, the player's touch at the piano, sympathetic vibrations of other strings, the makeup of *Klangfarbe* (a classification concept), *Klangschattierung* (variety of high, middle, and low tonal ranges), and the length of tones (sustaining quality).[90] In *Das Wesen* Riemann separates physical acoustics from aesthetic discussions as he investigates how tone and resonance impinge on pedaling. In this sense his analysis is echoed in Nicholas Cook's view of the scientific study of music as opposed to musicological discourse: the two categories are separate, but each must know something about the other.[91] Because Riemann realized that acoustical explorations only partially explain matters of tone in piano performance, he enveloped psychological as well as physiological elements in his development of piano technique. Riemann's passion to improve piano pedagogy was further expressed in his generously annotated second edition in 1913 of F. A. Steinhausen's analysis of piano technique.[92]

It was not long after the publication of *Das Pedal* in 1875 that several additional texts devoted entirely to piano pedaling appeared.[93] Most authors gave generous

[85] Schmitt, *Über die Kunst des Anschlags* (Vienna, 1894).
[86] Zellner, *Vorträge über Akustik*, 2 vols. (Vienna, 1892). The study features the harmonium.
[87] Riemann, *Populäre Darstellung der Akustik in Beziehung zur Musik: Im Anschluss an Hermann von Helmholtz' "Lehre von den Tonempfindungen"* (Brunswick, 1896).
[88] Ibid., foreword.
[89] Ludwig Riemann, *Das Wesen des Klavierklanges und seine Beziehungen zum Anschlag: Eine akustisch-ästhetische Untersuchung für Unterricht und Haus dargeboten* (Leipzig, 1911).
[90] Ibid., foreword, 5–6.
[91] Cook, *Music, Imagination and Culture* (New York, 1990), 243.
[92] Steinhausen, *Die physiologischen Fehler und die Umgestaltung der Klaviertechnik*, 2nd ed., ed. Ludwig Riemann (Leipzig, 1913). In blending ideas from acoustics and physiology, Steinhausen, a physician, took recourse in Helmholtz for analysis. For additional information on nineteenth-century pedagogy, see Myles W. Jackson, *Harmonious Triads: Physicists, Musicians, and Instrument Makers in Nineteenth-Century Germany* (Cambridge, Mass., 2006), 266–79.
[93] See Albert Lavignac, *L'école de la pédale du piano* (Paris, 1889); Georges Falkenberg, *Les pédales du piano* (Paris, 1891); John A. Preston, trans., *Guide to the Proper Use of the Pianoforte Pedals, with*

credit to Schmitt's pioneering work; however, except for Albert Lavignac and Leonid Kreutzer, comprehensive acoustical analysis was neglected. With a few exceptions, twentieth-century texts on pedaling disregard explicit acoustical aspects.[94] Laudatory references to Schmitt's work occasionally have appeared in texts on interpreting piano music. For example, in Jean Kleszynski's book on interpreting Chopin's piano literature, Schmitt's text is singled out as an excellent guide that throws great light on the execution of Chopin's piano works.[95]

Kreutzer in 1915 credited Schmitt for having provided the best exposition on pedaling up to that time.[96] He praised Schmitt for many novel ideas and benefited from Schmitt's work in creating his own systematic and comprehensive text using acoustics as a base, albeit with modifications and expansions of work done in the nineteenth century. Whereas Schmitt had favored a piano with the Viennese action (Bösendorfer), Kreutzer focused his attention on the twentieth-century piano based on the English/French action. He was critical of Schmitt's discussion on half-pedaling, since it could be carried out more easily on the nineteenth-century Bösendorfer than on the twentieth-century Steinway.

To my knowledge, no autonomous pedaling guide that features acoustical analysis was published between Kreutzer and the beginning of the twenty-first century. Helmut Brauss, professor emeritus at the University of Alberta, has linked the study of musical acoustics with the art of pedaling in a text originally published in German and translated into English.[97] With the advantage of early twenty-first-century technological advances explicated by the physicist Waldemar Maysenhölder, Brauss probes details of acoustics that apply to the beginnings and endings of pedaled sound and illuminates how pedaling affects dynamics and the perception of tuning. General principles are developed that are applicable to decision making. The discourse in many ways is revelatory. However, historical aspects of pedaling are left out entirely— a glaring omission in Brauss's work, according to his critics.

As pedagogues during the late nineteenth century became aware of the intricacies of and solutions to problems in artistic pedaling, so did composers. Vassily Safonov, teacher of Alexander Scriabin, emphasized watching the "foot" (i.e., pedal) in criticisms of his student's playing.[98] Pedal indications in scores of composers around the

---

Examples out of the Historical Concerts of Anton Rubinstein (Paris, 1897), and Teresa Carreño, *Possibilities of Tone Color by Artistic Use of Pedals: The Mechanism and Action of the Pedals of the Piano* (Cincinnati, 1919), reprinted together as *The Art of Piano Pedaling: Two Classic Guides* (Mineola, N.Y., 2003); Leonid Kreutzer, *Das normale Klavierpedal vom akustischen und aesthetischen Standpunkt* (Leipzig, 1915). Louis Köhler published *Der Klavier Pedalzug* (Berlin, 1882), using many of Schmitt's ideas, but without recognizing or giving him credit. Hugo Riemann makes this point in his *Vergleichende theoretisch-praktische Klavier-Schule*, 2nd ed. (Hamburg, 1890), pt. 1, 16.

[94] Joseph Banowitz, *The Pianist's Guide to Pedaling* (Bloomington, Ind., 1983), 11–3. Banowitz at first briefly touches on the importance of acoustical analysis, and occasionally it is implied in the text. See also Josef Gat, *Principles of Piano Technique* (Budapest, 1958). Gat's general text on piano technique offers an abbreviated acoustical orientation.

[95] Kleczynski, *The Works of Frederic Chopin and Their Proper Interpretation*, trans. from the 4th Polish ed. by Alfred Whittingham (London, 1892). Kleczynski adopted Schmitt's lined pedal notation in his examples. I thank Sandra Rosenblum for calling my attention to this reference.

[96] Kreutzer, *Normale Klavierpedal* (cit. n. 93), intro., 20, 24.

[97] Brauss, *Der singende Klavierton oder das "Wie" des Pedals* (Wilhelmshaven, 2003). Translated as *The Singing Piano Tone or the Artistry of Pedaling*, with an appendix by Waldemar Maysenhölder (Victoria, B.C., 2006).

[98] Quoted in Joseph Banowitz, "The Art of Pedaling," *European Piano Teachers Association Piano Journal*, no. 81 (2006): 15.

turn of the century are scarce and ambiguous. Long notated tones and articulation slurs can be taken as indicators of judicious damper pedaling in the piano scores of Scriabin, Max Reger, Ferruccio Busoni, and Arnold Schoenberg, among others. It would be helpful to explore in detail late nineteenth- and twentieth-century composers and their individual pedaling habits.[99]

CONCLUSION

It has been my intention in this brief sound tour to relate Hans Schmitt's pivotal, pioneering work on piano-pedal technique to the phenomena of artistic pedaling and the changing musical context. His emphasis on the intersection of music and musical acoustics charted the way to an appraisal of the control of sustained and articulated piano sound by means of the damper pedal. A number of pedagogues later discovered, extended, prescribed, and adapted Schmitt's innovative ideas to a continuing development of the kinship between acoustics and the practice of performance. This initiated an awakening to and a blending of objective (acoustic) science with the subjective (intuitive experiential) attitudes that created the past, which in turn shapes the way we know and experience such a relationship today. The process thrives on relentless remaking, since knowledgeable adaptation of a complex pedaling technique is in need of constant adjustment to the venue, the instrument, the music, the composer, and the historical milieu for comprehensive understanding and expression. Since all of these parameters are in flux, a prescribed (fixed) system of pedaling is not realistic. Without an understanding of the basic nature of the sounding body and its technical resources, an adaptable artistic orientation in particular contexts is wanting. Musicians were grateful for Schmitt's acoustical insights that finally gave pedaling technique a firm foundation. As a result, pianists listened more carefully and were aided in making knowledgeable decisions about how and when to use the pedal to enhance sound.

By influencing pedagogues to apply rigorous scientific analysis to musical sound, Helmholtz left an indelible footprint on the artistic production of pedaling. The study of acoustics brought about an awareness of sound feedback whereby the work of the scientist was made known beyond the scientific community. Indeed, nineteenth-century acoustics played a critical role in the creation of a new aesthetics and practice of piano pedaling. For Helmholtz, this would come as no surprise, as he saw the piano as simultaneously a musical and a scientific instrument. While much has been written by historians of science on Helmholtz's work on acoustics, we are finally beginning to see how his scientific contributions shaped piano playing. Indeed, even today, a worldwide network of inquiry by musicians and physicists continues to inform pianistic thought.[100]

[99] A brief introduction to such an analysis can be found in Hiebert, "Reflections on the Piano" (cit. n. 12). Rosenblum ("Pedaling the Piano" [cit. n. 14]) has touched on this issue.

[100] See, e.g., Nicholas J. Giordano Sr., *Physics of the Piano* (Oxford, 2010); Anders Askenfelt, ed., *Five Lectures on the Acoustics of the Piano* (Stockholm, 1990).

"A Towering Virtue of Necessity":
Interdisciplinarity and the Rise of Computer Music at Vietnam-Era Stanford

by Cyrus C. M. Mody* and Andrew J. Nelson†

ABSTRACT

Stanford, more than most American universities, transformed in the early Cold War into a research powerhouse tied to national security priorities. The budgetary and legitimacy crises that beset the military-industrial-academic research complex in the 1960s thus struck Stanford so deeply that many feared the university itself might not survive. We argue that these crises facilitated the rise of a new kind of interdisciplinarity at Stanford, as evidenced in particular by the founding of the university's computer music center. Focusing on the "multivocal technology" of computer music, we investigate the relationships between Stanford's broader institutional environment and the interactions among musicians, engineers, administrators, activists, and funders in order to explain the emergence of one of the most creative and profitable loci for Stanford's contributions to industry and the arts.

INTRODUCTION

Stanford University's CCRMA, or Center for Computer Research in Music and Acoustics, enjoys a reputation as a world-class computer music and sound research center. CCRMA faculty, staff, and students produce influential musical compositions, develop radical and impactful technologies, cultivate dense ties to other departments and external organizations, and manage an impressive patent portfolio that has yielded tens of millions of dollars in licensing revenue. While today such interdisciplinary cooperation between musicians and engineers may be heralded as a laudable—if still unusual—combination, such an outcome was hardly preordained at the time of CCRMA's early activities in the 1960s and 1970s.

On one level, this article addresses the simple question, Where did CCRMA come from? Our answer to that question, however, unravels a deeper set of issues, couched

* History Department, Rice University, MS #42, PO Box 1892, Houston, TX 77251-1892; Cyrus .Mody@rice.edu.

† Department of Management, Lundquist College of Business, University of Oregon, Eugene, OR 97403; ajnelson@uoregon.edu.

We gratefully acknowledge comments and suggestions from Christina Dunbar-Hester, Patrick Mc-Cray, Danny Crichton, and Joseph Klett, as well as from the editors, two anonymous reviewers, and other contributors to this volume. Mody's research is supported by the National Science Foundation via the Center for Nanotechnology in Society at the University of California, Santa Barbara, and the National Nanotechnology Infrastructure Network. Nelson's research is supported by the Ewing Marion Kauffman Foundation. Both authors contributed equally to this article.

© 2013 by The History of Science Society. All rights reserved. 0369-7827/11/2013-0013$10.00

in Vietnam War–era debates over the role of universities and their relationship to government or military funding. These debates, we argue, formed a milieu out of which interdisciplinary programs such as CCRMA could coalesce, acquire financial and political resources, and, ultimately, become self-sustaining and deeply integrated members of both a university like Stanford and a broad organizational field that combines commercial, academic, artistic, and other activities. We identify specific features of both the institutional environment at Stanford and the particular research conducted at CCRMA that, together, facilitated the center's emergence and growth. We focus specifically on pressures toward interdisciplinarity and "relevant" research, and argue that CCRMA researchers' expertise both in the "multivocal technology" of audio synthesis and processing and in novel forms of interdisciplinary collaboration explains why CCRMA emerged from (and thrived in) this milieu.

Music, science, and engineering were swirling around one another long before the late '60s, as documented by the other chapters in this volume. The core of our argument, however, is that the confluence of many different pressures brought on by the Southeast Asian conflict created an environment at Stanford in which these fields could effectively bind, react, and grow together. Our chapter thus emphasizes the institutional context surrounding the emergence and growth of computer music at Stanford and the ways in which institutional environments and particular developments at the intersection of music, science, and engineering came to be mutually reinforcing.

STANFORD'S TRANSFORMATION IN THE EARLY COLD WAR

The larger context that helped to facilitate CCRMA was the formation (and then, in the late 1960s, the sudden undermining) of the national science policy regime established in the early Cold War. More than almost any other university, Stanford capitalized on post–World War II changes in federal funding and the emergence of a military-industrial-academic complex to transform from a sleepy, provincial, undergraduate-focused institution into a world-leading research university. Yet that transformation came with a price. When the military-industrial-academic complex experienced crises of funding and legitimacy in the late 1960s, Stanford's dependence on that complex meant that it, in turn, saw steep budgetary declines and student unrest directed at the university's researchers and laboratories. To understand the environment that facilitated CCRMA's emergence, therefore, it is necessary to look back to the early Cold War.

Many hands contributed to Stanford's postwar transformation, but no one more than Frederick Terman.[1] When he became dean of engineering in 1946, Terman began remolding Stanford to "exploit" the "wonderful opportunity" provided by the Cold War expansion in federal research funding.[2] Terman largely followed the vision for science policy put forward by his mentor, Vannevar Bush, that came to be known as the "linear model of innovation."[3] This model presupposed that basic scientific re-

[1] Material on Terman is drawn largely from Stuart W. Leslie, *The Cold War and American Science: The Military-Industrial-Academic Complex at MIT and Stanford* (New York, 1993); Rebecca Lowen, *Creating the Cold War University: The Transformation of Stanford* (Berkeley, Calif., 1997); Eric J. Vettel, *Biotech: The Countercultural Origins of an Industry* (Philadelphia, 2006).

[2] Terman, quoted in Lowen, *Creating the Cold War University* (cit. n. 1), 96.

[3] There is a historiographic debate as to whether Bush should be given credit for the linear model, and even whether there was such a thing. See Benoît Godin, "The Linear Model of Innovation: The Historical Construction of an Analytical Framework," *Science, Technology and Human Values* 31

search, guided by curiosity and disciplinary conventions and ostensibly uninfluenced by economic gain or societal priorities, would yield generalized, fundamental knowledge that could be applied to specific, technological problems. Bush argued that several war-winning technologies had had their origins in curiosity-driven fields; most notably, the atomic bomb was in part made possible by unguided nuclear physics research that, before the war, had not seemed to have much relevance to military needs. The lesson Bush drew was that the Cold War state needed to invest large sums in basic research.[4]

Similarly, the lesson Terman took from his wartime experiences was "that the training of engineers was inadequate, that they didn't measure up to the needs of the war. . . . Most of the major advances in electronics were made by physicists . . . rather than by engineers."[5] Thus, back at Stanford, Terman revamped undergraduate engineering education to emphasize fundamental math and physics, and brought in star researchers who could attract federal funding and graduate students for basic engineering science research and who could bring more cutting-edge fundamental physics into engineering.

By prioritizing basic research, Terman also satisfied the Stanford family's directive that the university aid California industry. Faculty who engaged in applied research and development (R&D) of specific technologies risked competing with off-campus firms. By training students in basic research, however, Stanford's science and engineering departments generated both nonrival knowledge and personnel that flowed easily into the growing local electronics and aerospace sectors.

Terman used both carrot and stick to achieve his vision. In the traditional natural science departments, he used his administrative authority to override and isolate recalcitrant older faculty who were unwilling to take federal money or were too interested in teaching undergraduates, or whose research was not reductionist and fundamental enough.[6] He also pushed natural science departments to focus on subfields where fundamental research could easily serve high-tech industry and/or where federal funding was available. Geology, for instance, was steered toward petroleum geology, and physics toward high-energy accelerator research.[7]

In cases where change came too slowly, Terman created new hybrid departments, such as Applied Physics and Genetics,[8] and reinvented old ones (e.g., Metallurgy

(2006): 639–67; Philip Mirowski, *Science-Mart: Privatizing American Science* (Cambridge, Mass., 2010); David Edgerton, "The 'Linear Model' Did Not Exist: Reflections on the History and Historiography of Science and Research in Industry in the Twentieth Century," in *The Science-Industry Nexus: History, Policy, Implications*, ed. Karl Grandin, Nina Wormbs, and Sven Widmalm (Sagamore Beach, Mass., 2004), 31–57; Glen Ross Asner, "The Cold War and American Industrial Research" (PhD diss., Carnegie Mellon Univ., 2006). We are agnostic on this point. We are simply arguing that Bush articulated a justification for investment in basic research that hinged on the possibility that high-tech industries and militarily significant technologies would arise from that research. Over time, economists and policy makers elaborated that justification into what is now known as the linear model—and, over time, Terman elaborated his local implementation of Bush's vision at Stanford to follow suit.
[4] This is a tendentious interpretation of the importance of basic research in the Manhattan Project and other wartime programs, but one that was widely subscribed to at the time. See Rebecca Press Schwartz, "The Making of the History of the Atomic Bomb: Henry DeWolf Smyth and the Historiography of the Manhattan Project" (PhD diss., Princeton Univ., 2008).
[5] Terman, quoted in Leslie, *Cold War and American Science* (cit. n. 1), 54.
[6] Leslie, *Cold War and American Science*, 167; Vettel, *Biotech*, 53–65 (Both cit. n. 1).
[7] Lowen, *Creating the Cold War University*, 80; Leslie, *Cold War and American Science*, chap. 6 (Both cit. n. 1).
[8] Leslie, *Cold War and American Science*, 175–81; Vettel, *Biotech*, 65 (Both cit. n. 1).

became Materials Science[9]) in order to move methods from ostensibly more basic fields into more applied disciplines. He and his allies also created a series of nondepartmental research centers to serve as a bridge between university and industry, and between basic and applied research. First came the Microwave Laboratory in 1944, which early on functioned as the on-campus arm of a local start-up, Varian Associates.[10] Later centers included the Applied Electronics Laboratory, the Systems Techniques Laboratory, the Solid State Electronics Laboratory, and the Center for Materials Research. Though these centers did much fundamental research, their funding sources (and the most visible short-term applications of their work) were defense related, and the firms to which they had ties were major players in the military-industrial complex.

These nondepartmental research centers later provided an important model for CCRMA. For instance, several of these centers had industrial affiliates programs whereby firms could preview faculty research and recruit star students, a feature CCRMA later adopted. However, CCRMA was also molded by the backlash in the late '60s against two prominent characteristics of the early Cold War centers. First, though (as Stuart Leslie notes) these early centers were "interdisciplinary," the interdisciplinarity they embodied was often restrictive: methods and knowledge circulated among the physical and engineering sciences (sometimes with the expressed aim of making engineering more "scientific"), but there was little outreach to economics, law, music, English, political science, medicine, and so forth.[11] CCRMA, by contrast, was founded in—and exemplified—an era when many stakeholders placed a much higher value on more wide-ranging and egalitarian interdisciplinarity. Second, though these early centers did not avoid civilian funding and applications, their fortunes were largely tied to the national security state. Conversely, while CCRMA did not actively eschew national security funding and ties to the defense industry, it did reflect local and national calls for societal relevance by pursuing highly visible civilian technologies and ties to consumer products firms.

The center to which CCRMA was most closely related was the Stanford Artificial Intelligence Laboratory (SAIL). SAIL's roots can be traced to 1962, when John McCarthy arrived at Stanford and initiated an artificial intelligence (AI) project that built upon a similar project that he and Marvin Minsky organized at MIT in the late 1950s. McCarthy received financial support from the Pentagon's Advanced Research Projects Agency (ARPA) to fund a group of six people, establishing the Stanford Artificial Intelligence Project in 1963. With further support from ARPA, and bolstered by the formation of Stanford's computer science department in 1965, the "project" evolved into a "laboratory" and grew from fifteen people in 1965 to over one hundred by 1968. ARPA continued to provide anywhere from half to two-thirds of the funding in a given year, with NASA, the National Science Foundation (NSF), and the National Institute of Mental Health also providing significant support.[12]

---

[9] Lowen, *Creating the Cold War University*, 213; Leslie, *Cold War and American Science*, chap. 8 (Both cit. n. 1).

[10] Leslie, *Cold War and American Science* (cit. n. 1), chap. 6; Christophe Lécuyer, *Making Silicon Valley: Innovation and the Growth of High Tech, 1930–1970* (Cambridge, Mass., 2006), chap. 3.

[11] Leslie, "Playing the Education Game to Win: The Military and Interdisciplinary Research at Stanford," *Historical Studies in the Physical Sciences* 18 (1987): 55–88.

[12] Lester Earnest, ed., "Final Report: The First Ten Years of Artificial Intelligence Research at Stanford" (Stanford Artificial Intelligence Laboratory Memo AIM-228, July 1973); Bruce G. Buchanan,

A 1973 overview of the first ten years of the project reported that "the work of the Stanford Artificial Intelligence Project has been basic and applied research in artificial intelligence and closely related fields, including computer vision, speech recognition, mathematical theory of computation, and control of an artificial arm."[13] In other words, AI provided the underlying raison d'être, but actual applications were envisioned in a wide range of fields. The diversity of projects was matched by a diversity of political perspectives in the AI group, with McCarthy traversing from left to right over the course of the '60s while his deputy director, Les Earnest, moved in the opposite direction.[14] In both research outlook and political temperament, therefore, SAIL straddled a broad landscape just as the turmoil of the late '60s began. Its capacity to invite participants holding diverse political and disciplinary perspectives would be an important bequest to SAIL's eventual spin-off, CCRMA.

HISTORY OF THE STANFORD MUSIC DEPARTMENT

The Department of Music followed a rather different trajectory than Stanford's natural, social, and engineering science departments and their associated centers. In fact, Stanford did not have a music department until 1946, more than fifty-five years after the university's founding in 1891. This is not to say that music was not a major part of university life from the earliest days. For example, the Stanford Memorial Church, which occupies a central position on campus both geographically and (at least early on) socially, had an active music program from the start. Yet Stanford's musical activities, such as the band, orchestra, and various glee clubs, were largely initiated by students rather than faculty or the administration.[15] Formal courses were scarce and a degree program did not exist.

Through the university's first half century, Stanford administrators vacillated as to the place of music. In 1926, for instance, then-president Ray Lyman Wilbur wrote the board of trustees that "funds may be available for a course in harmony and composition as the basis for" the possible formation of "an adequate School of Music along the lines of the Yale School of Music,"[16] yet no such action was taken. In fact, it was alumni who moved forward by forming the Friends of Music at Stanford in the mid-1930s. Finally, in 1946, Stanford created a department of music, the last major American university to do so.[17]

The establishment of a department, however, did not ensure its quality and reputation. For example, Stanford remained unranked in the 1957 survey conducted by the

---

"Introduction to the COMTEX Microfiche Edition of Memos from the Stanford University Artificial Intelligence Laboratory," *AI Magazine* 4 (Winter 1983): 37–42.

[13] Earnest, "Final Report" (cit. n. 12), 2.

[14] John Markoff, "Optimism as Artificial Intelligence Pioneers Reunite," *New York Times,* 8 Dec. 2009, D4.

[15] Orrin Leslie Elliott, *Stanford University: The First Twenty-Five Years* (1937; repr., New York, 1977), 198–9.

[16] Wilbur to the Honorable Board of Trustees, Leland Stanford Junior University, memorandum, 9 Oct. 1926, Stanford University Archives, Stanford, Calif. (hereafter cited as SUA), SC27, box 17, folder 5, "Board of Trustees Supporting Documents."

[17] M. Tanner, "A Brief History of the Department of Music" (unpublished manuscript, Stanford, Calif., 1978); C. Smith, "Gift Enables Music Department to Construct Long-Needed Facility," *Stanford Daily*, 14 Jan. 1980, 1.

American Council on Education on quality of graduate faculty in music.[18] Moreover, Terman's reorientation of the university toward basic research may have run counter to the traditional emphasis on composition and performance—instruction-oriented subjects that were core foci of most academic music departments. Other humanistic departments, most notably Classics, may have offered Music a gruesome example of Terman's attitude toward what he deemed "impractical fields" that could not bring in external research funding or connect to high-tech industry: as provost, Terman stripped Classics of faculty lines, shrank its graduate program, and commanded that its remaining faculty teach only large, lower-level undergraduate courses.[19] The music department, therefore, was in a remarkably ambiguous and indeterminate position in the mid-1960s, particularly in comparison to its peer departments at other universities. Stanford's department was still emerging on the national scene, still searching out its areas of emphasis, and still uncertain about its place in Terman's university.[20] That uncertainty and immaturity made the music department ripe for organizational and technological experimentation just as the campus ferment of the Vietnam era began.

VIETNAM-ERA CRISES

The unrest that gripped Stanford in the late '60s and early '70s was—as on other campuses—a product of many deep divisions in American society. But Stanford was one of a smaller number of universities—such as MIT and Princeton—where the conduct of academic science and engineering was a focus of protest (often in conjunction with other issues, especially the war).[21] Science-based unrest at Stanford coalesced gradually, beginning in 1966 with debates about classified research on campus.[22] By the time the administration was forced to terminate its classified contracts in 1969, however, protesting students, reformist faculty, and influential politicians (such as Mike Mansfield and Edward Kennedy) had moved on to a much larger goal: the reconfiguration of the nation's—and hence also Stanford's—research enterprise to prioritize solving civilian social problems above generating fundamental knowledge or contributing to national security.

This coalition articulated—with many variations—a vision of academic science and engineering primarily funded by civilian agencies and emphasizing applied research deemed relevant to social issues such as pollution, energy, mass transit, public housing, and biomedicine. The types of basic research that had been the cornerstone of Terman's Stanford found little favor in the "relevance" agenda. Stanford's Student

[18] Kenneth D. Roose and Charles J. Andersen, *A Rating of Graduate Programs* (Washington, D.C., 1970), 48.

[19] Lowen, *Creating the Cold War University* (cit. n. 1), 159.

[20] Andrew J. Nelson, "Cacophony or Harmony? Multivocal Logics and Technology Licensing by the Stanford University Department of Music," *Industrial and Corporate Change* 14 (2005): 93–118.

[21] Vettel, *Biotech* (cit. n. 1), chaps. 5–6; Stuart W. Leslie, "Time of Troubles for the Special Laboratories," in *MIT: Moments of Decision*, ed. David Kaiser (Cambridge, Mass., 2011): 123–44; Matthew M. Wisnioski, "Inside 'the System': Engineers, Scientists, and the Boundaries of Social Protest in the Long 1960s," *History and Technology* 19 (2003): 313–33; Dorothy Nelkin, *The University and Military Research: Moral Politics at MIT* (Ithaca, N.Y., 1972).

[22] Our discussion of the controversies over classified research at Stanford and the Stanford Research Institute draws on Leslie, *Cold War and American Science* (cit. n. 1), chap. 9.

Radical Caucus, for example, dismissed basic research as "useless."[23] More subtle reformists, such as engineering graduate student and activist Stanton Glantz, argued that "the notion of 'basic research' often acts as a smokescreen to hide what we are doing . . . [and] to avoid facing the consequences" of academic research for the conflict in Southeast Asia.[24] Even faculty who had benefited from the early Cold War accumulation of a basic research stockpile, such as Robert Huggins (director of Stanford's ARPA-funded Center for Materials Research) were open to the idea that the time had come to apply that knowledge to "civilian technologies that have lain comparatively dormant in recent years, when primary attention was heavily concentrated upon . . . defense- and space-related matters."[25]

It is important to note that the deep divisions engendered by calls for relevance were not so much about whether academic research should benefit civil society, but how long it should take for that to happen, how directly that translation should be guided, and whether military funding hindered that translation. Researchers and administrators who were skeptical of the relevance agenda insisted that civilian technologies would result even from work funded by the military. Moderate reformers countered that academic scientists and engineers could and should quicken the pace of translation by seeking nonmilitary funding sources and being more attuned to research areas that would flow more easily into civilian applications. Radicals, meanwhile, believed urgent action was needed. Students for a Democratic Society and other campus activists painted researchers who took defense funding as complicit in the Southeast Asian conflict. In 1969, unrest over Stanford researchers' ties to the military came to a boil, with protesters picketing and chanting outside laboratories, setting fires and stealing classified information, pouring paint on President Kenneth Pitzer, staging a nine-day takeover of the Applied Electronics Laboratory (AEL) and the Systems Techniques Laboratory (STL), and more.[26]

RADICAL INTERDISCIPLINARITY AS A SOLUTION

The AEL/STL takeover and other confrontations contributed to CCRMA's founding in part by generating a pervasive anxiety about, as Glantz put it, "the possibility of violent confrontation between Stanford's technical and non-technical communities."[27] In response, many students, faculty, and administrators articulated the hope that greater cross-disciplinary understanding and collaboration would foster campus cohesion and prevent the breakdown of the institution. As President Richard Lyman put it after the AEL takeover,

> If we are in difficulties partly because our functions are many, and our focus can therefore never be single, it will do us no good to try to return to some simpler day. . . . Instead we ought to glory in the fact that some people are learning to appreciate Keats in one part of

[23] Student Radical Caucus, "Fire and Sandstone: The Last Radical Guide to Stanford," n.d. [almost certainly 1969], William Rambo Papers SC 132, ACCN 97-093, SUA, box 6, folder 10, "Student Unrest, 1968-2."

[24] Glantz, "Comments about Engineers for Engineering by an Engineer," *Grindstone: A Forum for Controversial Issues of Special Interest to the Engineering Community—Sponsored by the Student-Faculty Liaison Committee of the School of Engineering*, 30 Nov. 1970, 4–16, SUA, Arch 3009 The Grindstone.

[25] Quoted in Leslie, *Cold War and American Science* (cit. n. 1), 232.

[26] Ibid., chap. 9.

[27] Glantz, "Comments about Engineers" (cit. n. 24).

the campus, while others are solving problems of linear programming in another. Glory in it, and make a towering virtue of necessity by exposing the one group to the other, and each to a thousand further groups, at every available opportunity.[28]

The interdisciplinarity thought necessary to preserve Stanford and to quiet unrest was far more wide-ranging and egalitarian than that promoted earlier by Terman. Terman's interdisciplinarity had been a matter of regrounding engineering and social science disciplines in the ostensibly more fundamental knowledge of the natural sciences. Campus activists and faculty reformers of the Vietnam era, however, argued that the pressing social problems of the day were so complex as to require equal partnerships, not hierarchical relationships, ranging all across the social and natural sciences, engineering, and the humanities. As Stephen Kline, an engineer who co-founded Stanford's Values, Technology, and Society (VTS) program, put it, "The kinds of questions that do and should concern the students are: Do you build the SST [supersonic transport], and what is being done about smog? Questions of this sort cannot be seen clearly through the viewpoint of any single discipline."[29] Instead, Kline's VTS program offered a new approach, with "various combinations of scientists, engineers, philosophers, historians, anthropologists, psychologists, psychiatrists, sociologists, ethicists, and theologians—all working very closely together."[30]

Even skeptical faculty members and administrators in the Vietnam era commonly conflated "interdisciplinary" with "applied" or "relevant" research and/or painted discipline-based research as the opposite of "problem-oriented" research. As President Lyman put it in 1971, "No matter how earnestly the effort is made by scholars to mount collaborative attacks on social problems, . . . results are bound to be slow and halting. . . . It will take more than some marriages among academic disciplines."[31] Nevertheless, Lyman saw where things were headed: "If we succeed, as I trust we shall, in increasing the amount of multi-disciplinary, problem-oriented research that we do, this will happen in part because money is beginning to become available for such work from the Congress and from federal agencies."[32]

Thus, even skeptics of reform could be persuaded to embrace a more radical form of interdisciplinarity in the late '60s and early '70s because the rapidly changing federal funding situation offered strong incentives to do so. In part, there was simply less money to go around: federal nondefense R&D budgets declined by close to 30 percent from 1966 to 1976;[33] total federal R&D funding reached a peak of 3 percent of GDP in 1964 and declined steadily until the 1980s.[34] At the same time, civilian funders such as the NSF, the National Institutes of Health, and the Environmental Protection Agency began siphoning federal research funds away from national

[28] Quoted in "Defense Research Will Shift if Forced out of Universities," *Campus Report*, 23 Sept. 1970.
[29] "Values, Technology, and Society Included in Experimental Program," *Campus Report*, 19 May 1971.
[30] Ibid. Stanford later renamed the VTS program VTSS (Values, Technology, Science, and Society) and then STS (Science, Technology, and Society).
[31] "Revolt against Reason Can Have 'Ominous Consequences,'" *Campus Report*, 1 Jan. 1971.
[32] "Lyman Looks at Future: Toward a More Open University," *Campus Report*, 14 Apr. 1971.
[33] Intersociety Working Group, American Association for the Advancement of Science, *AAAS Report XXXIII: Research and Development, FY 2009* (Washington, D.C., 2009), 24.
[34] Homer A. Neal, Tobin L. Smith, and Jennifer B. McCormick, *Beyond Sputnik: U.S. Science Policy in the 21st Century* (Ann Arbor, Mich., 2008), 81.

security agencies, and shifted their emphasis toward applied research and engineering at the expense of basic science.

Given the unstable funding situation, Stanford researchers and administrators could see that the best strategy would be for individual faculty members to diversify their funding sources and reorient to Congress's new priorities. More radical interdisciplinarity was one way to appeal to a wider set of funders, while simultaneously dampening campus unrest. Thus, Stanford saw a dramatic mushrooming of interdisciplinary degree-granting centers in this era: a doubling in the number of such centers between 1968 and 1969, and a seven-times-higher rate of incorporation of such centers over the next twenty years than over the previous twenty. Most of these new centers were anchored in the humanities or social sciences, perhaps because, as Jamie Cohen-Cole has shown, an ideology of interdisciplinarity took hold of American social science in the '50s.[35] Indeed, when the university founded a Center for Interdisciplinary Research in 1972, it used an existing interdisciplinary social science center, the Institute for Public Policy Analysis, to nucleate a broadening of "the scope of interdisciplinary research activities to include engineering, the physical sciences, the professional schools and humanities, as well as economics and the social sciences."[36]

In the School of Engineering, meanwhile, many of the faculty housed in the extradepartmental research centers of the Terman era were, by the late '60s, searching for civilian applied projects, diversified funding sources, and (consequently) a broader type of interdisciplinarity. Stanford's electrical engineers—an important constituency for CCRMA's later explorations into computer music—were particularly prominent in turning toward interdisciplinary and applied, civilian projects such as biomedical technologies and aids for the handicapped. As John Linvill, chair of Electrical Engineering (EE), put it in a 1967 memo, "Stanford University can and should become more effective in studying and attacking the problems of today's society. Electrical Engineering, with its aim to bring technological tools to the solution of man's problems, is interested to join with other departments in working on these contemporary problems . . . [such as] environmental studies, urban problems, problems of developing countries, etc. . . . [which] cannot be attacked within a single discipline."[37] Few Stanford engineers altogether abandoned defense-funded projects, whether out of budgetary necessity or because they saw such work as both good science and good citizenship. Yet whatever their politics, many Stanford engineers were clearly moving in the late '60s toward work of the kind called for by antiwar reformers.

The situation in the late '60s and early '70s, then, was that the different parts of campus that would eventually contribute to CCRMA faced convergent pressures. For members of the still-unsettled music department, incentives for moving toward the Terman model by cooperating with engineers to bring in external basic-research

[35] Cohen-Cole, "Thinking about Thinking in Cold War America" (PhD diss., Princeton Univ., 2003).
[36] Nancy Donham, Stanford University News Service release on the Center for Interdisciplinary Research, 10 July 1972, SUA, Stanford University Sponsored Projects Office SC 344, box 14, folder "Center for Interdisciplinary Research (CIR), 1972–1974."
[37] John G. Linvill, executive head, Department of EE, to the Committee for the Study of Stanford's Educational Program, memorandum, "Re: Comments of Electrical Engineering on Stanford's Educational Programs and Objectives," 26 Jan. 1967, SUA, Calvin Quate Papers SC 347 (83-033)—1987 accession, box 3, binder "Electrical Engineering."

funding and to grow links with industry would have been apparent. Stanford's predominantly defense-funded engineers and computer scientists, meanwhile, could perceive incentives to modify the Terman model by cooperating with musicians and other humanists to demonstrate civilian applications. In such an environment, an emerging center that was interdisciplinary by nature, that addressed practical applications, that developed and leveraged technologies applicable in multiple settings, and that could garner external resources for these activities would suddenly find itself the embodiment of new organizational perspectives and priorities.

CCRMA'S ROOTS: "FOOLING AROUND" AND CHANCE ENCOUNTERS

John Chowning, CCRMA's cofounder and its most conspicuous initiator, arrived at Stanford in 1962 in order to pursue a doctoral degree in music composition. He had earlier studied composition in Paris, where he was exposed to—and intrigued by—electronic music. Upon his arrival at Stanford, with its newly created music department, however, he was dismayed to find that there were neither facilities for electronic music nor an interest in creating them.

Nevertheless, Chowning's interest in the field was known to others, including his fellow members of the Stanford Symphony Orchestra. In January 1964, one of these members passed him a copy of an article from *Science* that described how a computer could be used as a musical instrument. (The orchestra member's husband subscribed to *Science* because he was on the faculty at Stanford Medical School.) The article's author was Max Mathews, a researcher at Bell Telephone Laboratories. There, computer music developed from an attempt to create tones that would resemble speech but be more amenable to analysis. As Mathews and his colleague, John Pierce, put it,

> There is a very close analogy between the voice and a bowed-stringed instrument. . . . Concepts and analytical techniques developed in speech research should be useful in studying the sounds of bowed-stringed instruments, and in fact much of the computer-programming and all of the peripheral equipment used in this study were originally designed for speech research.[38]

Mathews's repositioning of the computer as a musical instrument thus reflects the discussion by John Tresch and Emily Dolan (in this volume) of "telos," or the ends to which instruments are used. Mathews's insight was to use music to simplify his data while at the same time importing a tool used for other purposes—the computer—into the world of music.

Critically for CCRMA, Mathews and Pierce were open to collaboration and conversation with musicians. In fact, Mathews, Pierce, and other Bell engineers collaborated with several artists in 1966 to produce a series of performance-art presentations, *9 Evenings: Theatre and Engineering*.[39] Mathews and Pierce were also closely

---

[38] Mathews, Joan E. Miller, Pierce, and James Tenney, "Computer Study of Violin Tones—Case 25952," Bell Labs technical memorandum, 15 Nov. 1965, John Pierce Papers, Huntington Library, San Marino, Calif. (hereafter cited as Pierce Papers), box 7, binder "Technical Memoranda 1956–1968."

[39] P. Miller, "The Engineer as Catalyst: Billy Kluver on Working with Artists," *IEEE Spectrum* 35 (July 1998): 20–9. More generally, Mathews, Pierce, and Chowning were leading exemplars of the heightened mutual regard and interaction among engineers and artists in the 1960s, particularly in

connected to Stanford through faculty members in applied physics and EE who had formerly worked at Bell Labs. Thus, Chowning was able to visit Mathews the summer after taking a computer programming course in the spring of 1964. Mathews provided Chowning with direction and, crucially, the Music IV computer program that Mathews had created to generate computer sound on Bell's IBM 7090.[40]

Computers were somewhat rare commodities at the time, however, and Chowning's search for one on which to implement Music IV led him to the rapidly expanding Stanford Artificial Intelligence Project. As he recalled:

> In autumn of '64 . . . I met Dave Poole. I had this box of [punch] cards from Max [Mathews]. We knew one another from the Stanford orchestra. He was a tuba player and I was a percussionist, so we were right next to one another. . . . Poole was an applied math major, maybe in his second or third year. He was a hacker. He was sort of on the periphery of the AI lab.[41]

With support from Poole, whose programming skills had become indispensable to the AI project, and from Les Earnest, the AI project's deputy director and the husband of a local music teacher, Chowning implemented Mathews's program on the AI project's PDP-1 computer later that year. Over the next several years, Chowning continued his work on computer-synthesized and manipulated sound, receiving his doctoral degree in 1966 and joining the music department faculty that same year. He had a major breakthrough in 1967, when some late-night "fooling around" (as Chowning himself describes it[42]) resulted in the discovery of frequency modulation synthesis, a technique that permits the creation of complex sounds with relatively few computations. Yamaha Corporation of Japan ultimately licensed the technique in 1975, and, after several more years of development by Yamaha and Stanford, it formed the basis of a suite of electronic musical instruments that was extremely profitable for both organizations. Once CCRMA was founded, its personnel built an active industrial affiliates program and an intellectual property portfolio of several dozen patents. Today, they contribute to a wide range of companies, many of which they started. The center itself is now one of the world's premier places for computer music and digital audio research.[43]

CCRMA's success had many sources. In part, it benefited from ties to firms such as AT&T and Yamaha and to the emergence of a global computer and electronic music community. CCRMA was also fostered in part by features of Stanford's music de-

---

New York (near which Bell Labs was located) and the Bay Area (including Stanford). See Matthew Wisnioski, *Engineers for Change: Competing Visions of Technology in 1960s America* (Cambridge, Mass., 2012); Andrew Pickering, *The Cybernetic Brain: Sketches of Another Future* (Chicago, 2011); Christina Dunbar-Hester, "Listening to Cybernetics: Music, Machines, and Nervous Systems, 1950–1980," *Sci. Tech. Hum. Val.* 35 (2010): 113–39.

[40] M. V. Mathews, "The Digital Computer as a Musical Instrument," *Science* 142 (1 Nov. 1963), 553–7. Chowning recalled the sequence of events in a 25 Mar. 2008 interview with Andrew Nelson in Palo Alto, Calif. Pieces of the story are captured in other accounts, too, including an interview of Chowning conducted by Vincent Plush on 31 May 1983 in Palo Alto as part of the Yale Oral History Project.

[41] Chowning, interview by Nelson (cit. n. 40).

[42] Chowning, interview by Plush (cit. n. 40).

[43] For other accounts of CCRMA's growth and the Yamaha relationship, see Nelson, "Cacophony or Harmony" (cit. n. 20); Tim Reiffenstein, "Codification, Patents and the Geography of Knowledge Transfer in the Electronic Musical Instrument Industry," *Canadian Geographer* 50 (2006): 298–318; Bob Johnstone, *We Were Burning* (New York, 1999).

partment. For example, the department's relative youth meant it had not developed a coherent internal identity around any one particular program. Rather, the music department at the time was experimenting with programs such as jazz studies, performance practice, and early dance, all of which were products of individual faculty members' initiative. Support from music department faculty members proved critical to Chowning's graduate studies, too. Under the chair system that was in place, graduate students could pursue any topic that their adviser supported, regardless of departmental interests or pressures. As Chowning recalled in 1982, "Not many universities would have allowed me the freedom to do what I've done."[44] The music department's policies thus enabled Chowning's unusual exploration of computer music.

CCRMA also benefited from a variety of fortuitous encounters and events, both inside and outside Stanford. These included Chowning's fellow orchestra member passing him Mathews's *Science* article, Mathews's willingness to share his computer program, Poole and Chowning's collocation in the Stanford symphony and Poole's subsequent assistance in implementing the computer code, and Chowning's late-night fooling around. Absent any of these, CCRMA's history might have unfolded quite differently. Yet these coincidences, as well as the felicitous characteristics and policies of the music department, had such an impact on CCRMA's development because of the larger institutional environment. That environment, as we have argued, was shaped by Vietnam-era tensions and budgetary conditions that incentivized practical applications and wide-ranging interdisciplinary collaboration.

INTERDISCIPLINARITY AND CCRMA'S TIE TO PRACTICAL APPLICATIONS

On one level, the very term "computer music" implies an interdisciplinary connection. Nevertheless, it is instructive to unpack the ways in which the activities at CCRMA reflected the more radical interdisciplinarity that emerged in the 1960s. This disciplinary diversity was evident, in part, in individuals such as Chowning himself. In writing of Chowning's promotion to full professor, for instance, the spring 1979 music department newsletter noted,

> Chowning's achievements in composition, teaching and research touch upon several disciplines. His research has been primarily in acoustics and psychoacoustics (the psychology of sound perception). He has developed computer techniques for creating the illusion of sound localization and movement through space, for artificial reverberation, and for synthesizing and generating any sound that can be produced by loudspeakers.[45]

Chowning and his associates were quick, however, to point out that no individual alone could master the wide variety of disciplines that contribute to computer music.

For instance, F. Richard Moore, a former CCRMA student who went on to build the computer music program at the University of California, San Diego, described the epistemic challenge of computer music this way in 1979:

> The complexity of the new technological tools . . . require[s] cooperative team efforts for their use. Musicians have traditionally worked in highly individualistic ways . . .

[44] Stanford Department of Music, "Computer Music Comes of Age," *Music at Stanford*, Oct. 1982. Included with the *Stanford Observer*, Oct. 1982, sec. 2.

[45] Stanford Department of Music, "Chowning Appointed to Full Professorship," *Music at Stanford*, Spring 1979, 1.

> [but] using computers to study music requires simultaneous attendance to . . . computer science, engineering, acoustics, psychoacoustics, and music. . . . It is unlikely (or at least rare) that individuals will possess all of the requisite skills and knowledge needed for effective use of computers in music making or study.[46]

Moore's reflections are not entirely new, of course: science and music share deep historical connections, as documented throughout this volume. The interdisciplinarity imagined and implemented by Chowning and Moore, however, involved multiple participants contributing unique—and equally valued—expertise, rather than individual scientist-musicians or scientists such as Helmholtz or Mach (on whom see, respectively, Kursell and Hui in this volume), who combined different forms of expertise in a single person.

Thus, CCRMA addressed the challenge of interdisciplinarity by bringing together individuals who were highly skilled in their particular disciplines but also reliant upon other individuals to fully realize their objectives. Rather than smothering individual creativity (which Moore cautioned that musicians would resist), CCRMA leveraged interdisciplinary teamwork to enable the full expression of individualistic goals. Contemporary descriptions of the center, in publications ranging from the Stanford campus newspaper to *Newsweek* to *Rolling Stone*, emphasized these interdisciplinary underpinnings. In fact, of the dozens of articles published on CCRMA in the 1970s, every one mentions the word "interdisciplinary," with most pieces dwelling at length on the range of disciplines found at the center.

The form of that interdisciplinarity was precisely fitted to the convergent incentives faced by different parts of the Stanford campus. As Chowning and two colleagues wrote in a 1974 grant proposal made jointly to the NSF and the National Endowment for the Arts (NEA),

> A major contribution to present and future music exists in the application of a rapidly developing computer technology to the art and science of music. The extraordinary results already obtained have occurred in those few instances where scientists and musicians have taken the opportunity to bring their respective skills to bear on problems of common interest in a rich interdisciplinary environment. It is an example of cooperation, but more, an expression of the freedom of intellect and invention, where creative minds from diverse disciplines have joined in a common goal to produce fundamental knowledge which must be the source for new music, and to produce works of art which reflect the scientific-technological riches of the present.[47]

That is, NSF funding for CCRMA would enable the center to expose musicians to the methods and values of scientists and engineers, allowing researchers to "produce fundamental knowledge" for the "science of music"—exactly the model established by Terman. At the same time, CCRMA would assist scientists and engineers in the "application of a rapidly developing computer technology to the art" of music "in a rich interdisciplinary environment"—exactly the kind of humanistic collaboration

[46] Moore, "The Nature of Music Research," *Directions: Center for Music Experiment and Related Research* (newsletter, University of California, San Diego), Dec. 1979, 1–2.

[47] John M. Chowning, Leland C. Smith, and Albert Cohen, "The Computer Music Facility: A New Musical Medium," proposal submitted to the NEA in conjunction with the proposal "Computer Simulation of Music Instrument Tones in Reverberant Spaces" submitted to the NSF, 18 June 1974, SUA, CCRMA ACCN 2001-262, box 7.

and civilian application that would bring diversified funding and approval from campus activists.

The structure and location of CCRMA within the broader university environment played a critical role in facilitating this disciplinary mixing. Current CCRMA director Chris Chafe, who arrived as a graduate student in the late 1970s, has argued that, on the one hand, CCRMA's administrative home in the music department is critical for attracting music personnel to the collaboration: "This is a cross-disciplinary, very artistic, technical, everybody-helping-each-other kind of environment. . . . Imagine the facility being located in a more technical department, what the barrier would be for musicians approaching. Would they feel they're free to come join the project if it were headed towards engineering?"[48] On the other hand, as Chowning recalled in 1982, CCRMA's early leaders "felt the separation [from the music department] was necessary because our work was different from the usual Music Department activity."[49]

In fact, CCRMA has always been physically separated from the rest of the music department, following the model of earlier semi-independent research centers: for example, the Center for Materials Research was aligned with, but not part of (or limited in its personnel to) the materials science and engineering department; the Integrated Circuits Lab was aligned with, but not part of or limited to the EE department. Chowning and other CCRMA leaders recognized that the trick was to be both separate from and connected to a variety of groups simultaneously. As Chowning remarked in 1975, "Every center tends to build a wall around itself. . . . We hope to provide intellectual ventilation as well as coordination."[50] CCRMA, therefore, thrived at an almost contradictory intersection between autonomy and authority, connectedness and isolation, individualism and teamwork. In a university environment struggling to find the appropriate organization for interdisciplinary activities, such ambiguity advantaged a group that could not quite be categorized and might, therefore, be all things to all people (or, at least, many things to many people).

At the center of CCRMA's multivalent milieu were the computer and associated sound-generating and processing algorithms. We label these tools "multivocal technologies" in reference to their ability to span both disciplinary boundaries and areas of applied interest. A multivocal technology exhibits generality and flexibility, and is subject to different interpretations depending on the group that employs or interacts with it. Our conceptualization of multivocal technologies differs from "boundary objects," though it certainly shares features in common with this concept. Fundamentally, boundary objects facilitate communication between diverse groups, enabling them to arrive at a shared understanding and to work together.[51] As is highlighted by the interdisciplinary context of CCRMA, a multivocal technology such as the computer can, like a boundary object, permit groups to work together; however, it does so in a way that is more reflective of Peter Galison's "trading zones"—where different

[48] Chafe, interview by Andrew Nelson, 7 Mar. 2002, Stanford, Calif.
[49] J. Bailie, "Computer Music: A Trend That's Caught Fire," *Stanford Daily*, 20 July 1982, 7.
[50] Paul Hertelendy, "Stanford's Musical 'Marriage,'" *Oakland Tribune*, 12 Jan. 1975, 20.
[51] Susan Leigh Star and James R. Griesemer, "Institutional Ecology, 'Translations' and Boundary Objects: Amateurs and Professionals in Berkeley's Museum of Vertebrate Zoology, 1907–39," *Social Studies of Science* 19 (1989): 387–420; Paul R. Carlile, "A Pragmatic View of Knowledge and Boundaries: Boundary Objects in New Product Development," *Organization Science* 13 (2002): 442–55; Beth A. Bechky, "Object Lessons: Workplace Artifacts as Representations of Occupational Jurisdiction," *American Journal of Sociology* 109 (2003): 720–52.

parties do not need to agree about the context of collaboration—than it is of boundary objects.[52]

More pointedly, our inspiration for multivocal technologies draws, in part, on John Padgett and Christopher Ansell's work on the Medici family in Renaissance Italy, which derived power from its ability to maintain separation between networks of political, economic, and social influence; the relationship illustrated by Padgett and Ansell is not so much focused on bringing together through commonality as it is on lying between by presenting different faces to different groups.[53] Thus, multivocal technologies not only embody a high level of interpretive flexibility, but also enable participants to exploit this interpretive flexibility to serve different ends depending on the pattern of relationships in play.[54]

That is, multivocality is intentional, even Machiavellian. Actors must choose to position themselves where they can sing different tunes to different audiences so that they can obtain resources from each audience. Other historians of technology have observed that computers and musical instruments have been useful "liminal entities" with which actors can mediate between different groups.[55] Particularly in the late 1960s, electronic musical instruments allowed such individuals to perform a "legitimacy exchange," whereby technologists accrued social capital from artists and vice versa.[56] Our study goes somewhat further, however, in showing how the multivocality of a technology (the computer) was amplified by innovation in the form of an organization (CCRMA).

The computer is not an inevitably multivocal technology, but it has been made multivocal in many different institutional contexts, including CCRMA. As Dick Moore wrote in 1979, "The computer may be viewed as a general purpose tool with the unprecedented function of extending our power of mind, as other tools extend our muscular or sensorial powers."[57] Chowning and others repeatedly emphasized the broad appeal of the computer as a research tool in gathering resources and accruing prestige for CCRMA. For example, the 1974 application to the NEA referenced above cited the research applications of the computer at Bell Labs, claiming, "Based on this past experience, the proposed [CCRMA] facility will maintain those attributes of generality, flexibility, and precision which have been of utmost importance in the research performed to date."[58]

As we have seen, Bell Labs researchers such as Mathews and Pierce employed the computer (almost literally) as a multivocal technology, not simply for generating and analyzing musical sounds, but also for building a diverse network of interlocutors interested in computer music: for example, Yale composer James Tenney, President Nixon's science adviser Edward David, and science fiction author Arthur C. Clarke. Pierce and Mathews in turn openly admired Chowning and CCRMA's similar use of

[52] Galison, *Image and Logic: A Material Culture of Microphysics* (Chicago, 1997).

[53] Padgett and Ansell, "Robust Action and the Rise of the Medici, 1400–1434," *Amer. J. Sociol.* 98 (1993): 1259–319.

[54] Wiebe Bijker, Thomas Hughes, and Trevor Pinch, eds., *The Social Construction of Technological Systems: New Directions in the Sociology and History of Technology* (Cambridge, Mass., 1987).

[55] Trevor Pinch and Frank Trocco, *Analog Days: The Invention and Impact of the Moog Synthesizer* (Cambridge, Mass., 2004).

[56] Fred Turner, *From Counterculture to Cyberculture: Stewart Brand, the Whole Earth Network, and the Rise of Digital Utopianism* (Chicago, 2006).

[57] Moore, "Nature of Music Research" (cit. n. 46), 2.

[58] Ibid., 3.

the computer as both a musical technology and the centerpiece of a heterogeneous network. As Pierce put it in letters of recommendation for Chowning,

> Most musicians shy away from the computer and need a great deal of help to use it effectively. Chowning not only knew how to make the computer work; he knew what to do with it. . . . In a day when many artists mouth science vainly, it is rare but extremely pleasing to find one, John Chowning, who understands and uses science.[59]

Or, more succinctly, "Chowning has contrived to show great originality and leadership. Starting from essentially nothing, he has brought diverse talents together into a field of common general interest."[60]

Endorsements from Bell Labs carried special weight at Stanford given Mathews's and Pierce's strong ties there. Mathews took a sabbatical in the Stanford Artificial Intelligence Project in 1969, and both he and Pierce eventually took faculty positions affiliated with CCRMA. Both men were also in regular contact with former Bell Labs team members at Stanford, including John Linvill (chair of EE) and Calvin Quate (in EE and applied physics, but also—on temporary assignment—an associate dean with oversight of Chowning's tenure case). Indeed, Quate sought Pierce's help in 1970 in securing philanthropic funding for his own answer to the relevance agenda, an acoustic microscope for biological research—in response to which Pierce teasingly asked whether the microscope was "fire resistant and bomb proof? Or, are things now quiet at Stanford?"[61]

Thus, CCRMA researchers' emphasis on generality allowed them to connect to broader networks that employed the same central tool and to draw legitimacy from their connection to those networks. Indeed, CCRMA researchers pushed the generality of their approach to extremes in search of funding and interlocutors. For example, a 1977 NSF grant application argued, "The work to be performed under this project is directed toward answering basic questions about the variety of forms in which information can flow between man and computer, and the factors which limit the rate of this information flow."[62] Here, music was the concrete context, but the use of a multivocal technology in that context enabled musicians to contribute to a highly abstract debate about the nature of information of the kind traditionally associated with physics, math, and EE.

The multivocality of the computer was indispensable in helping CCRMA's leaders satisfy the sometimes conflicting demands on faculty members at research universities in the 1970s. On the one hand, the computer aided CCRMA personnel in doing the kind of basic research that would establish their scientific credibility. As one reviewer of a 1977 NSF grant wrote, "In past proposals the National Science Foundation has had to consider whether it was supporting music as an art or whether it was supporting a scientific study. In this proposal, there is no question about the

[59] Pierce to Albert Cohen, Department of Music, Stanford, 21 Mar. 1978, Pierce Papers, box 5, folder "Misc. Correspondence H–Z, 1973–79."

[60] Pierce to Wolfgang Kuhn, Department of Music, Stanford, n.d., Pierce Papers, box 5, folder "Misc. Correspondence, H–Z, 1973–79."

[61] Pierce to Quate, 19 Oct. 1970, Pierce Papers, box 5, binder "Correspondence, External, August 1969–March 1971."

[62] John Chowning and Loren Rush, "An Interactive Graphical Environment for Musical, Acoustical, and Psychoacoustical Research," project summary for NSF grant no. MCS77-23743, 1977, SUA, CCRMA ACCN 2001-262, box 7.

focus of the effort. The work is entirely of a scientific nature; only the applications concern music."[63] Chowning himself reinforced this rhetoric. For example, in a 1982 student-newspaper story reporting that "since 1975, CCRMA has received over $1 million from the NSF to support its work," Chowning explained, "These grants have been based on the scientific aspects of our work in psycho-acoustics and signal processing."[64] That is, for some audiences, music provided the context of CCRMA's work, but the focus was on scientific advance.

On the other hand, some CCRMA personnel and external commentators recognized that music was not only a convenient context, but also a top beneficiary of research support. For example, another reviewer of the 1977 NSF grant emphasized that the musical implications of CCRMA's proposed research were exciting precisely because of their impact on a field of "practical" and "applied" importance (music):

> The timbre perception proposal is superb! It is a pattern for the kind of research that should be emphasized today. It is one of the too rare cases where scientific studies are making real contributions to an area of great applied importance. There are plenty of areas where good scientific methods are discovering interesting new things about the world, but often the new information does not have clear importance in applications. Likewise, there are plenty of important practical problems, but too often scientific methods can make only weak contributions to their solutions. The fundamental studies proposed here can produce information of enormous importance to music, both for immediate applications and for the far future.[65]

Six years later, a reviewer of another NSF grant also identified the arts not as a convenient context for fundamental research, but as a full participant in, and beneficiary of, CCRMA's brand of interdisciplinary, applied research:

> I cannot recommend this proposal more highly. . . . How many projects combine and bear fruit in fields as diverse as signal processing, artificial intelligence, and acoustics, while having profound implications in the arts? The benefits of such research to the music recording industry are clear.[66]

Thus, in the hands of CCRMA personnel, the computer allowed for continual feedback and alternation in the relationship between musical instrument and research. Not only was the musical instrument useful as a research tool—as in the accounts provided by Pesic and Kursell in this volume, as well as others—but insights about the research tool were made useful in a musical context as well.

Indeed, the blurring of distinctions among scientific research, technological development, and musical composition is reinforced by a peculiar feature of computer music. In traditional composition, the instruments are (relatively) fixed and the composition process focuses upon orchestration and the specific notes to be played by each instrument. In computer music, however, the instruments themselves need to be "composed" via computer code. The sonic possibilities are, therefore, unknown

[63] Reviewer comments on NSF proposal no. BNS-7722305, "Experiments in Timbre Perception," submitted by John Chowning and John Grey, 1 Sept. 1977, SUA, CCRMA ACCN 2001-262, box 7.
[64] Bailie, "Computer Music" (cit. n. 49).
[65] Reviewer comments on "Experiments in Timbre Perception" (cit. n. 63).
[66] Reviewer evaluation form for NSF proposal no. MCS-8214350, "An Intelligent System for the Knowledge-Driven Analysis of Acoustic Signals," submitted by John Chowning, 18 Feb. 1983, SUA, CCRMA ACCN 2001-262, box 7.

prior to—and become known through—composition. At CCRMA (as described by composer Michael McNabb, who earned his PhD there in the 1970s), the process of composing "instruments" and composing "music" was often iterative and involved the distributed cognition of an interdisciplinary team:

> It's really hard to even separate where all the creativity came from. When you talk about one piece of music, any one piece of music, there couldn't help but be at least half a dozen people involved besides the composer because we had to write our own software, the engineers had to build their own equipment, all this stuff. It was like you'd lose track: Did Julius [Smith, an electrical engineer] come up with this special [engineering] thing and we thought, "Well, that's cool. I'm going to use it in a piece." Or the other way around: "I want to do X" and they [people like Julius Smith] would say, "Well one way you could do that is like this."[67]

Compositional and research activities, and musical and technical activities, were inseparable. For this reason, many observers felt that CCRMA represented a new marriage between art and science. Each field benefited independently; science and engineering benefited art and, crucially, art provided a path forward for new scientific and technological developments.

By helping to blur these different domains, the computer facilitated CCRMA's reach beyond the ivory tower of research to engage the world of musical performance and appreciation directly. For example, in 1975, the *Stanford Daily* interviewed CCRMA personnel who downplayed "academic" activities and instead emphasized the practical applications of computer music:

> Fears that computer music will be "highly academic" are "clearly silly," Moorer stated. . . . "Computer music may be a 'laboratory thing' now but in five to 10 years," Moorer predicted, it will be a suitcase-size "performing stage popular instrument." The computer as a musical instrument could eventually "find its way into everyone's home," Grey added.[68]

Chowning, too, asserted (in 1979), "This is not an acoustical lab. We define our problems as 'real-world' problems; we don't use anechoic chambers [sound-absorbing rooms common in acoustics research environments] because nobody listens in one."[69]

This connection to "real-world" problems of performance allowed CCRMA to appeal for funding not just from civilian research-funding agencies such as the NSF, but also from arts organizations such as the NEA and the California Arts Council, and from commercial firms and industry groups. CCRMA's research tools were also multivocal enough to appeal to defense funding agencies, especially as time went on and defense R&D budgets bounced back (while the stigma associated with such funding waned). Indeed, by 1985, Stanford's Sponsored Projects Office was advising CCRMA to mine the intersection of civilian and national security funding: "NSF is mixed in with Navy and DARPA. . . . Helpful buzzwords on their interests include:

---

[67] McNabb, interview by Andrew Nelson, 17 Apr. 2009, San Francisco.

[68] Ann Amioka, "Computers Add New Dimension to Music," *Stanford Daily*, 5 Mar. 1975. Andy Moorer was an EE PhD student who would go on to develop the sound studio for LucasFilm and to found the firm Sonic Solutions. John Grey was a psychology doctoral candidate. It's worth noting that Grey's prediction was prescient: through LucasFilm, Moorer did bring computer music into every home, and the CCRMA-enabled Yamaha FM synthesis chip powered the first wave of sound cards in multimedia home computers.

[69] Michael Walsh, "Applying New Technology to an Ancient Art," *Palo Alto Times*, 12 Nov. 1979.

Conditioning sound waves, acoustics, signal processing—intelligibility problems, and signal processing—underwater acoustics."[70] The next year, CCRMA followed up by submitting a grant request to the Office of Naval Research. Similarly, in 1986 Pierce directed CCRMA's attention to a man from the National Security Agency "who expressed considerable interest in supporting work on a vocal-tract model for high-quality speech and song. The chances of success are 40–60%, even in these bad times."[71]

In an institutional environment that struggled with tensions and contradictions between basic and applied research, military and civilian applications, and disciplinary rigor versus interdisciplinary problem solving, CCRMA managed—via innovations in both organization and its multivocal research tools—to position itself as embodying all of these things simultaneously. That flexibility allowed CCRMA to secure resources from stakeholders with very different objectives and to nimbly reposition itself as the funding environment changed. The multivocality of CCRMA's tools, practices, and personnel also allowed it to connect into very different networks both inside and outside Stanford and thereby to accrue legitimacy and ameliorate conflict.

EARNING MERIT FOR CCRMA

That is, the formation of a constituency for computer music at Stanford, and its formalization under the umbrella of CCRMA, were facilitated by the fact that computer music offered something for everyone during a time of crisis. For idealists, it offered a more humanistic, holistic vision of science and engineering; for doves, the possibility of making military-industrial technologies relevant to civilian applications; for embattled administrators, a way to use an expansive form of interdisciplinarity to lubricate campus friction; and for cash-strapped, protest-besieged scientists and engineers, a way to solicit money from both military funders and civilian agencies, private philanthropies, and commercial firms. That diversity of funding in turn allowed CCRMA's members and (especially) Stanford's administrators to demonstrate to protesters that some of their demands were being met.

For CCRMA and electronic music to help those stakeholders achieve their aims, however, news of CCRMA's work had to be distributed widely. That is, if the budgetary and legitimacy crises of the late '60s and early '70s helped open the door for music to enter the laboratory at Stanford, those crises also required that computer music come back out of the laboratory and be heard by anyone who would listen. Trustees needed to be shown that Stanford faculty members were looking for funding wherever they could find it; campus activists needed to be shown that Stanford engineers were in the process of turning toward more civilian, socially relevant projects; administrators needed to be shown that CCRMA really could bring different sides of the campus together.

This theme played out wherever Stanford engineers sought expansive interdisciplinarity, civilian applications, and diversified funding in the late '60s and early '70s. For instance, the AEL takeover of 1969 prompted Stanford's School of Engineering

[70] Sponsored Projects Office to Patte Wood, CCRMA, handwritten memorandum, 1985, SUA, ACCN 2001-262, box 13.

[71] John R. Pierce to John Chowning, Patte Wood, and Julius Smith, memorandum, 15 Dec. 1986, SUA, ACCN-2001-262, box 4.

to prominently tout its faculty members' work on civilian-relevant applications such as biomedical sensors and aids for the handicapped—work that had previously been buried at the back of annual reports.[72] Similarly, the commingling of music and engineering played an outsize role in Stanford's promotion of itself (often, to itself) as an institution capable of overcoming Vietnam-era adversity.

For instance, President Lyman, in bemoaning the divisions tearing the university apart, made explicit gestures to music and computing as fields whose practitioners had far more in common than they might know or admit:

> It is an arrogant assumption of some humanists that no computer man reads Keats, and no electronics buff can dig Scarlatti. It is an arrogant counter-assumption of some technologists that no humanist has anything important to contribute to life in the technitronic age of the future. The university exists in part to attack such arrogant parochialisms.[73]

Similarly, researchers whose livelihoods were threatened by the protests against on-campus classified work eagerly embraced music as one of the civilian application areas benefiting from nominally secret research. As David Gray, a researcher in the AEL, wrote to Kenneth Arrow in 1966,

> That classified research will benefit the human race is apparent when one surveys the ever more numerous outgrowths of such research. To enumerate, there are SLAC [the Stanford Linear Accelerator Center], air traffic control, jet transportation, and isotopes for cancer research. . . . Musical instruments will be synthesized by a computer eventually placing an entire symphonic orchestra in the hands of a composer who will need only to communicate his composition to the synthesizer. He will then be able to listen and improve. At Stanford, research in encoding, in propagation of optical data processing, in integrated circuits, and in acoustic couplers are all relevant to such future development.[74]

When members of the divided university sought common ground, therefore, electronic music was one of a handful of guiding examples of how to do so.

For instance, when an ad hoc committee of faculty members was formed in 1971 to address "increased campus interest in research," it put together a research "exposition extending over the entire campus community. Interdisciplinary activities particularly are desired,"[75] such as

> the use of holography (three-dimensional laser holography) in displaying art objects, a continuous interactive opinion poll with a computer terminal, the use of electronic devices to help the blind and to measure medical data inside the body, and a presentation of computer-composed music.[76]

As a site of multivocal activity, the research that later became CCRMA's focus was an obvious fit for a public exhibition of Stanford faculty members' abilities to work

---

[72] Stanford Electronics Laboratories, *Stanford University Electronics Research Review* 11 (12 Aug. 1969), SUA, Collection 3120/4 STAN.

[73] Quoted in "Defense Research Will Shift" (cit. n. 28).

[74] Gray to Kenneth Arrow, memorandum, 2 July 1966, SUA, SC 132, ACCN 97-093, William Rambo Papers, folder 7, "Stud. Unrest, 1966-1."

[75] "Research Exposition Is Planned; Organizational Meeting Tomorrow," *Campus Report*, 9 Dec. 1970.

[76] "Research Exposition Is Planned," *Campus Report*, 27 Jan. 1971.

across disciplinary lines and to broaden the university's research portfolio beyond national security concerns.

Stanford administrators were particularly eager to have campus research reach out to the local communities of the San Francisco Peninsula, driven by pressure both from those local communities and from campus activists and reform-minded faculty who insisted that "as Stanford has grown in national prestige in the last twenty years, it has become simultaneously more and more isolated politically, educationally, and financially from its immediate community, neither serving it nor supported by it."[77] Few areas of research lent themselves better to overcoming such isolation than a performance art such as computer music. Indeed, CCRMA's leaders viewed CCRMA, and before it SAIL, as natural venues for connecting Stanford to its environs. A 1974 NSF funding application described the center's outreach efforts:

Publications and Performances
A normal function of the Center, indeed an obligation, will be the publication of results on a lay level as well as technical. Publications and tapes for performance could then be made available to large numbers of communities throughout the nation through the Executive Directors of the State Arts Councils.

Site Visits and Symposia
It is a normal circumstance to have a large number of visitors at the Stanford Artificial Intelligence Laboratory. Over the past years we have given demonstrations to groups ranging from school children to professionals in the field. With a relatively independent satellite system as proposed, we could significantly expand visits without causing inconvenience to the main laboratory. . . .
    As a part of the nation's Bicentennial, we propose that the Center at Stanford, an example of computer based interdisciplinary research and composition, organize a series of concerts of computer music, lecture-demonstrations, laboratory visits, and a symposium in the Spring and Summer of 1976. In this way, the public can share the excitement of applying its advanced technology to the discovery of new knowledge regarding sound and perception and to new means for the composition and performance of music.[78]

These efforts met with good success, too: by the early 1980s, CCRMA concerts had already attracted over a thousand attendees.[79]

OTHER FORMS OF TRANSLATION

Performances for local communities were by no means the only form of outreach envisioned for CCRMA. Chowning and other leaders organized their center to translate research out to society at large. Some of those conduits of translation were adapted from early Cold War centers such as the Microwave Lab; others emerged partly in response to the conditions of the early '70s.

One obvious CCRMA output was personnel. Students such as Dick Moore and Andy Moorer from CCRMA's early cohorts received offers from organizations, such

    [77] Flyer for "open meeting sponsored by FPAG (Faculty Political Action Group)," [ca. 18 Feb. 1971], SUA, Stanford University Sponsored Projects Office SC 344, ACCN 1987-130, box 1, folder 1, "Committee on Research 1970–71."
    [78] Chowning, Smith, and Cohen, "Computer Music Facility" (cit. n. 47), 5–6.
    [79] Xavier Serra and Patte Wood, eds., "Overview: Center for Computer Research in Music and Acoustics" (Stanford Department of Music Report Stan-M-44, March 1988); Julia Sommer, "Computer Music Audiences Mushroom," Stanford News Office press release, 28 June 1982.

as the Eastman School and the University of California, San Diego, that were looking to set up similar efforts in computer music. Musicians such as Györgi Ligeti stayed at CCRMA as composers in residence, then left to spread the word about computer music in general and Stanford's program in particular. Thus, through the production of personnel, CCRMA was able to export its organizational model. Indeed, that seems to have been one of Chowning's ambitions from the beginning. As the 1974 NSF proposal put it, CCRMA and its multivocal tools would serve "as a prototype for other and future systems."[80] Within the first few years of CCRMA's existence, the center had influenced programs in locations as far-flung as Paris and Hamburg.[81]

The production of personnel to export institutional models was, in some ways, an aim borrowed from semi-independent labs of the early Cold War such as Stanford's Center for Materials Research.[82] By the late '60s, however, leaders of Stanford's semi-independent research centers saw a need to produce something more than personnel. One path Stanford researchers increasingly took was to patent their research and seek firms to license those patents; some even founded start-ups to commercialize their research themselves. Music technologies played an outsize, if underappreciated, role in the exploratory commercialization of Stanford research in the '70s. For instance, one of Stanford's most active patenters in this era was Calvin Quate in applied physics and EE; among Quate's many patent disclosures was one from 1974 for an "electronic device for converting written music to audible sound."[83] CCRMA personnel, too, eagerly looked for commercial spin-offs from their research. Chowning, as noted, was an early and frequent patenter, and some of his intellectual property eventually became the most lucrative portion of Stanford's portfolio.

Commercial translations were not confined to patents and licenses, however. CCRMA also used its multivocal technology as a basis for cooperative research with industry, which the firms hoped might stimulate new uses (and hence markets) for their products. For example, the center submitted a 1979 grant application titled "Intelligent Systems for Music Analysis" under the NSF's university-industry cooperative research program, proposing to join with the Palo Alto–based machine-control firm Systems Control. Though Systems Control had no interest in music per se, it had much to gain from other applications of the work that happened to employ music as a context. Similarly, in 1975 James Angell (an EE faculty member) used money from Stanford's new Research Development Fund, "established to stimulate innovative research endeavors by both junior faculty and faculty who were redirecting their research interests" (i.e., faculty interested in adopting the relevant-research agenda), to support Moore's graduate work and to collaborate with Ed Taylor, an engineer at EPA Electronics. Moore, Taylor, and Angell then approached Intel, which told them "they would be happy to provide us with any amount of the componentry (especially integrated circuits) which they make that might be of use to us," presumably in hopes of finding new applications for their products.[84]

[80] Chowning, Smith, and Cohen, "Computer Music Facility" (cit. n. 47), 4.

[81] John Chowning, John Grey, James Moorer, Loren Rush, and Leland Smith, "Center for Computer Research in Music and Acoustics: Overview," 1 Dec. 1977.

[82] David Kaiser, "Cold War Requisitions, Scientific Manpower, and the Production of American Physicists after World War II," *Historical Studies in the Physical Sciences* 33 (2002): 131–59.

[83] Docket number S74-09, SUA, CCRMA ACCN 2001-262, box 7.

[84] Angell to Pat Devaney, "Re: RDF Report," 14 Jan. 1976, SUA, Richard Lyman Papers, SC 215 Series 3, box 23, folder "National Science Foundation, 3/1976–3/1977."

The most far-reaching of CCRMA's translations—at least in terms of physical distance—emerged from its collaborations with NASA. For its part, as early as 1962, NASA announced that it was "particularly desirous that the environment in which space research is conducted will be characterized by a multidisciplinary effort which draws upon creative minds from various branches of the sciences, technology, commerce and the arts."[85] Indeed, a 1972 NASA-Stanford agreement elicited proposals from "Art History, Genetics, Psychiatry, the Graduate School of Business, and Geology."[86] In that light, it is worth noting that there was an active Mars research group at the Stanford Artificial Intelligence Project, where CCRMA was based for its first years. In fact, Mike McNabb, one of CCRMA's earliest PhD students, composed a suite, *Music for Mars in 3-D*, tied to stereographic images from NASA's Mars Viking spacecraft. Another of McNabb's creations, *Invisible Cities*, was a ballet featuring an automated robot as a dancer and a computer-music soundtrack as the orchestra. McNabb's collaborators on the ballet included both the ODC/San Francisco Dance Company and the Veterans Administration Robotic Aid project, the latter of which was experimenting with how to employ robotics to assist injured Vietnam veterans.

CONCLUSION

As the other articles in this volume illustrate, music and science have a long common history. In this article, we have focused upon the organizational and institutional aspects of this relationship in a twentieth-century research university. Specifically, we have elaborated upon the ways in which Vietnam War–era debates at Stanford (and elsewhere) forged a new institutional environment that emphasized radical interdisciplinarity and relevant research. Although the activists and interests engaged in these discussions maintained significant heterogeneity, an emergent research center that was explicitly focused upon the intersection of music and science found both legitimacy and resources in this new environment. In turn, the work conducted at this center, in terms of both musical composition and technical advance, has had far-reaching impact in a variety of spheres.

Our analysis has focused particular attention on computer-music research methods and tools as multivocal technologies that span both disciplinary boundaries and areas of applied interest. In this way, the technologies themselves both reflected and facilitated the emergent institutional environment at Stanford. Even as Vietnam-era debates waned and campus interests turned firmly toward Silicon Valley entrepreneurship by the close of the century, the generality and flexibility of computer music allowed CCRMA members to orient toward start-up companies and new models of university intellectual property. Similarly, as the stigma attached to funding and projects associated with the military-industrial complex waned from the late '70s on-

---

[85] "Memorandum of Understanding between National Aeronautics and Space Administration and Leland Stanford Junior University Concerning Research Facilities Grant NsG(f) 2-62," 15 Dec. 1966, SUA, Richard Lyman Papers, SC 215 Series 1, box 125, folder "National Aeronautics and Space Administration."

[86] Jon Erickson, sponsored projects officer, to Harold Einhorn, associate vice president for academic programs, California State University, San Francisco, 22 July 1974, SUA, Richard Lyman Papers, SC 215 Series 1, box 125, folder "National Aeronautics and Space Administration."

ward, CCRMA employed its multivocal research tools to seek funding from the NSA and the Office of Naval Research.

Ultimately, our article sheds light upon the ways in which organizational and institutional arrangements can give rise to and underpin the environments in which music and science interact. From this perspective, Chowning's 1975 description of CCRMA's role—"to provide intellectual ventilation as well as coordination"[87]— might be equally apt to describe the role of institutional environments in structuring the relationship between music and science as a whole.

[87] Hertelendy, "Stanford's Musical 'Marriage'" (cit. n. 50).

Toward a New Organology:
Instruments of Music and Science

by John Tresch* and Emily I. Dolan†

ABSTRACT

The Renaissance genre of organological treatises inventoried the forms and functions of musical instruments. This article proposes an update and expansion of the organological tradition, examining the discourses and practices surrounding both musical and scientific instruments. Drawing on examples from many periods and genres, we aim to capture instruments' diverse ways of life. To that end we propose and describe a comparative "ethics of instruments": an analysis of instruments' material configurations, social and institutional locations, degrees of freedom, and teleologies. This perspective makes it possible to trace the intersecting and at times divergent histories of science and music: their shared material practices, aesthetic commitments, and attitudes toward technology, as well as their impact on understandings of human agency and the order of nature.

LOOKING BACKWARD FROM THE DIGITAL CONVERGENCE

This *Osiris* collection is part of a widespread and growing scholarly fascination for the connections between the sciences and the arts. While much of this work has focused on the visual arts, our editors have happily shifted their attention to sounds and music. The previous articles in the volume demonstrate through their diversity the rich paths of inquiry that are opened up by this focus, leading us from Young's correlations of light and sound and Helmholtz's experiments with *Klangfarbe* to the twentieth-century military-industrial university and to questions surrounding tuning, temperament, and the standardization of pitch. Our contribution picks up a theme that has loomed large over this collection in many guises; namely, that of instruments and instrumentality. In what follows, we attempt to think systematically about the very idea of the instrument and the central roles instruments play in science and music. The variable uses and interpretations of instruments are now firmly established as central topics in the history of science, and musicology appears to be heading in a similar direction, slowly overcoming the artificial divide that was created in the early twentieth century between the study of music (musicology) and the study of instruments (organology).[1] Our goal is to suggest ways of lacing together these par-

*Department of History and Sociology of Science, University of Pennsylvania, 249 S. 36th St., Philadelphia, PA 19104; jtresch@sas.upenn.edu.

† Department of Music, University of Pennsylvania, Music Building, 201 S. 34th St., Philadelphia, PA 19104; dolanei@sas.upenn.edu.

[1] According to organologist Wesley M. Oler, the term "organology" dates from Nicholas Bessaraboff's seminal work, *Ancient European Musical Instruments* (Cambridge, Mass., 1941): in a clear

© 2013 by The History of Science Society. All rights reserved. 0369-7827/11/2013-0014$10.00

allel developments, by focusing on the variable uses and modes of action of instruments in both fields.

One factor that has encouraged the growing interest in the technical infrastructures of music and science and the connections between them is the recognition that for the past two decades, just about anyone working in a field that requires communication—including scientists, visual artists, and musicians—must make use of a new set of tools: those provided by the computer. Media scholars, pundits, and precocious preteens have helped us recognize the fecundity of computers as both instruments and imaginative resources. While digital technology makes both artistic and scientific productions possible, it also provides metaphors and methods (networks and codes, bits of information crunched and transformed) that move across the alleged great divide between the two cultures. Further, the computer merges two aspects of technical instrumentation that are often seen as diametrically opposed. On one hand the computer appears as an autonomous, purely rational calculating machine, processing bits of abstracted information with inhuman rigor, speed, and accuracy. On the other, as digital media insert themselves into more and more aspects of our work, leisure, and social lives, computers seem to unite themselves fluidly and corporeally with human users, becoming emphatically "extensions of ourselves," as Marshall McLuhan characterized all media. Even at the level of hardware design, the "form factor" of digital technology makes these devices appear as supple, sticky interfaces, forming an ergonomic skin connecting us to the world.

The sudden ubiquity of this diversified but integrated interface explains some of the urgency with which both the history of science and music studies are turning to studies of instruments: What is new, many ask, and what was ever thus, in humans' engagement with instruments? The current proliferation of digital technologies has also altered our perspective on those objects that the digital replaces or imitates. All of the new forms of musical production and consumption—iTunes, iPods, YouTube, and digital editing software—reconfigure the relationship between machines, instruments, and their traditional functions. Take for example Apple's popular Logic Pro program, which allows its users the ability to control every aspect of producing music, "from the first note to the final master."[2] Musicians can record, edit, and mix music, with access to thousands of sampled instruments, pedal boards, filters, amplifiers, loops, and spaces (i.e., acoustic profiles of particular locations, from cathedrals to wine cellars). Its encyclopedic functionality blurs the distinction between physical instruments and already synthesized material.

But Logic Pro doesn't only imitate the sounds of specific instruments. Apple promises that the user can "play models of the Hammond B3 organ, the Hohner Clavinet D6, and the Fender Rhodes, Wurlitzer, and Hohner electric pianos—with all the character and quirks of the originals."[3] Its interface seeks to reproduce the physical

reference to Michael Praetorius's term *organographia* (see n. 6 below), Bessaraboff used the term to refer to "the scientific and engineering aspects of musical instruments." See Oler, "Definition of Organology," *Galpin Society Journal* 23 (1970): 170–4. The Galpin Society, dedicated to the study of instruments and still flourishing today, was founded in the same decade that Bessaraboff's study was published. On the history of scientific instruments, see Albert van Helden and Thomas L. Hankins, eds., *Instruments*, vol. 9 of *Osiris*, 2nd ser. (1994).

[2] "Logic Pro," Apple, http://www.apple.com/logicpro/top-features/ (accessed 13 Apr. 2013).

[3] Ibid. Logic Pro is one of a number of forms of software displaying this kind of "technostalgia." The new Peter Vogel CMI iPad app, for example, allows the user to recreate the experience of working with the original Fairlight Computer Musical Instrument sampler, right down to setting the fail rate of the

characteristics of mixers, pedal boards, and drum machines: the user can still turn virtual knobs and dials and press virtual buttons—a feature that is surely as much a concession to nostalgia as it is a useful way of bridging the gap between physical device and software. Logic Pro transforms its embedded objects. Within this world, a guitar is no longer a physical prosthesis for the performer, liberating forms of artistic expression while simultaneously circumscribing the performer's range through its technical specifications and limitations. In Logic Pro, the instrument becomes synonymous with its effects; it becomes, as it were, purely aesthetic—a particular texture, a timbre, as well as a cultural resonance that can be conjured up with a few clicks. We see and hear the original instrument, as a material object, in new ways. The metamorphosis from physical objects to digital plug-ins draws attention to the historicity of Logic Pro's new and old instruments; instruments have life histories, and multifaceted and changeable personalities.

Similar observations about the re-mediation of scientific instruments are inescapable. In contemporary sciences, computer-based models and simulations of processes at subatomic, molecular, physiological, cerebral, geological, or astrophysical scales illustrate a growing power of digital imitation; unprecedently massive new data sets are compiled and subjected to automatic analysis; new fields, such as bioinformatics, come into being thanks to comparatively easy access to vast computing power.[4] Ultimately, every aspect of laboratory life—from experiment, visualization, and data storage and processing to communication among lab members, including scheduling, collaboration, and publishing—has been transformed by the presence of computers. The tools that made up the previous infrastructure and material backdrop of the lab have been replicated, subtly altered, and amalgamated into layered and networked instruments of instruments.

The fact that the arts and the sciences now share so many aspects of their technical infrastructure results in isomorphic logics, shared working strategies, and common imaginaries between them. It has also made it easier to recognize earlier convergences between the fields and to find in them compelling anticipations of our own moment, leading to reflection on the ways in which instruments have facilitated intersections between the fine arts and the sciences, as well as the different ways they have been understood in relation to humans. While we could build on a growing body of work that juxtaposes the sciences with the visual arts, the instruments of science and music in many ways form a more intriguing pair.[5] While visual art objects are records of past work with paint, pencil, canvas, or stone, for there to be music at all, the instrument has to be currently at work (at least, that was the case until the arrival

---

floppy-disk reader. On the historical performance and re-creation of electroacoustic technologies, see Joseph Auner, "Wanted Dead and Alive: Historical Performance Practice and Electro-acoustic Music from Abbey Road to IRCAM," in *Communicating about Music: A Festschrift for Jane Bernstein*, ed. Roberta Marvin and Craig Monson (Rochester, forthcoming).

[4] In addition to the explosion of writings seen since 2012 on "big data" (e.g., Steve Lohr, "The Age of Big Data," Sunday Review, *New York Times*, 11 Feb. 2012), see the influential statements of Jim Gray in *The Fourth Paradigm: Data-Intensive Scientific Discovery*, ed. Tony Hey, Stewart Tansley, and Kristen Tolle (Redmond, Wash., 2009); for critical discussion, see Chris Kelty, Lilly Irani, and Nick Seaver, eds., *Crowds and Clouds*, issue no. 2 of *Limn*, http://limn.it/issue/02/ (accessed 29 Sept. 2012).

[5] For studies of correlations between visual and acoustic media in nineteenth- and twentieth-century arts and sciences, see Mara Mills and John Tresch, eds., *Audio/Visual*, vol. 43 of *Grey Room* (Spring 2011).

of automatic recording). Further, from Pythagoras onward there has been a supposition of unity between science and music. At the same time, there is reason to think of these fields as complementary: musical instruments express the inner states of the composer or the performer, moving outward from the mind to the world, while scientific instruments bring external states of the world into the consciousness of observers, moving from the world to the mind. Given the current acute sense of the ways in which new instruments of communication and representation are changing our modes of thinking, arguing, and perceiving, the time is ripe for trying to get historical perspective on these changes, by means of a comparative study of instruments in different fields.

Part of our inspiration comes from Renaissance texts such as Sebastian Virdung's *Musica getutscht und ausgezogen* (1511) and Michael Praetorius's *Syntagma Musicum* (1618).[6] These organographical treatises catalogued a wide range of musical instruments in a systematic fashion, organizing them into families according to their construction and use. The same classifying impulse lies behind the present article, but our focus is broader in that we also include scientific instruments and attend to historical differences. We are aiming at an analysis—and perhaps a new taxonomy— of the ways people (primarily in the West) have understood instruments' action and their bearing on humans. In what follows, we will sketch some of the features that might define a systematic study of the natures, uses, degrees of agency, and ends of instruments in different fields and at different times: a new organology.

If all we were after were a historical inventory of instruments, the first task would be to assemble materials. In the case of music, we would consider taxonomies, organologies, orchestration treatises, collection guides, and museum inventories. For science, we would turn to similar instrumental compendia: laboratory manuals, *cabinets de physique*, and kits for calibrating instruments and setting up laboratories, as well as the correspondence exchanged between different observatories and labs about coordinating and standardizing equipment. But we are aiming at something more. We want to think about instruments as actors or tools with variable ranges of activity, with changing constructions and definitions, and with different locations in both technical and social formations. We want to ask, What aspects of instruments have been variable (or have been seen to be), and what were the consequences of that variation? What were the larger arrangements of technology, social roles, and elements of the natural world (breath, electricity, the air, metal, biological specimens) into which particular instruments were woven? How was their action understood: Were they neutral vehicles for human intention or external nature, or did they transmute or modify the impulses they carried along? What larger projects, goals, or conceptions of either the arts or sciences, or both, did they help to articulate?

To get at these differences and similarities, we will take a lesson from a prominent focus of attention in recent history of science: the different ethical ideals at work in scientific research. In parallel with studies of "epistemic virtues" and the ethics of the knowing subject, we propose as a thought experiment—one that will surely provide more questions than answers—a comparative study of the ethics of instruments.

---

[6] Virdung's work is the earliest printed treatise on instruments; for more information, see Beth Bullard's translation, *Musica getutscht: A Treatise on Musical Instruments* (Cambridge, 1993). The term "organology" derives in part from the second volume of Praetorius's *Syntagma Musicum*, entitled *De Organographia*, which was devoted to descriptions of musical instruments.

THE ETHICS OF INSTRUMENTS

Instruments are integrated in diverse ways with human activities; they also influence understandings of human conduct and freedom. In thinking about instruments as having an ethical dimension, our approach dovetails with recent attempts to characterize moments in the history of science by the epistemic virtues that have guided the pursuit of natural knowledge.[7] Ethics, in such works, turns out to be important for epistemology: knowledge appears not merely as a set of ideas or even practices, but as a form of life, with distinct ideals, moral codes, activities, and understandings of the self.[8] This broadened conception of ethics as a form of life resonates with comparative studies and histories of the self (e.g., by Marcel Mauss, or more recently by Charles Taylor and Jerrold Siegel).[9] It also takes inspiration from the later works of Michel Foucault.

In the works published toward the end of his life, Foucault reframed ethics as essentially concerned with the self's relation to the self. He saw this relation as consisting of four dimensions: an *ethical substance*, or the part of the self understood to be addressed by ethics, including the relevant domains of activity (economic exchange, food, sex, vocation, etc.); a *mode of subjection*, or the relation of the subject to explicit codes of conduct, rules, and obligations; the *ethical work*, or the activities through which the subject is constituted; and finally the *telos*, or the ends toward which this activity is directed.[10] Foucault laid particular emphasis on the idea of ethics as an aestheticization of existence: in his interpretation of texts from ancient Greece and the Hellenistic period, he saw the care of the self (in ascetic practices such as self-examination and journal keeping) as guided by the ideal of creating a beautiful life. In late interviews and occasional pieces he presented more recent ethical constellations, such as Baudelaire's theorization of dandyism in the 1860s and attempts to refashion friendship as the basis of gay life in the 1970s, as modern updates of the view of ethics as an aesthetics of existence. Ethics was less concerned with moral proscriptions than the practices and ideals through which one constituted oneself as a free subject and as a work of art.[11]

Yet Foucault's conception of the ethical telos of ancient philosophy has been criticized as too narrow. According to Pierre Hadot, his focus on aesthetics obscured the

---

[7] See, e.g., Peter Galison and Lorraine Daston, *Objectivity* (New York, 2007); Matthew Jones, *The Good Life in the Scientific Revolution* (Chicago, 2004).

[8] For Daston and Galison, ethical conduct is "a way of being in the world, for a group or individuals" (*Objectivity* [cit. n. 7], 40).

[9] Mauss, "A Category of the Human Mind: The Notion of Person; The Notion of Self," in *The Category of the Person: Anthropology, Philosophy, History*, ed. Michael Carrithers, Steven Collins, and Steven Lukes (Cambridge, 1985), 1–25; Taylor, *Sources of the Self: The Making of the Modern Identity* (Cambridge, Mass., 1989); Siegel, *The Idea of the Self: Thought and Experience in Western Europe since the Seventeenth Century* (Cambridge, 2005).

[10] Michel Foucault, *The History of Sexuality*, vol. 2, *The Use of Pleasure*, trans. Robert Hurley (New York, 1990).

[11] "The idea of morality as obedience to a code of rules is now disappearing, has already disappeared. And to this absence of morality corresponds, must correspond, the search for an aesthetics of existence." Michel Foucault, "An Aesthetics of Existence," in *Politics, Philosophy, Culture: Interviews and Other Writings*, ed. Lawrence Kritzman (New York, 1988), 49; see also his interview entitled "Friendship as a Way of Life" in *Foucault Live: Interviews, 1961–1984* (New York, 1996), 203–7, and his essay "What Is Enlightenment?" in *The Foucault Reader*, ed. Paul Rabinow (New York, 1984), 32–50. For a more recent, hilarious theorization of dandyism, see Lord Breaulove Swells Whimsy, *The Affected Provincial's Companion*, vol. 1 (New York, 2006).

Epicurean and Stoic understanding of ascetic practices as means of situating oneself in the universal order and accessing a "cosmic consciousness" as embodied in the figure of the sage.[12] While Foucault reframed ethics as a transhistorical dandyism, Platonic and Stoic self-fashioning aimed instead at placing the individual into harmony with the order of the universe—which is not necessarily the same thing. For ancient philosophy, ethics was closely tied to cosmology.[13] We might further add the obvious point that in all standard accounts, ethics bears upon one's conduct toward family, friends, allies, and enemies; it involves the self's relation not only to the self, but to others. Following Hadot, then, we need to expand Foucault's conception of ethics to include further contexts of ethical conceptualization and activity: the self's relation to the cosmos or nature, and the self's relation to other selves.

Modified in these ways, Foucault's later works provide a useful, multidimensional framework for comparing ethical systems. But what if we were to go one step further, and apply this framework not only to humans—in their capacities as active and knowing subjects—but to the instruments with which humans engage as they create knowledge and other cultural products? What if we shift our gaze from the relations between self and self, and self and nature (as traditionally explored in ethics and epistemology), to consider the variable relations between selves and instruments, tools, and machines? What then comes to light is a framework for studying the historical variations in an ethics of instruments—a history that parallels and intersects that of ethics and of the knowing subject. A comparative ethics of scientific instruments could also be juxtaposed with the relations of selves and instruments found in fields other than science; for instance, in music.

Such a project requires us to reckon with the many different ways in which instruments, tools, and machines have been understood. We need to expand our view beyond the standard notion of the tool as utilitarian and passive, and beyond the ideal of the machine as embodying inhuman precision and standardization—the uniform, predictable, sharp-edged ideal underlying "mechanical objectivity."[14] Across time and in different contexts, instruments and machines have changed in their material configuration, their mode of activity, their relations to other objects and people, and their aims. These changes have had consequences for how humans understand themselves. Upending the view of technical progress as increasing domination over nature, for instance, Thoreau famously said, "We do not ride on the railroad, it rides upon us." Are tools understood as granting us mastery, or are they seen as reducing us to cogs in what Lewis Mumford called the "megamachine"?[15] Or, alternatively, are they seen as establishing a complicated balance, weaving us into the fabric of a second nature? The answers to these questions will depend on the era, the instrument, and the field.

To give precision and reach to the notion of an ethics of instruments, we propose the following analytical categories. Though they are inspired by the four axes that

[12] See Hadot, *Philosophy as a Way of Life* (Oxford, 1995); discussion in Arnold Davidson, "Ethics as Ascetics: Foucault, the History of Ethics, and Ancient Thought," in *The Cambridge Companion to Foucault*, 2nd ed., ed. Gary Gutting (Cambridge, 2005): 123–48.

[13] In *The Good Life* (cit. n. 7) Matthew Jones shows how this was also the case for early modern natural philosophers.

[14] See Daston and Galison, *Objectivity* (cit. n. 7).

[15] Henry David Thoreau, *Walden; or, Life in the Woods* (1854), in *A Week on the Concord and Merrimack Rivers, Walden, The Maine Woods, Cape Cod* (New York, 1985), 396; Mumford, *Technics and Human Development*, vol. 1 of *The Myth of the Machine* (New York, 1971).

made up Foucault's analysis of the self's relation to the self, we have tweaked and tuned them in order to apply them to instruments. An instrument's ethics, we suggest, may be analyzed according to the following categories:

1. The *material disposition* of the instrument: the nature and configuration of its elements, and the materials and parts that make it up. Also, and perhaps most important, this disposition is defined by which parts are seen as necessary to make the object an instrument of a certain type, and which may be varied to alter its specific action. (This corresponds to what Foucault called the ethical substance, that part of the self made an object for moral reflection.)

2. The instrument's *mode of mediation*: whether its action is considered to be autonomous or passive, modifying or transparent, hidden or visible (corresponding to Foucault's ethical activity, the work conducted to make oneself a subject of ethics).

3. The *map of mediations* of which the instrument is a part. Such maps, joining together a number of distinct elements, may be rather complex: in music they include air, sound, composers, players, other instruments, and listeners, as well as orchestration treatises and rules of composition; in the sciences they include the phenomena being investigated, the observer or experimenter, and other elements in the experimental system, as well as rules of method, laboratory protocols, scientific institutions, and patterns of moving between observation and generalization. (This category relates to what Foucault calls the mode of subjection, or the subject's relation to rules or obligations.)

4. The *telos* of an instrument's activity, or its ends (Foucault uses the same term to describe the goal of ethical work). What is the nature of the enterprise within which the instrument is deployed; what are its social contexts and uses, and the social, economic, and political relations they express, reinforce, or perhaps modify? At the level of telos we might also want to bring in broader conceptions of the goals attributed to instruments (much as Hadot suggested in his critique of Foucault mentioned above): not the instrument's relation to itself but its relation to its users and those exposed to its products, as well as its impact on the entire collective. Furthermore, we might consider the relationship an instrument is seen to entertain with the natural order, with the cosmos as a whole.

This approach means that we will be applying concepts to nonhuman objects that are usually attributed to humans. Exploring the different forms and degrees of agency attributed to instruments suggests that the qualities of sentience, activity, and intention might not always belong only to humans but also to objects often classed as inanimate, including machines and instruments.[16] Yet, as Bruno Latour has argued, an ambivalence between human agency and the agency of machines is a common theme in considerations of technological inventions of all kinds: "The label 'inhuman' applied to techniques simply overlooks translation mechanisms and the many choices that exist for figuring or defiguring, personifying or abstracting, embodying or disem-

---

[16] Such a conception builds on recent interrogations of the liveliness of matter; e.g., Bill Brown, "Thing Theory," *Critical Inquiry* 28, no. 1 (2001): 1–22; Jane Bennett, *Vibrant Matter: A Political Ecology of Things* (Durham, N.C., 2010); Manuel De Landa, *A Thousand Years of Non-linear History* (Cambridge, Mass., 2000).

bodying actors. When we say that [technologies] are 'mere automatisms,' we project as much as when we say they are 'loving creatures'; the only difference is that the latter is an anthropomorphism and the former a technomorphism."[17] If instruments are frequently accused of making humans act mechanically, why should we not take seriously instruments' oft-noted lifelike capacities?

The following pages elaborate and illustrate the four categories we have proposed as steps toward a new taxonomy—a classificatory scheme for ordering the long series of scientific instruments and the long series of musical instruments. Our examples are drawn from medieval and early modern settings as well as from the Enlightenment, romanticism, industrial modernity, and postmodernity. If this kaleidoscopic view is somewhat disorienting, so much the better: we hope that our instrument-focused approach to the conjunctions of music and science will allow readers to see (and hear) things from a new angle. At the very least, the organological framework we develop should be a provocation to thinking about distinct aspects of the instruments at work in both of these fields. In addition, we see this analysis as providing the grounds for the construction of parallel historical series for scientific and musical instruments, and, eventually, for a comparison between them. In other words, we hope to address such questions as, What do the ways in which people built, used, and thought about musical instruments in a given period tell us about the scientific instruments in the same period, and vice versa? Where do historically specific ways of thinking about scientific instruments converge with those for musical instruments? More broadly, what do those interactions tell us about the changing relationships between science and the arts, or about the changing relationships among humans, nature, and technology? While answers to such questions must lie in the future, throughout the following discussion of the four categories that make up an ethics of instruments—material disposition, mode of mediation, map of mediations, and telos—certain suggestions along these lines should be apparent, and more will be sketched in the conclusion.

MATERIAL DISPOSITION: MICROSCOPE, TELESCOPE, KEYBOARD

In examining the material disposition of an instrument, we ask, What is the instrument made of, and how are its components arranged? Its configuration of parts might change, even as the same name is applied to it. Take for example the microscope. From early designs even preceding Robert Hooke's, microscopes involved lenses and mirrors. Their aim was to enlarge visual access to the very small, and they worked by deviating rays of light to magnify the image of an object as it reached the eye. In contrast, the electron microscope, which first appeared in the 1930s, projects a ray of electrons through a thin slice of the object being observed; these are then captured on photographic film. Because electrons have much shorter wavelengths than the photons of visible light, they grasp much finer textures. Similarly, the scanning electron microscope sends a highly focused beam across an object's surface; as the electrons' energy is transformed, it is recorded as the basis for a reconstruction of spatial features. Though these instruments share a name, the electron microscope—

[17] Latour, "Where Are the Missing Masses? The Sociology of a Few Mundane Artifacts," in *Shaping Technology, Building Society: Studies in Sociotechnical Change*, ed. Wiebe Bijker and John Law (Cambridge, Mass., 1992), 225–58, on 241.

as well as nanotechnology's defining instrument, the scanning tunneling microscope (or STM)—are constructed and function according to principles quite distinct from the lenses and luminous optics of the instruments originally called "microscopes."[18]

The telescope has undergone a comparable evolution.[19] From Roger Bacon's thirteenth-century explorations with "seeing stones" to Newton's reflecting telescope, the optics of natural light and vision were put to work in making very distant objects close, again by bending rays of light to expand the visible image. Improvements in lenses and increases in scale continued apace. In the twentieth century, new telescopes appeared that no longer used visible light as their medium. In 1931, radio waves emitted from the Milky Way were detected, and the radio telescope, with its characteristic disk pointed at the sky, became a defining feature of the modern observatory. Infrared rays are tracked by telescopes usually placed on mountaintops, and ultraviolet and gamma-ray telescopes are attached to satellites; some, like the Hubble, travel untethered through space. Although these instruments often result in visual images—colored and shaded according to the conventions of romantic landscape painting—these are post hoc constructions, visual mappings of the vast quantities of data generated by automatic sweeps of the sky, given a wide berth for selection and modification by their creators.[20]

Beyond their smaller scale, cost, energy inputs, and technical complexity, the underlying physical processes at work in the early microscope and telescope were completely different from those of their more recent successors: direct ocular observation versus automatic collection of electromagnetic radiation or capture of electron beams and their subsequent statistical analysis.[21] Microscopes and telescopes each offer an example of an instrument that undergoes profound transformations at the level of its material disposition—what it is made of, which aspects of the external world it engages with, how it functions materially—while remaining an instrument of the same kind.

In contrast, the material disposition of an instrument, or some key part of it, might remain basically constant over time, while the instrument, as it combines with other elements, becomes an entirely new tool used for different functions. The keyboard, for instance, offers an example of an enduring interface that has been attached to a wide range of technical apparatuses. The basic concept—an instrument controlled by a series of levers operated by fingertip pressure—is ancient. Within music, it has been part of the organ, harpsichord, clavichord, piano, and synthesizer, as well as shorter-lived inventions whose names are largely forgotten today. Medieval key-

---

[18] Jutta Schickore, *The Microscope and the Eye: A History of Reflections, 1740–1870* (Chicago, 2007); William Croft, *Under the Microscope: A Brief History of Microscopy* (Singapore, 2006); Cyrus Mody, *Instrumental Community: Probe Microscopy and the Path to Nanotechnology* (Cambridge, Mass., 2011).

[19] Henry King, *The History of the Telescope* (New York, 2011); Alison Morrison-Low, Sven Dupré, Stephen Johnston, and Giorgio Strano, eds., *From Earth-Bound to Satellite: Telescope, Skills and Networks* (Leiden, 2011).

[20] Elizabeth Kessler, *Picturing the Cosmos: Hubble Space Telescope Images and the Astronomical Sublime* (Minneapolis, 2012).

[21] In this respect, astronomy resembles the many other fields of science that have been transformed in the last decade and a half by their heavy reliance on automatic data collection and algorithmic data mining. See the discussion in the case of meteorology and more generally in Paul Edwards, *A Vast Machine: Computer Models, Climate Data, and the Politics of Global Warming* (Cambridge, Mass., 2011), and, more programmatically, in Hey, Tansley, and Tolle, *Fourth Paradigm* (cit. n. 4).

boards were, by and large, diatonic—that is, they had seven or eight pitches to an octave. The keyboard layout we know today in the modern piano, which comprises the twelve pitches of the chromatic scale, has existed since the fifteenth century.

Throughout history, various inventors have attempted to alter the traditional design, creating keyboards that move away from C major as the center, use additional keys that allow microtonal music, or are arranged in different configurations understood to allow for easier transposition or more ergonomic playing, but such efforts have remained experimental. The interface of the keyboard has structured in profound ways inventors' conceptions of new instruments and machines. The late eighteenth and early nineteenth centuries witnessed a flurry of activity aimed at inventing an instrument to capture an idealized "voice of nature"—a sweet, nuanced sound that would resemble human singing. Benjamin Franklin's glass harmonica (invented in 1761), a series of tuned glass bowls, played directly with moistened fingertips and threaded on a rod that the performer turned with a foot treadle, spurred this activity. The shortcomings of Franklin's mechanism—its inability to play rapid passages, sluggish response time, and nasty habit of making performers sick—inspired later inventors to combine the glass harmonica's ethereal tones with a more manageable interface. And that interface was, for most inventors, the standard keyboard: from Ernst Florens Friedrich Chladni's clavicylinder to the harmonichord of Johann Gottfried and Johann Friedrich Kaufmann, instruments of this period contained wildly differing mechanisms, all accessed by means of a keyboard. It was precisely the presence of the keyboard that made these instruments accessible to the public imagination.[22] Indeed, for an instrument inventor *not* to use a keyboard for a complex new instrument required some justification. Synthesizer pioneer Don Buchla, for example, rejected using a keyboard interface for his 1970 Music Box. "A keyboard is dictatorial," he argued. "When you've got a black and white keyboard there it's hard to play anything but keyboard music. And when there's not a black and white keyboard you get into the knobs and the wires and the interconnections and the timbres, and you get involved in many other aspects of the music, and it's a far more experimental way."[23] Trevor Pinch and Frank Trocco have shown how Robert Moog also initially resisted fitting his synthesizer with a keyboard until his collaborator Herb Deutsch convinced him to reconsider.[24]

But the keyboard helped to structure nonmusical instruments as well. In the early 1840s, the interface of the first machines for mechanically composing and distributing type was directly modeled on the piano keyboard; many of these devices were called "pianotypes." While this name made the musical roots of these devices explicit, the lever systems of many early typewriters did not look much like pianos, arranging keys in circular, spherical, and semicircular layouts. However, Samuel W. Francis based the action of his typewriter on that of the mid-nineteenth-century piano, and the Remington typewriter followed this lead. Even when the keyboard did

[22] On these inventions, see Emily I. Dolan, "E. T. A. Hoffmann and the Ethereal Technologies of 'Nature Music,'" *Eighteenth-Century Music* 5, no. 1 (2008): 7–26; Myles Jackson discusses the scientific context of some of these instruments in *Harmonious Triads: Physicians, Musicians, and Instrument Makers in Nineteenth-Century Germany* (Cambridge, Mass., 2006).

[23] Quoted in Trevor Pinch and Frank Trocco, *Analog Days: The Invention and Impact of the Moog Synthesizer* (Cambridge, Mass., 2002), 44.

[24] Ibid., 59.

not directly replicate the layout of a piano, skills could transfer from one domain to the other. In a late nineteenth-century history of the still rapidly developing machine, P. G. Hubert reported that

> the most elaborate typewriter ever constructed was that made a few months ago for little Joseph Hofmann, the boy pianist. . . . After less than two hours' practice, he wrote letters in several languages to friends in different parts of the world, who might not yet know what a typewriter is.[25]

Hofmann (1876–1957) was a child prodigy and one of the great pianists of the twentieth century; his purported ability to master typing in two hours testified to his general virtuosic dexterity. But the connection existed even for nonvirtuosi. Those skilled at the piano were better positioned to develop dexterity on writing machines, a fact with important social consequences: because the pianotype was less physically taxing than previous techniques of text composition, women could take over the craft at lower wages—replacing skilled male workers.[26] Likewise, Kittler has argued that the resemblance between piano and typewriter enabled women to transition from leisured domestic music making into paid secretarial work. This connection was clear to those witnessing the birth of the female clerk in the nineteenth century. Writing in 1895, two German economists remarked:

> Today, the *typist* has evolved into a kind of type: she is generally very high in demand and is the ruling queen in this domain not only in America but in Germany as well. It may come as a surprise to find a practical use for what has become a veritable plague across the country, namely piano lessons for young girls: the resultant dexterity is very useful for the operation of the typewriter. Rapid typing on it can be achieved only through the dexterous use of *all fingers*.[27]

The history of the keyboard helps raise a series of questions about instruments' material dispositions. What conditions are necessary for a new instrument to succeed? How are the material elements of one instrument reterritorialized onto new instruments? Likewise, the history of the microscope and telescope lead us to ask, What must stay the same for an instrument to remain the same kind of thing over time? How much variation is possible within a single family of instruments, and along which dimensions? At what point have the underlying mechanics and mode of function shifted such that we can no longer call the instrument by the same name as its precursors?

MODE OF MEDIATION: WHO'S PLAYING WHOM?

In January 2011, the renowned guitarist and composer Marc Ribot held a month-long residency at Le Poisson Rouge in Manhattan. Those in attendance watched Ribot transition from charismatic virtuoso, single-handedly commanding attention, to thoughtful ensemble member, self-effacingly attuned to the needs of the group. Even

[25] Hubert, "The Typewriter; Its Growth and Uses," *North American Review* 146, no. 379 (1888) [unpaginated].

[26] François Jarrige, "Le mauvais genre de la machine," *Revue d'histoire moderne et contemporaine* 54 (2007): 193–221.

[27] Quoted in Friedrich A. Kittler, *Gramophone, Film, Typewriter*, trans. Geoffrey Winthrop-Young and Michael Wutz (Stanford, Calif., 1999): 194–5.

more striking, however, was his relationship to his instrument: at times the guitar appeared as a seamless extension of his body, utterly under his control; other times, he made the separation between performer and instrument clear, cradling the guitar in his arms like he was embracing another being. He often seemed to meet the instrument halfway, grappling with the guitar's own agency. This fluidity calls attention to the different modes of mediation that may be at work when a user interacts with an instrument. Any history of instruments must also account for their changing forms of agency and visibility. Do we understand a given instrument within a given context as passive and obedient to the hand of the user, or does it appear as active, occasionally beyond the user's complete control? How much does the instrument control the user, and vice versa?

Different historical periods and cultural settings have offered starkly contrasting answers to these questions. In the case of the late Enlightenment and the romantic era, intriguing connections between the agency of musical and scientific instruments abounded. The early part of the career of the famed astronomer William Herschel (1738–1822) was spent amid not telescopes but musical instruments. Born in Hanover, he worked as an oboist and violinist for the Hanover Guards. Abandoning the post and its harsh conditions, he performed and composed music outside London, eventually securing a position as organist at the Octagon Chapel in Bath in 1766. He composed over two dozen symphonies, numerous concertos, chamber music, music for keyboard, and vocal works. A versatile instrumentalist, in January 1787 he performed concertos in Bath for both oboe and violin and a sonata for the harpsichord. The autograph manuscripts of his oboe concertos contain an unusual number of nuanced and precise performance indications.

During the 1770s, Herschel's astronomical interests grew. He began to construct his own telescopes and stands, meticulously grinding his own lenses; he collaborated with his sister Caroline, whose childhood encounter with typhus left her unmarriageable in the eyes of her parents.[28] In 1781, Herschel realized that the sidereal object he had thought was perhaps a comet was actually a planet. He named the planet Georgium Sidus; only later (perhaps unfortunately) was it renamed Uranus. After this discovery, King George awarded Herschel an annual salary, allowing him to concentrate entirely on astronomy. Yet elements of his musical career persisted. Connecting his musical and astronomical endeavors was the notion of practice. In a letter from 1782, Herschel wrote:

> I do not suppose there are many persons who could ever find a star with my power of 6,450, much less keep it, if they found it. Seeing is in some respects an art, which must be learnt. To make a person see with such a power is nearly the same as if I were asked to make him play one of Handel's fugues on the organ. Many a night have I been practising to see, and it would be strange if one did not acquire a certain dexterity by such constant practice.[29]

Looking through a telescope required the same kind of dedicated practice as the performance of fugues at a keyboard. Herschel's skill in constructing instruments rein-

---

[28] See Michael Hoskin, *Discoverers of the Universe: William and Caroline Herschel* (Princeton, N.J., 2011).

[29] Herschel to Dr. William Watson, 7 Jan. 1782, in Constance Lubbock, *The Herschel Chronicle: The Life-Story of William Herschel and His Sister Caroline Herschel* (New York, 1933), 99–101, on 101.

forced his artful seeing. His telescopes were neither wholly active nor entirely passive instruments: each new and larger telescope that he constructed altered what he
saw and how he understood what he saw (a fact crucial to his understanding of nebulae as progressively developing cloudlike material).[30]

Herschel's view of instrumental work as an intimate, skillful collaboration was
theorized in quasi-biological terms a generation later by the Prussian polymath
Alexander von Humboldt, who saw scientific instruments as outgrowths of human
capacities:

> The creation of new organs (instruments of observation) increases the intellectual and
> not infrequently the physical powers of man. More rapid than light, the closed electric
> current conveys thought and will to the remotest distance. Forces, whose silent operation
> in elementary nature, and in the delicate cells of organic tissues, still escape our sense,
> will, when recognized, employed, and awakened to higher activity, at some future time
> enter within the sphere of the endless chain of means which enable man to subject to his
> control separate domains of nature, and to approximate to a more animated recognition
> of the Universe as a Whole.[31]

The instrument merges with and extends our senses, and unites us in thought and in
action with the cosmos. Both Herschel and Humboldt presented the use of instruments as a "dance of agency" akin to that which Andrew Pickering describes as "the
mangle of practice."[32]

Yet the reciprocal agency at work in the mechanical romanticism of Herschel and
Humboldt is certainly not the only way in which the activity of instruments has been
understood. One classic understanding is of the instrument as entirely passive. In
the case of music, one thinks of instruments perfectly responsive to the impulses of
performers. Science maintains the dream of the "transparent" instrument, one that
renders, in a known and entirely predictable way, the givens of nature. In the late
eighteenth century, the balance, as used across diverse sciences (in the form of scales,
calorimeters, and economic balance sheets), implied an equivalence of inputs and
outputs, with no interference by the instrument.[33] The nineteenth-century scientific
ideal of "mechanical objectivity," which focused on devices that bypassed the pecu-
liarities of human intentions and will, likewise assumed that machines would passively transmit the world out there, a mediation that in no way altered the phenomenon being observed.

At the opposite extreme we find expressions of concern that the unsought agency
of an instrument will upset the delicate equilibrium between observer and world: the
history of twentieth-century science evidences constant worry about the artifacts and
noise produced by observing machines, from bubble chambers to functional magnetic resonance imaging machines (fMRIs)—traces not of the object being studied but of the experimental apparatus itself. In music, the recurring figure of a self-

---

[30] See Simon Schaffer, "Herschel in Bedlam: Natural History and Stellar Astronomy," *British Journal for the History of Science* 13, no. 3 (1980): 211–39.

[31] Humboldt, *Cosmos: A Sketch of a Physical Description of the Universe*, vol. 2, trans. E. C. Otté
(London, 1849), 742. Humboldt presented instruments as active organs throughout *Cosmos*; he described the telescope as having "exercised an influence similar to some great and sudden event" (739).

[32] Pickering, *The Mangle of Practice: Time, Agency, and Science* (Chicago, 1995).

[33] Norton Wise, "Mediations: Enlightenment Balancing Acts; or, The Technology of Rationalism,"
in *World Changes: Thomas Kuhn and the Nature of Science*, ed. P. Horwich (Cambridge, Mass.,
1993), 207–56.

playing instrument frequently involves diabolical deals, magical flutes, and uncanny hunting horns; the negative associations of this myth are active today in denunciations of lip-synched performances as soulless and robotic.[34] Daft Punk—a French electronic dance music duo who perform in helmets making them look like robots—make intriguing sport of this topos. Their 2006/7 *Alive* tour featured synchronized light patterns, beats, and vocoder phrases emanating from a giant illuminated pyramid, a spectacle whose preprogrammed nature in no way diminished the crowds' delight. Charisma and spontaneity were paradoxically presented as being on the side of the machine, provoking an ironic recognition that all participants in the event — performers, audience members, and technical apparatus—were "human after all," as in the title of Daft Punk's 2005 album.

In performance, agency may also shift among different instruments working together. In a jazz session's call-and-response and turn taking, one instrumentalist moves forward as soloist, improvising freely, while the other members of the combo maintain a steady background; the soloist retreats and a member of the background comes forward to work a new melodic line. The cuing required for taking turns involves an intersubjectivity that is all the more complex the more an ensemble tends toward the outer limits of free jazz, where all improvise at once.[35] In a scientific experiment, the management of agency is also at a premium. The ideal of a statistical experiment, for instance, involves control over the agency of all elements except one: the independent variable. This widely held conception of the mode of mediation in the elements of an experimental system (with multiple controlled variables, a single independent variable, and a single dependent variable) arises from the conjunction of early nineteenth-century writings on scientific method with late nineteenth- and early twentieth-century statistics. A new organology would explore the historicity of this and related formulations of instruments' modes of mediation, examining the prevailing understandings and evaluations of instruments' degree of agency in a given period and field.

MAP OF MEDIATIONS: EXPERIMENTAL SYSTEMS, ORCHESTRATION

So far, we've considered the elements that make up an instrument and their different possible arrangements; we've thought about whether instruments are seen as acting under their own impulses, or if they are entirely dependent on an outside agent, be it the human using them or some other force. A third dimension along which we can analyze and compare instruments in action is the location they occupy with regard to other elements: other instruments, the range of users involved with them, their objects, their audiences. In parallel with Foucault's notion of a mode of subjection—the subject's relation to rules and external obligations—for an instrument, this is a map of mediations.

[34] See, e.g., E. T. A. Hoffmann's short story "Der Sandmann" (1816) as translated by E. F. Bleiler in *The Best Tales of Hoffmann* (New York, 1979), 183–214. The story involves Olimpia, a piano-playing and deceivingly lifelike automaton who leads to the horrific downfall of the protagonist; Lisa Morales, "Ashlee Simpson and That Lip-Syncing Feeling," *Washington Post*, 26 Oct. 2004.

[35] Ornette Coleman, *Free Jazz: A Collective Improvisation*, Atlantic SD 1364, 1961, 33⅓ rpm; John Litweiler, *The Freedom Principle: Jazz after 1958* (New York, 1984). See also Arnold Davidson's keynote lecture, "Improvisation as Ethical Form," for the symposium "Improvisation and Ethics: A Conversation," held at Columbia University 13 Nov. 2008. The talk is available through Jazz Studies Online: http://jazzstudiesonline.org/?q=node/987 (accessed 29 Sept. 2012).

The history of science of the past thirty years has made the map of mediations articulated by an instrument a central topic. The ways in which points of a map may be adjusted, black-boxed, and revised are signaled clearly by Hans Jörg Rheinberger's "experimental system," which he defines as a "basic unit of experimental activity combining local, technical, instrumental, institutional, social, and epistemic aspects."[36] Many of the field's terminological innovations gesture in the same direction: the Foucauldian *dispositif* or apparatus, the "experimental assemblage," the "network." A handful of such systems have become paradigmatic. For instance, in seventeenth-century England, well-heeled gentlemen of the Royal Society gathered to observe facts produced by carefully regulated instruments, including Robert Boyle's air-pump, in experiments in Gresham College choreographed by Boyle. The scenography separated visible work performed by Boyle and other natural philosophers from the labor of "invisible technicians," the "lowly mechanics" who prepared the rooms, built and maintained much of the equipment, and assured the successful conduct of experiments. The diagram of the elements linked by the air pump thus included humans—the fellows of the Royal Society; the mechanics; the elements brought into contact with it (barometers, mercury thermometers, candles, and doomed doves); the architectural setting; and the extension of this time and space into other times and spaces through the virtual witnessing made possible by the reporting in the *Proceedings of the Royal Society*.[37] The instrument can be seen as the cornerstone of all these relationships. Similarly celebrated is the laboratory of Louis Pasteur, with its exemplary experiment involving two swan-necked flasks containing beef broth. After the swan neck was broken off of one of them and exposed to the air, tiny organisms grew in that solution, but not in the flask whose neck remained unbroken. The demonstration and explication of these flasks helped make Pasteur's lab "an obligatory point of passage" linking hygienists, the state, farmers, vintners, physicians, and thirsty schoolchildren.[38] Likewise, the arrangement of Franciscus Cornelis Donders's laboratory to correlate reaction times to cognitive events formed a map that articulated "partial objects derived from the experimenter and the experimental subject (eyes, hands, voices, etc.), more or less isolated organs (hearts, lungs, muscles, nerves, etc.), energy sources, styli, sooted paper, tables, notes, and publications."[39] As these familiar examples show, a map of mediations is not static but rather a chart of movement, a symphonic or choreographic transcription.

In music, two terms exist for this kind of complex organization: instrumentation and orchestration. The former describes the act of distributing musical material to instruments generally; the latter term is used in the specific context of the particular assemblage we know as an orchestra. The instrumentation or orchestration of a particular piece of music depends upon numerous factors, ranging from the aesthetic to

---

[36] Rheinberger, *Toward a History of Epistemic Things: Synthesizing Proteins in the Test Tube* (Stanford, Calif., 2007), 238. Rheinberger's sense of the importance of procedural rules in an experimental system, as well as his division between "technical objects" (elements that remain fixed) and "epistemic things" (those that are subject to variation) suggests the close connection between a map of mediations and the specific mode of mediation of its elements.

[37] Steve Shapin and Simon Schaffer, *Leviathan and the Air-Pump: Hobbes, Boyle, and the Experimental Life*, 2nd ed. (Princeton, N.J., 2011).

[38] Bruno Latour, *The Pasteurization of France*, trans. Alan Sheridan and John Law (Cambridge, Mass., 1988).

[39] Henning Schmidgen, "The Donders Machine: Matter, Signs, and Time in a Physiological Experiment, ca. 1865," *Configurations* 13, no. 2 (2005): 211–56, on 211.

the institutional. That is to say, the fact that we can speak of "orchestration" implies not just the existence of orchestras but also standardization in what those ensembles are. The term itself was first used in the nineteenth century and reflected the consolidation of the modern orchestra as a musical body, an institution, and a concept (roughly contemporary with an emerging standardization of scientific laboratories and globally distributed observatories).[40] This transformation radically altered how composers could interact with instruments and the ways in which they wrote for them. In the late seventeenth century, wind instruments frequently doubled already existing string parts; in the early eighteenth century it was not uncommon for composers to write a generic upper line that either a flute or oboe could play. Over the course of the eighteenth century, the ensemble expanded to include more instruments and became relatively uniform across geographical regions. This stability meant that composers writing for the orchestra could write instrument-specific parts: parts for flutes, clarinets, and oboes would take advantage of the unique qualities of those particular instruments. By the early nineteenth century, it was expected that a piece of music for an orchestra would engage with the different voices of the ensemble in sensitive and effective ways. This involved knowing the technical capacities and limitations of each instrument and the effects that they could create when combined with each other. With this new way of engaging with the orchestra came a new kind of compositional aid: the instrumentation treatise. These documents were distinctly different from the organological treatises of the previous centuries, since they addressed both the collective practice of writing for instrumental combinations as well as the specific qualities of individual instruments. Earlier treatises on instruments, by contrast, had sought to classify and categorize instruments, often emphasizing their similarities rather than their differences.

Though instrumentation treatises existed in the eighteenth and early nineteenth centuries, these have been largely overshadowed by Hector Berlioz's *Grand traité d'instrumentation et d'orchestration moderne*.[41] Berlioz, in lively language, instructs his reader about the personality and characters of each instrument and offers examples of their most effective uses, drawn from contemporaries such as Giacomo Meyerbeer as well as his precursors (Gluck, Beethoven, etc.). But crucially, Berlioz also discusses the art of combining instrumental sonorities, imagining what an ideal orchestra would look like (the answer involves over eight hundred instrumentalists and singers) and what sort of powerful and otherworldly sonic combinations could be achieved by such an ensemble. Just as important as understanding the individual instrumental characteristics and their combinations was knowing how to control them. A large ensemble risked devolving into mere noise, without the intervention of an all-powerful composer-conductor who could coordinate the myriad parts of the ensemble (Berlioz famously likened the orchestra to an enormous keyboard instrument upon which the conductor played, and fantasized about synchronizing players via electrical telegraph). But Berlioz's conception of orchestral music went beyond instruments and the conductor: equally crucial were the space in which the orchestra

[40] On the history of the orchestra as an institution, see John Spitzer and Neal Zaslaw, *The Birth of the Orchestra* (Oxford, 2004). On the history of the idea of orchestration, see Emily I. Dolan, "Haydn's Creation and the Work of the Orchestra," *19th-Century Music* 34, no. 1 (2010): 3–38.

[41] Originally published in Paris, 1844; revised 1855; translated by Hugh Macdonald as *Berlioz's Orchestration Treatise: A Translation and Commentary* (Cambridge, 2002).

played (outdoor orchestral music, he argued, didn't really exist) and the arrangement of the players in relation to the audience.

Berlioz's treatise therefore alerts us to questions we might ask about instrumental collectives in both music and science: How do different instruments within a group relate to each other? How are they coordinated? What sort of spaces do they require? What institutions support the ensemble? Since the advent of electronic music in the 1920s, many issues surrounding instrumentation and orchestration have transmuted into questions concerning the arrangement of studios and experimental music labs; now, on top of asking about the relationship between individual instruments, we can ask additional questions: How is a particular studio constituted? What is the relationship between people developing new technologies and those producing music with those technologies? How do scientific and musical practices merge and distinguish themselves? In their article in this volume, Cyrus Mody and Andrew Nelson explore precisely these sorts of questions; indeed, their study of the institutional and political contexts of the Center for Computer Research in Music and Acoustics at Stanford traces a map of mediations very much in the sense we are proposing.

THE ENDS OF INSTRUMENTS

To what ends are instruments used? This question has already lurked behind many of the above discussions. The answer may seem obvious in the case of musical instruments: a musical instrument is intended to play music. But even beyond the vast range of uses and meanings of musical performance, it's also true that musical instruments have at various points been employed to nonmusical ends. Stepping beyond our limited Western focus for a moment, consider the *shakuhachi* flute, made from bamboo root. An instrument capable of extreme nuance and subtle shading, the shakuhachi flute is associated with a wide musical repertoire. In the Edo period, the instrument was used by the *komuso* monks, in the Fuke sect of Zen Buddhism, who played it as a means of breathing meditation. But these flutes served another purpose as well: shakuhachi flutes were also taken up by *ronin*, samurai warriors who lacked masters and were therefore forbidden to wear swords. Many became—or disguised themselves as—monks and carried and played flutes: the hefty, club-like flute in their hands became a useful and discreet weapon.

Even if we restrict our investigation of musical instruments to more traditional musical performances, we can still find a variety of intended ends. One use of trombones in baroque-era sacred music was as an instrument of coordination. The trombone could double vocal parts and thereby help the singers keep their pitch. In the late Enlightenment, we find a new use: in Mozart's *Don Giovanni*, the trombone enters as the sound of sacred authority, signaling the ghostly presence of Il Commendatore, whom the lascivious Don kills at the beginning of the opera. The trombone's meaning in Mozart's opera relied on its earlier function within the church; it was strongly associated with a sacred context, even if within that context it was not used to signal divine might.

Comparably, one historian of scientific instruments has jokingly diagnosed what he calls "the astrolabe syndrome": the difficulty of finding interesting things to say about the instruments most commonly displayed in science museums.[42] But as far

---

[42] Jean François Gauvin, "The Astrolabe Syndrome," *JFG: Carrière et vie professionelle* (blog), 9 Mar. 2009, http://jfgauvin2008.wordpress.com/2009/03/09/the-astrolabe-syndrome/.

as an instrument's telos goes—the overall projects within which it is inscribed, the ways in which it articulates humans' relations to each other and to the cosmos—astrolabes could hardly be more compelling. The fact that there are so many of them, making up a significant proportion of many collections, underlines the wide availability of this instrument. In the early modern period, astrolabes were of use to travelers, often on board ships driven by the tangle of motivations science historians have come to know well: the acquisition of commodities and specie, knowledge of the vast variety in the flora and fauna of diverse regions of the world, enhancement of the standing of the sovereign sponsor, the glorification of God and the spreading of his word, the extension of empire through trade, conquest, and settlement. Further, a great number of the historical astrolabes in science museums are of Islamic origin. Asking about *their* telos leads to consideration of Islamic theories of the four winds; the importance of natural philosophy as a means of worship; and the place of astrology in medieval Islamic courts. More vividly, the number and distribution of small, portable astrolabes, and the prominent markings on them that indicated the *Qibla*—the direction of Mecca—show them as tools of social coordination across vast distances. Five times a day, in a wave across the planet that follows the movement of the sun across the sky, the faithful conduct their prayers, aligned to a single point in space, thanks to this handy tool.[43]

Similarly, tracing the different projects within which an instrument is deployed over time may illuminate large-scale structural transformations in the aims of musical composition and performance. The organ, for instance, has a long-standing association with churches, cathedrals, and the liturgy; in the seventeenth century, the Jesuit natural philosopher Marin Mersenne investigated its mechanical properties as part of a project uniting worship with empirical and rational study: "Organs were the reification of Mersenne's universal harmony, a harmony juxtaposing the spiritual and the worldly, the music of pure consonances with the levers, gears and bellows of a mechanical device. To worship God while listening to the music of an organ or to discover God's natural creations by means of the latter's mechanical parts was not that incongruous to someone like Mersenne."[44] In the nineteenth century, Meyerbeer heightened the impact of his opera *Robert le diable* by employing an organ in the opera hall, a potentially risqué displacement from worship to entertainment.[45] A similar journey was traced in the twentieth century, when the Hammond organ, developed as a less expensive and smaller replacement for church organs, drifted from gospel to blues and jazz, due in large part to "The Incredible" Jimmy Smith's use of it on Blue Note albums including *The Sermon!* of 1958. The Hammond's overdriven sound powered Bob Dylan's momentous switch from folk to electric rock in 1965's "Like a Rolling Stone"; it became an evocative, vibratory mainstay of psychedelic

[43] Sarah Schechner Genuth, "Astrolabes: A Cross-cultural and Social Perspective," introduction to *Western Astrolabes*, by Roderick Webster and Marjorie Webster (Chicago, 1998); David King, *Astronomy in the Service of Islam* (Hampshire, 1993); "An Islamic Astrolabe," on the Starry Messenger website of the Department of History and Philosophy of Science, Cambridge University, http://www.hps.cam.ac.uk/starry/isaslabe.html (accessed 4 Apr. 2013).

[44] Jean-François Gauvin, "Music, Machines, and Theology (II)," *JFG: Carrière et vie professionelle* (blog), 15 Mar. 2009, http://jfgauvin2008.wordpress.com/2009/03/15/music-machines-and-theology-ii/.

[45] Emily I. Dolan and John Tresch, "'A Sublime Invasion': Meyerbeer, Balzac, and the Opera Machine," *Opera Quarterly* 27, no. 1 (2011): 4–31.

rock of the 1960s and '70s, and returned in the 1980s and '90s as a favored texture in acid jazz and hip-hop, an aural nod to soulful precursors.

Scientific instruments also serve many ends. They may be used in exploratory basic research or routine industrial production; like the thermometer, insulin meter, or calculator, they may be part of everyday domestic life. These differing modes of use make them part of different maps of mediation and connect them to different ends. In the eighteenth century, the instruments associated with Newtonian mechanics, including telescopes, Atwood machines, and electrical machines, were used to investigate properties of nature; they could also publicly demonstrate the truths of natural philosophy. These distinct modes and maps, however, shared a common telos: confirming Newtonian laws of nature as the basis of God's stable universe, and establishing natural philosophers as privileged interpreters of both divine and temporal laws.[46] In the nineteenth century, one could consider the role played by telescopes—John Herschel's in South Africa or Lord Rosse's in Parsonstown—in debates about natural theology, evolution, and multiple worlds, including the famous "moon hoax." Later in the century, French popular astronomer Camille Flammarion's telescope was a gateway to wonder for a mass audience. By the twentieth century the complex instruments of the observatory were put to work, worldwide, for national security, the space race, and the search for extraterrestrial life.[47] Whether we discuss church organs or telescopes, tracing instruments' changing aims, and the ways they articulated humans' relations to nature and to technology, may offer a revealing, perhaps counterintuitive shorthand for the technical and cultural histories of music and science over several centuries.

CONCLUDING INTERACTIONS

Exploring the history of music and science along the four dimensions we have sketched—material disposition, mode of mediation, map of mediations, and telos— aims to make visible the changes as well as the continuities in the ways in which instruments have been used and understood. In our scattershot examples, we have encountered several moments where the history of science and the history of music intersected. The ultimate aim of this study, of course, would be to make visible the historical patterns of these intersections. While we are still far short of an organological précis (or epic) of this sort, a handful of inflection points seem particularly promising for further study.

For starters, we might consider the long-standing close relation between music, astronomy, arithmetic, and geometry inscribed in the medieval quadrivium, united by Pythagorean conceptions of form and relation. At the same time, we could explore the variable relationship between the sciences of the quadrivium as abstract theories and the practices that accompanied them: these liberal arts concerned with form were traditionally set above and apart from the lowlier mechanical arts that concerned instruments and machines. At what points did the status of mathematical instruments and musical instruments change, and how did these changes correlate with

---

[46] Simon Schaffer, "Machine Philosophy: Demonstration Devices in Georgian Mechanics," in Van Helden and Hankins, *Instruments* (cit. n. 1), 157–82, on 160.

[47] See Morrison-Low et al., *From Earth-Bound to Satellite* (cit. n. 19).

changes in the theoretical sciences? For instance, a defining feature of early modern science was the shift of natural philosophy from a primarily contemplative domain to a field dependent on instruments and experimentation.[48] Might there be interesting comparisons between, on one hand, the relation of music as liberal art to music as performance, and, on the other, the relation of the sciences of arithmetic, geometry, and astronomy formerly associated with the quadrivium to the technical practices of mixed mathematics and experimental natural philosophy? How did the rising status of the mechanical arts in natural philosophy—as in Galileo's prominent use of the telescope—affect conceptualizations of musical instruments in early modern music theory?

Moving forward in time, we have elsewhere examined continuities and breaks between Enlightenment and romantic-era uses and representations of both scientific and musical instruments.[49] Along with several contributors to this volume, we might consider the intersection of the musical and scientific instruments in Helmholtz's laboratory, his studies of the ear and the eye, and his artificial production of stimuli to affect them. Focusing on the same period, however, we would further have to consider the ways in which understandings of music and science bifurcated. Baudelaire's famous attack in 1859 on photography as anathema to true art—which he believed to be concerned only with the imagination—was arguably embodied in Wagner's choice at Bayreuth to make the orchestra invisible in front of the stage: audiences might be distracted from the spiritual and idealist transports of his music and spectacle if they were forced to witness the instruments that made them possible. Later, the futurists' love of technology led them to break from musical idealism as radically as possible, with Luigi Russolo's *Art of Noise* manifesto: "We certainly possess nowadays over a thousand different machines, among whose thousand different noises we can distinguish. With the endless multiplication of machinery, one day we will be able to distinguish among ten, twenty or thirty thousand different noises. We will not have to imitate these noises but rather to combine them according to our artistic fantasy."[50] This ambition—and this new alliance between music and other technologies—continued into musique concrète. As we suggested at the beginning of this article, the tendency seems to have reached a new plateau in contemporary music: the MIDI sampler and subsequent music-composition programs that reproduce the waveforms of all existing musical instruments transform incidental sounds captured from any setting and produced by any means into musical instruments. The electronic archive of musical sounds becomes the potentially infinite supply source for cut-and-paste compositional practices, at the same time as scientific discovery is reconceived, across disciplines, as the automatic, algorithmic mining of vast, delocalized databases.

Such would be some of the turning points signaled in a longer story that could be

---

[48] Peter Dear, "What Is History of Science the History *Of*? Early Modern Roots of the Ideology of Modern Science," *Isis* 96 (2005): 390–406; see also Lissa L. Roberts, Simon Schaffer, and Peter Dear, eds., *The Mindful Hand: Inquiry and Invention from the Late Renaissance to Early Industrialisation* (Amsterdam, 2007).

[49] Emily I. Dolan, *The Orchestra Revolution: Haydn and the Technologies of Timbre* (Cambridge, 2013); John Tresch, *The Romantic Machine: Utopian Science and Technology after Napoleon* (Chicago, 2012).

[50] Russolo, *The Art of Noise: Futurist Manifesto, 1913*, trans. Robert Filliou (New York, 2004), 12.

woven out of the materials furnished by a more complete organology of music and science. We conclude with a more modest tale, one from a period when the techniques of beautiful sound and natural knowledge were also closely entwined and promised new experiences and a new framing of the world. Back in the 1770s, one of the founders of musicology, Charles Burney, made a number of trips around Europe to collect material for his famed four-volume *History of Music*. He visited church organs, encountered street music, went to the opera, and heard a variety of orchestras perform. At the same time, he carried out acoustical experiments, testing the properties of the echo in the Villa Simonetta on the outskirts of Milan; he sought out and admired a grand orrery built by Matthew Hahn; and he visited the Jesuit scholar and astronomer Roger Boscovich and the physicist Laura Bassi, both of whom demonstrated for him their instruments and machines, much to Burney's delight.

For Burney, to understand music *and* physics, opera *and* astronomy, one had to have firsthand experience with the instruments involved in their execution and performance. Burney further declared that with more time he would have gone beyond musicology to create a journal dedicated to "the present state of arts and sciences, in general."[51] A new organology could underwrite such a project, allowing it to embrace a wider history, curating, concretizing, and animating it with the multifarious objects that reside (though often far apart) in museum collections. Now, at a moment when all sounds and phenomena are portrayed as collapsing into a single medium of easily convertible data, music scholars and historians of science can gain by retrieving some of Burney's avid, eye- and ear-witness attention to the concrete spaces, routines of practice, community formations, and creative lines of flight in both musical and scientific forms of life. In exploring what we have been calling the ethical specificities of the instruments that hold such arrangements together, they can begin once more to measure out the discords and harmonies between these fields.

[51] Charles Burney, *The Present State of Music in Germany, the Netherlands, and United Provinces*, 2 vols. (London, 1775), 2:332.

Notes on Contributors

Emily I. Dolan is Associate Professor of Music at the University of Pennsylvania. Her first book, *The Orchestral Revolution: Haydn and the Technologies of Timbre* (Cambridge, 2013), explored the birth of a new form of attention to sonority, uncovering the intimate relationship between the development of modern musical aesthetics and the emergence of orchestration. She is currently working on a project on the instrumentality of music that delves into ideas of instrument invention and classification, historiography, and the birth of modern organology.

Daniel Gethmann is Assistant Professor at the Institute of Architectural Theory, History of Art, and Cultural Studies at Graz University of Technology. His research focuses on the fields of auditory culture, media theory, and the history and theory of cultural techniques. He is the author of *Das Narvik-Projekt: Film und Krieg* (Bonn, 1998) and *Die Übertragung der Stimme: Vor- und Frühgeschichte des Sprechens im Radio* (Zurich, 2006).

Elfrieda Hiebert (1921–2012) was Director of the Chamber Music Program at Mather House, Harvard University, from 1976 to 2005. She also taught piano to Harvard students and continued coaching chamber music until 2009. As a Fred Fellow at the University of Wisconsin–Madison, she was awarded a PhD in Musicology with an additional concentration in chamber music performance mentored by Eva Badura-Skoda and Rudolf Kolisch.

Julia Kursell is Professor of Historical Musicology at the University of Amsterdam. She previously worked as a research scholar at the Max Planck Institute for the History of Science. She is the author of *Schallkunst: Eine Literaturgeschichte der Musik in der russischen Avantgarde* (Vienna, 2003) as well as numerous articles on the history of the auditory physiology and psychology of music, the history of auditory media, and twentieth-century composition. Currently, she is completing a book manuscript on Hermann von Helmholtz's experimental musicology.

Cyrus C. M. Mody is Assistant Professor of History at Rice University. His first book is *Instrumental Community: Probe Microscopy and the Path to Nanotechnology* (Cambridge, Mass., 2011). He is currently working on a second monograph, "The Long Arm of Moore's Law: Microelectronics and American Science, 1965–2005."

Andrew Nelson is Assistant Professor of Management and Bramsen Faculty Fellow in Innovation, Entrepreneurship, and Sustainability at the University of Oregon. He received his PhD from Stanford. His research focuses on the diffusion and commercialization of university-based research in computer music, green chemistry, and biotechnology.

Peter Pesic is Tutor and Musician in Residence at St. John's College in Santa Fe, New Mexico. He is the author of several books, including *Abel's Proof: An Essay on the Sources and Meaning of Mathematical Unsolvability* (Cambridge, Mass., 2003), and *Sky in a Bottle* (Cambridge, Mass., 2005), and has edited works by Planck, Maxwell, Gauss, and Weyl for Dover Publications and Princeton University Press. He has been awarded the Peano Prize and is a Fellow of the American Association for the Advancement of Science and of the American Physical Society, as well as a Guggenheim Fellow.

Sonja Petersen is Academic Researcher and Teacher in the Department of History, Section for the History of the Impact of Technology, University of Stuttgart. She received her PhD in 2010 in History of Technology at Darmstadt University of Technology and worked from 2010 to 2011 as Guest and Visiting Professor in the Department of Product Design at the Offenbach Academy of Art and Design. Her current research interests are the history of private energy consumption and the meaning of writing and drawing in the creation and acquisition of knowledge.

Armin Schäfer is Professor of German Literature and History of Media Culture in the Department of Cultural and Social Sciences at the University of Hagen.

Henning Schmidgen is Professor of Media Aesthetics at the University of Regensburg, Germany. He just finished a comprehensive account of physiological and psychological time experiments in the nineteenth and early twentieth centuries. His recent books include *Die Helmholtz-Kurven: Auf der Spur der verlorenen Zeit* (Berlin, 2009) and *Bruno Latour: Zur Einführung* (Hamburg, 2011).

Bernhard Siegert is Gerd Bucerius Professor for the History and Theory of Cultural Techniques at Bauhaus University, Weimar, Germany, and, since 2008, one of the two directors of the International Research Center for Cultural Technologies and Media Philosophy in Weimar. He has published numerous books and articles in the fields of literary studies, media theory and media history, the history of science, and art history, and is coeditor of the *Archiv für Mediengeschichte* and the international journal *Zeitschrift für Medien- und Kulturforschung*.

John Tresch is Associate Professor in History and Sociology of Science at the University of Pennsylvania. His book *The Romantic Machine: Utopian Science and Technology after Napoleon* (Chicago, 2012) showed how new devices associated with steam, electricity, light, and life were woven into projects for reorganizing both society and the cosmos in the years before the Revolution of 1848. He is currently working on a book on nineteenth-century American science and the inventions and discoveries of the author and West Point–trained engineer Edgar Allan Poe.

Axel Volmar is Postdoctoral Research and Teaching Fellow in the Department for Media Studies at Siegen University, Germany. He recently finished his doctoral dissertation, which reconstructs the auditory culture of science since 1800. Volmar has published various articles in the history of science, sound studies, and media studies, and is the editor and coeditor of several collected volumes on auditory culture and the temporality of media technologies, including *Zeitkritische Medien* (Berlin, 2009), *Das geschulte Ohr: Eine Kulturgeschichte der Sonifikation* (with Andi Schoon; Bielefeld, 2012), and *Auditive Medienkulturen: Techniken des Hörens und Praktiken der Klanggestaltung* (with Jens Schröter; Bielefeld, forthcoming).

Roland Wittje is Lecturer in History of Science at the University of Regensburg, Germany. He has published on the history of nineteenth- and twentieth-century physics, including nuclear physics in Norway, the history of teaching science, and university heritage, and is coeditor of *Learning by Doing: Experiments and Instruments in the History of Science Teaching* (with Peter Heering; Stuttgart, 2011). His current research project focuses on the transformation of acoustics from the nineteenth century to the interwar period.

Index

accommodation studies, 123–4, 129
acoustic radio astronomy, 65–6
acoustics: development in Great War, 55–7; equivalent-circuit diagram, 41, 53, 57, 61; Hansing studies, 220–4; Helmholtz's contributions, 238–41; Koenig, 224–5; oscillations of piano strings, 219–21; piano making, 212–27; in piano pedaling, 238–41, 244–50; Schwingungsforschung research, 51–8; as scientific practice, 82–3; transverse oscillations, 220–1
Adler, Guido, 241
Advanced Research Projects Agency (ARPA): Ruina testimony on seismology, 97–8; SAIL, 257; on Speeth's method, 95, 97–100; support of Stanford, 257, 260
AEG Forschungsinstitut, 53
alternating current transformer, 50–1, 56
astrolabes, 294–5
AT&T, 58, 59, 113, 264
auditory detection method, 81
auditory processing research, 95, 98

Barnett, Miles A.F., 78
bells: acoustic color, 114–8; acoustics of sounds, 110–4; cultural semiotics of, 107–10; Helmholtz on, 200–1; and measuring time, 180; strike note, 113, 117–8; use in warfare, 106
Bell Telephone Laboratories, 58; human-speech synthesizer, 59; interferences investigations, 65–6
Berger, Carl Georg, 225–6
Berlin Institut für Schwingungsforschung, 61
Bethe, Hans, 92–3
Burney, Charles, 298

Cage, John, 162–3
camera silenta, 165–6, 167, 188
camera silentissima, 186
Center for Computer Research in Music and Acoustics: music research, 270–2; and NASA, 276; roots of success, 263–365
Chladni, Ernst Florens Friedrich: vibration experiments, 26, 36, 197–8
circuit diagrams, 42–4
clocks: synchronizing via telegraph, 175–7, 181–3
Cold War: nuclear arms race, 84–5; prompting seismology studies, 89–92; spurring research, 80–1
Cruft Laboratory, 163, 164

Daft Punk, 291
Darwinians, 133, 134, 135
Derrida, Jacques, 75–6, 77
differentiated string set, 227–31
digital sound technologies, 63, 279–80

Dombois, Florian, 82
Doppler effect, 124

ear: anatomy, 127, 134; functioning, 29, 30, 34, 45, 46, 89–90, 99, 114, 139, 155–6, 188, 193, 198, 201; mechanics, 124–9
Eccles, William Henry, 57–8
Eitz, Carl, 137–8
electrical engineering: rise of, 50; at Stanford, 262
electricity: electric arc, 52–3; in warfare, 44
electroacoustics, development of, 40, 44
electromagnetic transducer, 46–7, 56–7
equivalent-circuit: advance of, 56–61, 61–2; diagrams, 41–2; 63; in electroacoustics, 41–2; explaining AC currents, 50–1; representational, 42–3; use, 53, 56–7, 60–3
exhaustion, 158–60

fatigue, body, 157–8
FitzGerald, George Francis, 48–9
Fourier's theorem, 194, 195, 196, 198, 199
Frantti and Levereault, 99–100

Grotrian, Willi, 222–3, 227

Hainisch, Victor, 227–31
Hansing, Siegfried: differentiated string set, 227–31; piano acoustics, 212–4, 223–4
harmonics: anharmonic bells, 111–3, 114; Hansing's work on, 219; harmonic analyzer, 118; musical notation and, 108; Schouten's experiments, 114; Smith's work on, 23; telegraph, 5; theory, 196–7; Young's work on, 20–3, 30
Hartmann-Kempf, Robert, 51
Harvard University: anechoic room, 163–5; Münsterberg laboratory, 177–8
Heaviside, Oliver, 49–50; Heaviside layer, 78; spatial-theoretical concept, 77
Heinrich Hertz Institut, 53, 58, 61
Helmholtz, Hermann von: contributions to piano making, 214–5, 223, 227, 229, 253; experiments, 169, 170; harmonics studies, 113–4; interference studies, 228; Mach's reworking of, 130, 132; measuring time, 168–9, 170; musical background, 238–9; musikalische Klangfarbe, 191–3; physical acoustics, 41, 198–200; physical acoustics research, 44–5; prolonging sound studies, 206–8; on pure intervals, 139–40; resonance experiments, 200–2; Schmitt sourcing, 245, 246, 248; sensations of tone, 45–8, 214, 239–41; sign theory of vision, 122, 124; sound local to ear, 206, 208; tone color, 191–3; voltage-source equivalent, 47–8; vowels as set of Klangfarben, 202–5, 209

301

SUGGESTIONS FOR CONTRIBUTORS TO OSIRIS

OSIRIS is devoted to thematic issues, conceived and compiled by guest editors who submit volume proposals for review by the OSIRIS Editorial Board in advance of the annual meeting of the History of Science Society in November. For information on proposal submission, please write to the Editor at osiris@etal.uri.edu.

1. Manuscripts should be submitted electronically in Rich Text Format using Times New Roman font, 12 point, and double-spaced throughout, including quotations and notes. Notes should be in the form of footnotes, also in 12 point and double-spaced. The manuscript style should follow *The Chicago Manual of Style*, 16th ed.

2. Bibliographic information should be given in the footnotes (not parenthetically in the text), numbered using Arabic numerals. The footnote number should appear as superscript. "Pp." and "p." are not used for page references.

   a. References to books should include the author's full name; complete title of book in *italics*; place of publication; date of publication, including the original date when a reprint is being cited; and, if required, number of the particular page cited (if a direct quote is used, the word "on" should precede the page number). *Example*:

     [1] Mary Lindemann, *Medicine and Society in Early Modern Europe* (Cambridge, 1999), 119.

   b. References to articles in periodicals or edited volumes should include the author's name; title of article in quotes; title of periodical or volume in *italics*; volume number in Arabic numerals; year in parentheses; page numbers of article; and, if required, number of the particular page cited. Journal titles are spelled out in full on the first citation and abbreviated subsequently according to the journal abbreviations listed in *Isis Current Bibliography*. *Example*:

     [2] Lynn K. Nyhart, "Civic and Economic Zoology in Nineteenth-Century Germany: The 'Living Communities' of Karl Möbius," *Isis* 89 (1999): 605–30, on 611.

   c. All citations are given in full in the first reference. For succeeding citations, use an abbreviated version of the title with the author's last name. *Example*:

     [3] Nyhart, "Civic and Economic Zoology" (cit. n. 2), 612.

3. Special characters and mathematical and scientific symbols should be entered electronically.

4. A small number of illustrations, including graphs and tables, may be used in each volume. Hard copies should accompany electronic images. Images must meet the specifications of The University of Chicago Press "Artwork General Guidelines" available from the Editor.

5. Manuscripts are submitted to OSIRIS with the understanding that upon publication copyright will be transferred to the History of Science Society. That understanding precludes consideration of material that has been previously published or submitted or accepted for publication elsewhere, in whole or in part. OSIRIS is a journal of first publication.

OSIRIS (ISSN 0369-7827) is published once a year.

Single copies are $33.00.

Address subscriptions, single issue orders, claims for missing issues, and advertising inquiries to *Osiris*, The University of Chicago Press, Journals Division, PO Box 37005, Chicago, IL 60637.

Postmaster: Send address changes to *Osiris*, The University of Chicago Press, Journals Division, PO Box 37005, Chicago, IL 60637.

OSIRIS is indexed in major scientific and historical indexing services, including *Biological Abstracts*, *Current Contexts*, *Historical Abstracts*, and *America: History and Life*.

Copyright © 2013 by the History of Science Society, Inc. All rights reserved. The paper in this publication meets the requirements of ANSI standard Z39.48-1984 (Permanence of Paper). ⊚

Paperback edition, ISBN 978-0-226-02939-9

Osiris

A RESEARCH JOURNAL DEVOTED TO THE HISTORY OF SCIENCE AND ITS CULTURAL INFLUENCES

A PUBLICATION OF THE HISTORY OF SCIENCE SOCIETY

EDITOR
ANDREA RUSNOCK
University of Rhode Island

COPY EDITOR
RACHEL KAMINS
Washington, DC

PAST EDITOR
KATHRYN OLESKO
Georgetown University

PROOFREADER
JENNIFER PAXTON
The Catholic University of America

OSIRIS EDITORIAL BOARD

FA-TI FAN
State University of New York, Binghamton

BERNARD LIGHTMAN
York University
EX OFFICIO

ILANA LÖWY
Institut National de la Santé et de la Recherche Medicale, Paris

W. PATRICK MCCRAY
University of California, Santa Barbara

TARA NUMMEDAL
Brown University

SUMAN SETH
Cornell University

HSS COMMITTEE ON PUBLICATIONS

BRUCE HUNT
University of Texas, Austin
CHAIR

JOSEPH DAUBEN
The Graduate Center, City University of New York
SECRETARY

SORAYA DE CHADAREVIAN
University of California, Los Angeles

FLORENCE HSIA
University of Wisconsin, Madison

MICHAEL GORDIN
Princeton University

ANGELA CREAGER
Princeton University
EX OFFICIO

EDITORIAL OFFICE
DEPARTMENT OF HISTORY
80 UPPER COLLEGE ROAD, SUITE 3
UNIVERSITY OF RHODE ISLAND
KINGSTON, RI 02881 USA
osiris@etal.uri.edu